Preludes to U.S. Space-Launch Vehicle Technology

UNIVERSITY PRESS OF FLORIDA

Florida A&M University, Tallahassee
Florida Atlantic University, Boca Raton
Florida Gulf Coast University, Ft. Myers
Florida International University, Miami
Florida State University, Tallahassee
New College of Florida, Sarasota
University of Central Florida, Orlando
University of Florida, Gainesville
University of North Florida, Jacksonville
University of South Florida, Tampa
University of West Florida, Pensacola

UNIVERSITY PRESS OF FLORIDA

Gainesville · Tallahassee · Tampa · Boca Raton · Pensacola
Orlando · Miami · Jacksonville · Ft. Myers · Sarasota

Preludes to U.S. Space-Launch Vehicle Technology

Goddard Rockets to Minuteman III

J. D. Hunley

13 12 11 10 09 08 6 5 4 3 2 1

Library of Congress Cataloging-in-Publication Data
Hunley, J. D., 1941–
Preludes to U.S. space-launch vehicle technology :
Goddard rockets to Minuteman III / J. D. Hunley.
v. cm.
Includes bibliographical references and index.
ISBN 978-0-8130-3177-4 (alk. paper)
1. Launch vehicles (Astronautics)—United States—History—20th century.
2. Rocketry—United States—History—20th century.
3. Minuteman (Missile)—History—20th century. I. Title.
TL785.8.L3H86 2008
629.470973—dc22 2007042543

The University Press of Florida is the scholarly publishing agency
for the State University System of Florida, comprising Florida A&M
University, Florida Atlantic University, Florida Gulf Coast University,
Florida International University, Florida State University, New College
of Florida, University of Central Florida, University of Florida,
University of North Florida, University of South Florida,
and University of West Florida.

University Press of Florida
15 Northwest 15th Street
Gainesville, FL 32611-2079
http://www.upf.com

This book is dedicated to Barnet R. Adelman, Wilbur Andrepont, Charles Bartley, Robert C. Corley, Daniel Dembrow, Ross Felix, Robert L. Geisler, Edward N. Hall, Charles Henderson, Kenneth W. Iliff, Karl Klager, Franklin H. Knemeyer, Grayson Merrill, Ray Miller, Edward W. Price, Milton W. Rosen, Ed Saltzman, Ronald L. Simmons, Ernst Stuhlinger, H. L. Thackwell, and Robert C. Truax.

Contents

Preface and Acknowledgments

This book and its sequel, *U.S. Space-Launch Vehicle Technology: Viking to Space Shuttle,*[1] address a significant gap in the literature about access to space. There are numerous and quite excellent works covering various aspects of missile and space-launch-vehicle development and some general accounts. But there is no study that traces in a detailed and systematic way how the technology evolved from its beginnings with Robert Goddard and with the German V-2 missile to the end of the cold war.

Another problem with the existing literature is the lack of agreement among sources about specifics. From measurements of length and diameter to those of thrust and accuracy, sources differ. These two books cannot claim to resolve the differences, but they acknowledge them in endnotes and indicate which sources seem most credible.

I first began working on these histories in 1992 when I undertook a much more modest, monographic study comparing the contributions to U.S. launch vehicle technology of the Wernher von Braun group that developed the V-2 in Germany and then immigrated to America after World War II with those of the group around Theodore von Kármán and Frank Malina at the Jet Propulsion Laboratory near Pasadena, California. I quickly found that the literature would not permit such a comparison without a much broader assessment of rocket technology. As a result I wrote the account that follows.

Because the material to be covered is so broad as well as technical, what I had originally conceived as a single volume had to be divided into two. The present book covers primarily missile development, because many of the launch vehicles borrowed technology and whole rocket stages from missiles, most of which developed before the launch vehicles. In a couple of cases, to provide continuity of coverage, early uses of missiles as components of launch vehicles are discussed in this book. Then *Viking to Space Shuttle* steps back in time to pick up development of Viking and Vanguard, which preceded most of early missile production. The second book then continues with the Thor-Delta, Delta, Atlas, Scout, Saturn, and Titan space launch vehicles, concluding with treatment of the space shuttle. Both books are written in such a way that they can be understood by a general audience, but I hope they will also prove useful to scholars, engineers, and others who already possess an extensive knowledge of some of the material covered.

Although the two books constitute a continuous whole, intended to be read as such, some readers primarily interested in either missiles or launch vehicles may want to read just one of the volumes. They will find much that is new in each. But readers primarily interested in missiles should be aware that I have not written a complete history of even the ballistic missiles that *Goddard Rockets to Minuteman III* mainly covers, while providing only limited treatment of some tactical missiles. Since my focus is on the technologies that contributed to launch-vehicle development, I spend comparatively little time with the business end of missiles, the warheads. Instead, I concentrate on propulsion, structures, and guidance and control technologies because these carried over—though often with considerable modification—to launch vehicles.

By the same token, those primarily interested in launch vehicle technology will find that they have missed out on much of the history if they start with *Viking to Space Shuttle*. It is obviously impossible to understand the Thor-Delta family of launch vehicles without knowledge of the Thor missile, or the Atlas launch vehicles without knowing how the Atlas missile developed. Links between the Titan I and II missiles and the Titan launch vehicles are equally important. Even the Scout and Saturn launch vehicles borrowed much technology from missiles. And the solid-rocket motors for the Titan III and IV as well as the solid-rocket boosters for the space shuttle borrowed much technology from the Polaris and Minuteman missiles.

In researching and writing both volumes, I have incurred many debts of gratitude to an enormous number of people. I acknowledge many of them in endnotes, but unfortunately I can no longer remember everyone who assisted me in a great variety of ways. I owe particular gratitude to Roger Launius, who, as my boss at the NASA History Office, first encouraged me to begin studying the subject of rocket technology and has provided unfailing support in more ways than I can recount. (He also was wise enough to discourage me from attempting to cover the entire gamut of American missile and launch-vehicle technology, but I was not smart enough to follow that advice.) Lee Saegesser and Jane Odom, archivists at the NASA History Office, both provided extensive support, as have the rest of the archival staff there, including Colin Fries and John Hargenraether as well as their predecessors, Jennifer and Bill Skeritt.

Mike Neufeld shared his research on the V-2 and arranged for me to study the captured files on Peenemünde at the Smithsonian National Air and

Space Museum. In 2001–2, I was fortunate enough to become the Ramsey Fellow at that institution. I am very grateful to NASM for granting me the fellowship and thus allowing me to complete my research for this book. Mike kindly read several chapters and offered me his criticism. Many of his colleagues at the museum, along with the archival staff, were also extremely helpful. I had valuable discussions with John Anderson, Tom Crouch, David DeVorkin, and Mike Neufeld about aspects of my work. Also very helpful were a host of archivists, librarians, curators, docents, and volunteers including Marilyn Graskowiak, Dan Hagedorn, Gregg Herken, Peter Jakab, Mark Kahn, Daniel Lednicer, Brian Nicklas, George Schnitzer, Leah Smith, Paul Silbermann, Larry Wilson, Frank Winter, and Howard S. Wolko.

Chapters 3, 5, and 9 of this volume plus chapters 6 and 7 of *Viking to Space Shuttle* contain material I published earlier in chapter 6 of *To Reach the High Frontier: A History of U.S. Launch Vehicles*, edited by Roger D. Launius and Dennis R. Jenkins (2002). The material in the present book incorporates much research done since I wrote that chapter, and it is organized differently. But I am grateful to Mack McCormick, rights manager at the University Press of Kentucky, for confirming my right to reuse the material that appeared in the earlier version.

Many people read earlier chapters of this book and provided suggestions for improvement. They include Matt Bille, Roger Bilstein, John Bluth, Trong Bui, Virginia Dawson, David DeVorkin, Ross Felix, Mike Gorn, Pat Johnson, John Lonnquest, Ray Miller, Fred Ordway, David Stumpf, Frank Winter, and Jim Young. Persons who furnished documents or other source materials that would have been difficult to locate without their assistance include Nadine Andreassen, Liz Babcock, Scott Carlin, Robert Corley, Dwayne Day, Bill Elliott, Robert L. Geisler, Robert Gordon, Edward Hall, Charles Henderson, Dennis Jenkins, Karl Klager, John Lonnquest, Ray Miller, Tom Moore, Jacob Neufeld, Fred Ordway, Ed Price, Ray Puffer, Karen Schaffer, Ronald Simmons, Ernst Stuhlinger, Ernie Sutton, Robert Truax, P. D. Umholtz, and Ben Zibit.

Librarians, historians, and archivists at many institutions assisted my research in a variety of ways. They include Air Force Historical Research Agency archivist Archangelo Difante; Air Force Space and Missile Systems Center historian Harry Waldron and archivist Teresa Pleasant; Air Force Flight Test Center historians Jim Young and Ray Puffer; China Lake historian Leroy Doig; Clark University Coordinator of Archives & Special Collections Dorothy E. Mosakowski; Bill Doty at the National Archives, Laguna Niguel, California; Dryden Flight Research Center librarian Barbara Rogers;

and JPL archivists John Bluth, Barbara Carter, Julie Cooper, and Margo E. Young. Several reference librarians at the Library of Congress should be added to the list, but I don't have their names.

Thanks are due to everyone who consented to interviews (included in endnote references and lists of sources). They not only provided their time, and often editorial comments on the transcribed interviews, but agreed to permit me to use the information in the interviews. In addition, many people discussed technical issues with me or provided other assistance, such as help with photographs. These include Ranney Adams, Wil Andrepont, Stan Backlund, Rod Bogue, Al Bowers, George Bradley, Mark Cleary (who went to great lengths to provide me with nineteen photographs), Robert Corley, Daniel Dembrow, Robert L. Geisler, Mike Gorn, Mark L. Grills, John Guilmartin, Burrell Hays, Charles Henderson, J. G. Hill, Cheryl Hunley, Michael Hunley, Ken Iliff, Fred Johnsen, Karl Klager, Franklin Knemeyer, Tony Landis, Niilo Lund, Jerry McKee, Ray Miller, Tom Moore, Sarah Parke, Ed Price, Milton Rosen, Jim Ross, Bill Schnare, Carla Thomas, Woodward Waesche, Herman Wayland, and Paul Willoughby. To all of the people above, I offer my thanks for their generous help.

I would like to express my special appreciation to Jacob Neufeld and Robert L. Geisler for their helpful comments on drafts of this book. Both of them also assisted me in my research. Jack Neufeld kindly sent me a copy of his interview with Col. Ed Hall and a copy of a recent book published by the Air Force History and Museum Program. His publications about rocketry and Gen. Bernard Schriever have been indispensable sources for what I have written about Air Force programs. Bob Geisler has shared several sources with me as well as his knowledge about missile development. He also chaired a session at an AIAA Joint Propulsion Conference in 1999 at which he invited me to present a paper. At that conference I was able to obtain several seminal papers from fellow presenters that were key sources for parts of this history. Both gentlemen were also refreshingly forthcoming in allowing the University Press of Florida to share their names with me, which is how I am able to thank them personally instead of as anonymous referees.

I am deeply indebted to Ann Marlowe for her exceptionally diligent copyediting, Jacqueline Kinghorn Brown for her terrific help as in-house editor, editor-in-chief John W. Byram for advice and help on many matters, and everyone else at the University Press of Florida for their contributions to the book. But I especially thank Ann and almost think her name belongs on the title page along with mine.

Last but not least, I thank my agent and friend Neil Soderstrom for his encouragement, editorial advice, fruitful suggestions, help in finding a publisher for both books, and, above all, friendship through thick and thin in the face of many demands on his time that should have taken precedence over helping me. He has made me look at these books in a different way than would have occurred to me on my own. They owe much to him and his generosity of time, hard work, and spirit.

There are undoubtedly many other individuals who have assisted my research and writing over the years but whose help or names have been erased by the passage of time (and old age) from my active memory bank. I can only apologize for the oversight and say a generic "thank you very much." None of the people acknowledged here bears any responsibility for the details and interpretations that appear in these two books. For them, I alone am responsible. But I hope these generous individuals will approve of the way I have used the suggestions, comments, materials, and information they provided or helped me to find.

Introduction

Although black-powder rockets had been around for centuries, it was not until 1926 that American physicist and rocket developer Robert H. Goddard launched the first known liquid-propellant rocket. Despite this auspicious beginning, not until the mid-1950s did the United States begin to invest significant resources in rocket development. Already by the end of January 1958 the United States had launched its first satellite, and within a generation it had developed a series of missiles and launch vehicles of enormous power and sophistication. The Atlas, Titan, Scout, Delta, Saturn, and space shuttle launched a huge number of satellites and other spacecraft that revolutionized our understanding of the universe, including our own planet, and brought events and reporting from all parts of the world into the American living room with unprecedented speed. How could the United States have advanced so rapidly from the relatively primitive rocket technology available on a small scale in the mid-1950s to the almost routine access to space available by the 1980s?

This book and its sequel, *U.S. Space-Launch Vehicle Technology: Viking to Space Shuttle*, attempt to answer that question and to trace the convoluted technological trajectory from Goddard's imaginative but problem-prone early rockets to the huge Saturn V and the complex space shuttle, among other launch vehicles. The history of these vehicles has been punctuated by failures on the path to overall success. But on the whole, the achievements have been remarkable.

Perhaps most remarkable have been the unique features of the space shuttles. As the United States approaches the end of shuttle flights in 2010, it is appropriate to reflect that in some ways this astonishing but troubled launch vehicle and spacecraft was the culmination of the development process discussed in these two volumes. It represented a bold dream of converting previously expendable missile and launch vehicle technologies into a reusable source of routine access to and return from space, analogous to airliners and large cargo aircraft for the nearer skies. In one sense the effort was a failure, since the Air Force continues to rely on expendable launch vehicles and NASA is retreating, under budgetary and safety pressures, from reusability to a concept akin to the Saturn launch vehicles of the Apollo era.

In another sense, however, the crew and cargo launch vehicles Ares I and V (discussed in the final chapter of *Viking to Space Shuttle*) are themselves a legacy of the shuttles, since they will use shuttle experience and technology as part of the basis upon which to build a more affordable and safer way to return to the Moon and even go to Mars.

It is also worthwhile to recognize that many of the achievements of the space shuttles would have been extraordinarily difficult to accomplish without the unique features built into the shuttles. To give but one example, after space shuttle *Discovery* launched the Hubble Space Telescope in April 1990, it quickly became apparent that the enormous promise of this astronomical instrument was marred by a small but critical flaw in its primary mirror. Following a partial correction by computer enhancement, a planned routine repair mission turned into a rescue mission in December 1993 in which the huge telescope was recaptured in the payload bay of space shuttle *Endeavour*, outfitted with a corrective mechanism for the optics of the primary mirror, and serviced in other ways to allow the scientific instrument to function as originally intended.

Some 1,200 women and men were involved in orchestrating, designing, practicing for, and carrying out this delicate and complex repair mission, which could hardly have been performed by an expendable launch vehicle coupled with any other existing spacecraft. *Endeavour*'s astronauts used five spacewalks to install the device using additional mirrors to correct Hubble's optics as well as to replace failed gyroscopes and the Wide Field/Planetary Camera. They then installed equipment to improve the telescope's failing computer memory, among other things. Although Hubble had been providing important new scientific data even before the rescue mission, afterwards it began to live up to and even exceed the performance that astronomers had expected from it, including provision of the first solid evidence for the existence of black holes (regions in space of intense gravitational force) and spectacular images that graced the pages of newspapers and even appeared on a cover of *Newsweek*.

Literally millions of people had watched as the shuttle astronauts performed their repairs, and the entire team responsible for the mission received the 1993 Robert J. Collier Trophy from the National Aeronautic Association "for outstanding leadership, intrepidity, and the renewal of public faith in America's space program by the successful orbital recovery and repair of the Hubble Space Telescope." This and other almost equally astonishing achievements showed the unique value of the space shuttle as the fruition of a comparatively short but intense period of development of Space-Launch capabilities.[1]

Although the focus of these two books, *Viking to Space Shuttle* as well as *Goddard Rockets to Minuteman III*, is on technology used by launch vehicles, which permitted space exploration such as that carried out by Hubble, *Goddard Rockets to Minuteman III* is mostly about missiles and can be read by itself as a history of missile technology. Missiles follow trajectories aimed at places on Earth instead of the heavens; they carry warheads instead of satellites or spacecraft. But especially in the area of propulsion, they use much the same technology as launch vehicles. In fact, many launch vehicles have been converted missiles. Others have borrowed stages from missiles.

One major irony stands out in this process. While the complexity and sophistication of missiles and launch vehicles gave birth to the expression "rocket science," careful study of the vehicles' development reveals many instances in which the designers and operators encountered problems they did not fully understand. They frequently had to resort to trial-and-error fixes to make their rockets perform as intended. Although data about and understanding of the advancing technologies continually increased, each large jump in scale and performance introduced new difficulties. Rocketry was, and is, as much an art as a science, fitting the description of engineering—as distinguished from science—provided by Edwin Layton, Walter Vincenti, and Eugene Ferguson, among others. (Besides engineering as art, these scholars also emphasized engineering's focus on doing rather than knowing, on design of artifacts rather than understanding the universe, and on making technological decisions in the absence of clear understanding—all features that distinguish engineering from science.)[2]

This is not to say that science and scientists did not contribute to rocket technology. For example, Ronald L. Simmons earned a B.A. in chemistry at the University of Kansas in 1952 and went on to work for thirty-three years as a propulsion and explosives chemist with the Hercules Powder Company, a year with Rocketdyne, and thirteen years with the U.S. Navy at Indian Head, Maryland. Among other projects, he worked on upper stages for Polaris, Minuteman, Poseidon, and Trident.

In 2002 he wrote, "I consider myself to be a chemist . . . even though my work experience has been a lot of engineering. I really believe the titles are arbitrary, though I consider myself a scientist rather than an engineer." He added in relation to the issue of rocket engineering versus rocket science, "'Tis amazing how much we don't know or understand, yet we launch large rockets routinely . . . and successfully . . . that is when we pay attention to details and don't let the schedule be the driving factor. . . . By and large, I believe that we understand enuff to be successful . . . yet may not understand why." Although he spent much of his career working with double-base propellants—those using nitrocellulose (NC) and nitroglycerin—he admitted, "There is much no one understands about nitrocellulose (and black powder for that matter) in spite of the fact that NC has been known since 1846 and black powder since before 1300!"[3]

Chronologically, the two books follow the development of American rocket technology through the end of the cold war in 1989–91. Chapter 2 of *Goddard Rockets to Minuteman III* covers the German V-2 because it became one of the foundation stones for U.S. rocket technology. Many of the V-2's developers immigrated to the United States after the end of World War II. They became the nucleus of the later NASA Marshall Space Flight Center. Under the leadership of Wernher von Braun, many of these Germans (along with hundreds of Americans) oversaw the development of the Saturn launch vehicles that lifted twelve astronauts on their journey to the Moon in the Apollo program.

Viking to Space Shuttle ends about 1990–91 with the close of the cold war because, after that, launch vehicle development began a new chapter. Funding became more spartan, and the United States began borrowing technology from the Russians, who had competed with American missile and launch vehicle technology during the Soviet era.

Most readers of this book will have watched launches of the space shuttle or other launch vehicles on television. For those less familiar with the fundamentals of rocketry, this may help: Missiles and other rockets lift off from Earth through the thrust created by the burning of propellants (fuel and oxidizer).[4] This combustion creates expanding exhaust products, mostly gaseous, that pass through a nozzle at the back of the rocket. The nozzle contains a narrow throat and an exit cone that cause the gases to accelerate, thereby increasing thrust. The ideal angle for the exit cone depends on the altitude and pressure at which it will operate, with different angles needed at sea level than at higher altitudes where the atmosphere is thinner and the outside (ambient) pressure is lower.

Rockets in the period covered by this book typically used multiple stages to accelerate the vehicle all the way to its designed speed. When the propellants from one stage became exhausted, that stage would drop off the stack, so that as succeeding stages took over, there was less weight to be propelled to higher speeds. Multiple stages also permitted using exit cones of varying angles for optimal acceleration at different altitudes.

Most propellants required an ignition device to begin combustion, but hypergolic fuels and oxidizers ignited upon contact, dispensing with the need for an igniter. These types of propellants typically had less propulsive power than the extremely cold (cryogenic) liquid oxygen and liquid hydrogen, but they required less special handling than their cryogenic counterparts. Liquid oxygen and liquid hydrogen would boil off if not loaded just before launch, so they needed a lot more preparation time before a launch could occur. Hypergolic propellants, by contrast, could be stored in propellant tanks for comparatively long periods, allowing almost instant launch upon command. This was an especial advantage for missiles, and for spacecraft launches that had narrow "windows" of time, when the desired trajectory was lined up with the launch location only for a short period as Earth rotated and circled the Sun.

Solid-propellant missiles and rockets also enjoyed rapid-launch capabilities. They were much simpler and usually less heavy than liquid-propellant rockets because the fuel and oxidizer filled the combustion chamber without a need for propellant tanks, high pressure or pumps to force the propellant into the chamber, extensive plumbing, and other complications. Typically, technicians loaded a solid propellant into a combustion chamber with thin metal or a composite structure as the case, insulation between the case and the propellant, and a cavity in the middle where an igniter started combustion. Engineers designed the internal cavity to provide optimal thrust, with more exposed propellant surface providing more instant thrust and a smaller amount of surface providing less initial thrust. The propellant burned from the inside toward the case, with the insulation protecting the case as the propellant burned outward. The disadvantage of solids was the difficulty of stopping and restarting combustion, which could be done with valves in the case of liquids. Thus, for launch vehicles, solids usually appeared as initial stages, called stage 0, to provide maximum thrust for the initial escape from Earth's gravitation field or as upper stages (although the Scout remained a fully solid-propellant, multistage launch vehicle from 1960 to 1994).

Liquid propellants found more frequent use for the core stages, usually stage 1 and often stage 2, of launch vehicles such as the Atlas, Titan, Delta, and space shuttle. They also served in upper stages that needed to be stopped and restarted in orbit for insertion of satellites and spacecraft into particular orbits or trajectories. But the process of injecting fuels and oxidizers into the combustion chamber proved to be fraught with problems. For reasons that have been difficult for engineers to understand, mixing the two types of propellants in the needed proportions frequently resulted in oscillations that could destroy the combustion chamber. Known as combustion instability, this severe problem only gradually yielded to solutions, as each scaling up of a particular type of engine usually caused new problems that required their own specific solutions.

Solid propellants also experienced combustion instability. Problems with solids were somewhat different from those with liquids. But as with liquids, the solutions required much research and, often, trial and error before they could be solved, or at least ameliorated.[5]

Besides propulsion systems, rockets required structures that would withstand the high heats of combustion, intense dynamic pressures as the vehicles accelerated through the atmosphere, shock waves as they passed through the speed of sound (referred to as Mach 1), and aerothermodynamic heating from friction while traveling at high speeds through the atmosphere. Because weight slowed acceleration to orbital speeds and altitudes, structural issues required much research to find lighter materials that would still withstand the rigors of launch. Engineers gradually found new materials that were strong, heat resistant, light, and, if possible, affordable.

Another field of research was aerodynamics. Missiles and launch vehicles needed to have as little drag (friction from the atmosphere that slowed flight and increased temperatures) as possible. They also had to be steerable by means of vanes, canards, moveable fins, vernier (auxiliary) and attitude-control rockets, fluids injected into the exhaust stream, and/or gimballed (rotated) engines or nozzles.

Associated with these types of control devices were various guidance and control systems incorporating computers programmed to adjust steering and keep the missile or launch vehicle on course. Such systems varied greatly in design and weight. They involved increasingly sophisticated programming of the computers. But they were essential to the success of both missiles and launch vehicles.[6]

Chapter 1 of *Goddard Rockets to Minuteman III* introduces the two rocket pioneers who had the greatest influence on American missiles and launch vehicles, the American physicist and rocket experimenter Robert H. Goddard and the Romanian-German rocket theorist Hermann Oberth. Both were fascinating characters with highly inventive minds. Although Goddard's innovations foreshadowed many later rocket technologies, his failure to publish many details of his research and development during his lifetime limited his influence. Oberth published his more theoretical conceptions in greater detail and had real influence on Wernher von Braun and other Germans who developed the V-2 missile before and during World War II and then immigrated to the United States. Through them, Oberth arguably had greater influence on U.S. missile and launch-vehicle development than did Goddard. But it can also be argued that they had a synergistic effect, with Goddard providing an example of how to develop rockets, at least to a point, while Oberth provided more theoretical details about rocket development in sources that he published early enough for them to be consulted by early rocket developers.

Although the V-2 was only one of many influences on American rocket technology, it was important, if sometimes overrated. Chapter 2 of *Goddard Rockets to Minuteman III* discusses the development of this missile and provides the technical information needed for later analysis of ways in which the V-2 was and *was not* a stepping-stone for American rocketry. Chapter 3 covers rocket development in the United States before, during, and shortly after World War II at what became the Jet Propulsion Laboratory (JPL) near Pasadena, California. Chapter 4 discusses other American rocket efforts from 1930 to 1954, culminating in a joining of German and JPL rocket technologies in the Bumper WAC project. Meanwhile, other efforts during and after World War II yielded important solid-propellant innovations that paralleled those at JPL, which had worked on both liquid and solid propellants. These developments at JPL form the subject of chapter 5.

Chapter 6 of *Goddard Rockets to Minuteman III* covers the Redstone missile and its modification into the first stage of the Juno I launch vehicle that placed the first U.S. satellite in orbit on January 31, 1958. The upper stages of the Juno I employed JPL technology, which again blended with that of Wernher von Braun's team in Alabama as they had on the Bumper WAC. The Redstone itself combined German and American contributions to rocketry, including some from the Air Force's Navaho missile. Chapters 7 through 9 cover the Atlas, Thor, Jupiter, Titan I and II, Polaris, and Minuteman missiles that produced still other technologies and separate stages used on later launch vehicles. Without the cold war and the developments it prompted, these contributors to launch vehicle technology would have evolved far more slowly they did.

Chapter 1 of *Viking to Space Shuttle* covers the Viking and Vanguard programs, separate American rocket efforts that contributed importantly to missile and launch vehicle technology, including the use of gimbals for steering. Chapters 2 through 7 then discuss the uses of missile technology in development of the Delta, Atlas, Scout, Saturn, Titan, and space shuttle launch vehicles. Development of these vehicles, and of the missiles that preceded or were contemporaneous with them, was by no means problem-free. Besides cold-war threats and resultant funding, factors in the evolution of rocket technology in the United States included the efforts of both technical rocket engineers and others who were often less intimately involved with technical matters than with engineering the social aspects of missile and launch-vehicle development by promoting that cause in Congress, the Pentagon, and the media (as, previously, the V-2 was advocated to the Nazi regime in Germany). Without the promotional efforts of these so-called heterogeneous engineers, even the cold war might not have been enough to overcome the inertia that stood in the way of complex and expensive development, punctuated by many well-publicized failures in the early years.

Other factors in the rapid development of U.S. rocketry included both rivalry between military services (and the rocket firms that supported them) and, at the same time, a high degree of cooperation and sharing of knowledge by the competitors. A wide range of disciplines was required to design and develop rockets, calling for unselfish collaboration among competitors in solving problems that occurred during tests and operational launches. Universities also played a role in this process, as did a growing technical literature. Relatedly, the movement of personnel between firms, carrying technical knowledge from one project to another, professional networks, and federal intellectual property arrangements all helped promote and transfer innovation, leading to increasingly powerful and sophisticated launch vehicles that placed satellites in orbit and sent spacecraft on missions to explore our solar system and beyond.

A final contributing factor to the rapid and ultimately successful development of launch vehicle technology was a variety of management systems that helped to integrate efforts on the various systems in missiles and rockets, to keep them on schedule, and to promote configuration and cost control. All of these factors and others are discussed in the chapters that follow in both books.

Both books are organized essentially by project. As the dates in the chapter titles suggest, there was a great deal of overlapping in time between projects. Since these projects borrowed technologies from one another, there is an inherent problem with presenting technical materials in such a way that readers not highly familiar with the history of rocket technology can easily follow the story. The problem is compounded by the fact that different systems on a given missile or rocket were developed simultaneously. Thus it is impossible to follow a strictly chronological path in the narrative. Even if that could be done, the result would hardly be comprehensible. To assist the reader, I have included a chronology of key events or achievements and a glossary of technical terms and acronyms in this book, so that if a chronological detail or the meaning of a term or acronym is forgotten, these appendices may be consulted.

Some readers may be disappointed that I have not included in this history of technology a discussion of its ethical dimensions. It is widely known, for example, that some of the Germans who worked on the V-2, including von Braun, were members of the Nazi Party and were implicated to various degrees in the use of concentration-camp labor for the production of the missile at an underground facility named Mittelwerk. These issues have remained controversial. Historian Michael J. Neufeld has explored them and related concerns at some length in an article that lays out the case against von Braun in particular. He presents a balanced case against von Braun, pointing out the many ambiguities, complexities, and gaps in the evidence. Although he comes down on the accusatory side of the argument, to his credit he admits that much of the evidence is inconclusive.[7] In a subsequent exchange with von Braun's distinguished colleague Ernst Stuhlinger, Neufeld states, "Von Braun made a Faustian bargain with the German Army and National Socialist Regime in order to pursue his long-term dream of exploring space."[8]

While I agree that in some sense von Braun's was a Faustian bargain, one could also ask whether building missiles for other countries, even though they are not reprehensible in the same ways that Nazi Germany was, is not to some degree a Faustian bargain. During the long cold war, nuclear-tipped missiles were never fired in anger, and the argument can be made with considerable validity that they in fact served the cause of peace through deterrence or a "balance of terror." But there is always the possibility they will someday be fired at a perceived enemy, unleashing unprecedented destruction of life. While well aware of these ethical issues, I am not prepared to address them categorically. They are matters about which reasonable human beings can differ profoundly, and I remain somewhat agnostic about them. I neither condemn von Braun and other designers and developers of missiles nor sing their praises. I merely record their contributions to the technologies that led to launch vehicles. Because of the complexity and ambiguity of the issues involved and because these two books are already very long, I hope readers will forgive me for not dwelling on moral and ethical concerns in these two volumes.

1

The Beginnings

Goddard and Oberth, 1926–1945

The modern history of rocketry in the United States begins, in different ways, with American physicist/rocket developer Robert H. Goddard (1882–1945) and Transylvanian-German rocket theorist Hermann Oberth (1894–1989). These two visionaries are correctly regarded as among the three preeminent pioneers in the early development of rocketry and spaceflight in various parts of the world. The other member of this worldwide trio, Russian theorist Konstantin E. Tsiolkovsky (1857–1935), influenced U.S. rocketry only indirectly if at all, after the end of the cold war, when Russian rocket technology became available for American use. Although Tsiolkovsky's work preceded that of Goddard and Oberth, and the Russian calculated such crucial parameters as orbital velocity (17,900 mph) and escape velocity (25,050 mph) for spacecraft, his works did not circulate widely even in the Soviet Union and were not translated into Western languages until the 1940s after Goddard and Oberth had independently arrived at their own theories, which were similar to Tsiolkovsky's.[1]

Goddard and Oberth both significantly influenced rocket enthusiasts who succeeded them. They shared some inclinations and goals. And they confronted quite similar (often negative) reactions to their early publications. But their procedures contrasted markedly. Goddard, the quintessential lone inventor, pursued a pattern of secrecy throughout most of his life. Although his example inspired others, his secretiveness hindered the United States from developing missiles and rockets as rapidly as it might have if he had devoted his real abilities—which at least bordered on genius—to the cooperative development needed for such complex devices.

Oberth, on the other hand, openly published most of the details of his more theoretical findings and then contributed to their popularization in Germany. He was not a citizen of Germany at the time but belonged to that country by ethnic and cultural background despite his birth in Transylvania, then part of the Hungarian portion of the Austro-Hungarian Empire. His early efforts were significantly responsible for launching a spaceflight movement that led directly to and influenced the A-4 (V-2) rocket of World War II fame, developed principally at Peenemünde on the German Baltic coast. Then, through the emigration of Wernher von Braun and his rocket team from Germany to the United States after the war, Oberth contributed indirectly to American missile and spaceflight development. As is well known, on July 16, 1969, a Saturn V rocket engineered under the direction of von Braun blasted off on the Apollo 11 mission. Four days later the first humans landed on the Moon. Curiously, this fulfilled a prophecy not only of President John F. Kennedy in May 1961 but also of Oberth's maternal grandfather, Friedrich Krasser, who had predicted in July 1869 that people would land on the Moon in one hundred years.[2]

Why and how could a German from Transylvania possibly have a greater effect upon the U.S. space program than Goddard, the first person in the world known to launch a liquid-propellant rocket, a revered figure called by such grandiose titles as Father of Modern Rocketry?[3]

Goddard

Robert Goddard was born on October 5, 1882, in Worcester, Massachusetts, an industrial city about forty miles west of Boston. His father, Nahum, was something of an inventor. Nahum encouraged Robert's bent toward experimentation and invention. He also supplied the boy with a microscope, a telescope, and a subscription to *Scientific American*. And he emphasized to his son that it was better to work independently than for someone else. Despite this, the elder Goddard initially worked for a manufacturer of machine knives and then married the boss's daughter. The Goddards moved to Roxbury, a middle-class suburb of Boston, when Nahum found work there with another manufacturer of machine knives, and soon he and another employee purchased the firm. However, the family moved back to Worcester after Nahum's wife was diagnosed in 1898 with tuberculosis. Robert's grandmother took over the raising of her frail grandson. She helped give him the self-confidence he would need for rocket development.

Worcester prided itself on its mechanical reputation. According to Goddard biographer Milton Lehman, its local heroes included cotton-gin inventor Eli Whitney and Ichabod Washburn, who developed the techniques for drawing wire from steel. In Lehman's view, the "self-protective and ingrown" proclivities of Worcester contrasted with the "academic and sometimes outgoing" spirit of Boston. The two conflicting influences fostered competing strains in the young Goddard that characterized him the rest of his life—an inclination to protect and prove his ideas in the Worcester manner as opposed to transmitting them freely in the style of Boston.[4]

Toward the end of his grade school years Goddard suffered a variety of ailments described as colds, pleurisy, and bronchitis. In all probability, they marked the beginnings of tuberculosis. Although it was not diagnosed until 1913, Robert suffered from this disease most of his life. Because of illness, by the time he entered Boston English High School in 1898 and signed up for the general science curriculum, he had fallen behind other students his age. Recurrent sickness kept him out of school most of the term. He compensated with self-education, reading books on the atmosphere, electricity, chemistry, and chemical analysis, as well as on magic.[5]

While attending Boston English, Goddard took a course in algebra but did not distinguish himself. For the next two years illness kept him out of school altogether, although he continued his self-education. Among other things, he read Cassell's *Popular Educator,* which his father had given him. He learned there of Newton's laws of motion. At home in Worcester, he continued to read *Scientific American* as well as textbooks borrowed from the public library.[6]

Thus prepared, the eighteen-year-old Robert Goddard entered the new South High School at Worcester in 1901. He determined to "shine" in physics and math, now important to him because of a transforming experience. At the age of seventeen back on October 19, 1899, as he relates the story, he had climbed a tall cherry tree behind the barn of the Goddard family house in Worcester. He had been reading H. G. Wells's *War of the Worlds* and Jules Verne's *Journey from the Earth to the Moon.* Up in the tree, he imagined "how wonderful it would be to make some device which had even the *possibility* of ascending to Mars." This seemingly idle daydream had a profound influence upon him. "I was a different boy," he wrote, "when I descended the tree from when I ascended, for existence at last seemed very purposive."[7]

Almost yearly after this event, Goddard referred to October 19 as Anniversary Day in his diary. When in Worcester thereafter, he would revisit the tree.[8] In school, he reported, he found it easy to excel in physics but had to struggle against his "previous distaste" for mathematics. He nevertheless managed to lead the class in math.[9]

He graduated from South High in June 1904 at the age of twenty-one "with highest honors." The young man went on to earn a B.S. in general science from Worcester Polytechnic Institute in 1908 and a Ph.D. in physics at Clark University, also in Worcester, Massachusetts, in 1911.[10] Available information about Goddard and Clark University suggests that he got an excellent education in physics there.[11] In Arthur Gordon Webster, who held a Ph.D. from the University of Berlin, the small school had an excellent physics and mathematics professor trained by the eminent Hermann von Helmholtz. Webster's lectures in mathematical physics were significant in advancing education in American physics, which in his day was typically more experimental than mathematical. Held to be a forceful and lucid lecturer, he provided a solid grounding in theoretical and experimental physics to nearly thirty Ph.D.'s, including Goddard.[12]

Goddard seems to have begun serious development of rockets on February 9, 1909, as a graduate student at Clark, when he performed his first experiment on the exhaust velocity of a propellant. He was already in the habit of jotting thoughts on rocketry in a notebook, such as an entry for September 6, 1906, suggesting the idea of ion propulsion that led in 1915 to one of his eventual 214 patents.[13] However, it was not until he was a research fellow at Princeton during 1912–13 that he developed the rocket theories published in his famous paper "A Method of Reaching Extreme Altitudes" (1919/1920). He had written the original version in 1914 and revised it in the light of experiments he performed in 1915–16.[14]

By this time he had become an assistant professor of physics at Clark, later a full professor (1919), then head of the department and director of the physical laboratories (1923).[15] Evidently aware as early as 1909–10 that liquid propellants were more efficient than solid propellants, he penned notes with references to a "general theory of [a] hydrogen and oxygen rocket" in 1910. But Goddard found hydrogen and oxygen difficult to obtain, so his experiments in 1915–16 were limited to smokeless powder, which posed the "least experimental difficulty."[16]

Using high-heat smokeless powder as a propellant, a steel combustion chamber to permit high operating pressures, and a de Laval nozzle (named for Swedish engineer Carl de Laval, who in 1889 had designed it to convert thermal into kinetic energy with high efficiency), Goddard was able to achieve an energy efficiency of 64.5 percent. This compared to the 2 percent he had measured for common firework-type rockets. (He calculated these efficiencies from the energy theoretically available in the powder compared with the exhaust velocities, measured with a ballistic pendulum.) Finally, he developed an apparatus that permitted him to fire a rocket in a vacuum. This demonstrated conclusively that a rocket could operate without reacting against the air, as would be required at high altitudes where the air density was thin. He showed with this device that the recoil of the rocket was "the result of an actual jet of gas, and was not due to reaction against the air,"[17] as many otherwise knowledgeable people at the time supposed.

Goddard had earned a salary of $1,000 per academic year as an assistant professor at Clark until 1916–17, when the president of the institution offered him an increase in pay to $1,500 "provided that your health permits you to carry full work."[18] Even with this pay increase, Goddard believed that he had "reached the limit" of the work on rocketry that he could do "single-handed, both because of expense, and also because further work will require more than one man's time," as he wrote to the president of the Smithsonian Institution in September 1916. Goddard asked whether the president—Charles Doolittle Walcott, whose title in fact was secretary—could recommend a source of funding for his work "upon a method of raising recording apparatus to altitudes exceeding the limit for sounding balloons," and indicated that a mass of one pound could be elevated as high as 232 miles by a rocket weighing 89.6 pounds. He emphasized the importance of his efforts to science, especially meteorology. He also referred to potential military uses, although he thought at this time that these were limited and would constitute a "loss to science." Characteristically, he urged that his research not be made public. He even stated his fear that if his device were not explained "in a manner satisfactory to myself," public opinion might "be influenced against the method even before it has been given a fair presentation."[19]

Two things are noteworthy about Goddard's letter to the Smithsonian. First, his inclination toward secrecy is apparent. Second, although his cherry-tree epiphany showed an interest in spaceflight, here he spoke only of the more plausible goal of reaching extreme altitudes for meteorological and other research. This marked a characteristic difference between the staid scientist from New England and the less cautious Oberth, the German who was later to write boldly of spaceflight even though doing so was no more respected in scientific circles in Germany than in the United States. Yet Goddard was just as interested in spaceflight as Oberth.[20]

In 1920 Goddard submitted a report to the Smithsonian on using a rocket to investigate space in a spirit similar to Oberth's later writings, and in it he even urged an appeal for public support.[21] However, this report remained unpublished, as did an earlier one written in 1918 discussing such topics as the use of atomic energy and expeditions to the "regions of thickly distributed stars."[22] In short, although Goddard retained an interest in spaceflight throughout his life, he didn't publish much on the subject for reasons he stated in a letter of 1940 on the issue. There he suggested that the matter

Table 1.1. Grants to Goddard, Exclusive of Military Support

Source	Years Used	Amount
Smithsonian Inst., Hodgkins Fund	1917–21	$5,000
Clark University	1921	2,500
Clark University	1922	1,000
American Assn. for the Advancement of Science	1924	190
Smithsonian Inst., Cottrell Fund	1924–29	5,000
Smithsonian Inst., Cottrell Fund	1929	2,500 [a]
Carnegie Institution of Washington	1929–30	5,000
Daniel Guggenheim	1930–31	25,000
Daniel Guggenheim	1931–32	25,000
Smithsonian Inst., Hodgkins Fund	1932	250
Guggenheim Foundation	1933	2,500
Guggenheim Foundation	1934–35	18,000
Guggenheim Foundation	1935–36	18,000
Guggenheim Foundation	1936–37	20,000
Guggenheim Foundation	1937–38	20,000
Guggenheim Foundation	1938–39	20,000
Guggenheim Foundation	1939–40	20,000
Guggenheim Foundation	1940–41	20,000 [b]
Total		$209,940

Source: Goddard Papers, 190, 224, 233, 277, 322–23, 451, 469, 470–71, 472–73, 524, 528, 531, 663, 680, 726, 744, 832, 838, 851, 874, 930, 1028, 1061, 1168, 1264, 1352, 1423, 1433, 1443, 1466, 1502, and esp. 1557.

a. Cf. Goddard to Abbot, *Goddard Papers*, 663, and Merriam to Goddard, *Goddard Papers*, 726, suggesting that the amount was actually $5,000 rather than the $2,500 shown in *Goddard Papers*, 1557n, and entered here.

b. A Summary of Expenditures for 1940–1941 in *Goddard Papers*, 1423, shows an additional $3,000 as a special 1940–41 Guggenheim grant that is not included in the note on page 1557. Also not included is a special Guggenheim grant in September 1941 for $10,000 that was repaid on December 27, 1943. Perhaps the $3,000 was likewise repaid, but this is not explained in the *Papers*.

would not be "scientifically respectable" until a rocket had "risen far into, or above, the atmosphere; and it is to this end that I am devoting all my energies at present."[23]

Meanwhile, Goddard's initial appeal to the Smithsonian in 1916 produced a response from the secretary, who indicated an interest in "a number of problems" Goddard's method might solve. Walcott asked how much his device and experiments would cost. Goddard replied that he did not think he could reach heights of 100–200 miles "within a time as short as one year for less than $5,000." The result was a grant in 1917 from the Thomas George Hodgkins Fund for that amount.[24] This constituted but the first of many such grants, summarized in table 1.1. They made Goddard one of the best-funded researchers in the United States before World War II. (To give an idea of the magnitude of Goddard's funding in twenty-first-century terms, $5,000 in 1919 would have been worth almost $65,000 in 2006, while $20,000 in 1940 equaled over $280,600 in 2006.[25])

Before Goddard could make much further progress with his rocket development, World War I intervened. He then reached an agreement with the U.S. Army Signal Corps to develop rockets for military purposes. During 1917–18, first in Worcester and then in Pasadena, California, on land and in workshops belonging to the Mount Wilson Solar Observatory, Goddard and a number of assistants and machinists developed single-charge, recoilless, and multiple-charge rockets before the armistice on November 11, 1918, ended the war. The Army, with the assistance of one of Goddard's associates, Clarence N. Hickman, ultimately developed the recoilless device together with a shaped charge into the famous bazooka armor-piercing weapon used in World War II. (Hickman later earned a Ph.D. from Clark University and ultimately became chairman of the rocket section of the National Defense Research Committee during World War II, so this was one instance where Goddard influenced later developments in the field of rocketry.)[26]

Meanwhile, having wound up his wartime research efforts, Goddard proposed that the Smithsonian publish his findings about rockets. Characteristically, he apparently did so only under a threat by Professor Webster that he, Webster, would publish a paper on rockets if Goddard didn't.[27] The result was "A Method of Reaching Extreme Altitudes." It appeared as volume 71, number 2, of the *Smithsonian Miscellaneous Collections* for 1919, actually published in January 1920. In it Goddard set forth the problem as raising a "recording apparatus beyond the range for sounding balloons (about 20 miles)" and stated that he had found an "approximate method" to be necessary "in solving this problem, in order to avoid an unsolved problem in the calculus of variations," a method of proceeding more in keeping with engineering than science.[28]

His solution showed "that surprisingly small initial masses would be necessary" to launch an apparatus weighing one pound to "any desired altitude" so long as the rocket ejected gases at a high velocity and the great majority of its mass consisted of propellant material that would steadily diminish in weight as it burned. He went on to describe the experiments already summarized above and to explain the scientific importance of investigating the upper atmosphere. To launch a rocket this high required complete combustion of the propellants. They had to burn a "little at a time" in a combustion chamber that was small but strong," permitting high chamber pressures that would then exhaust through a smooth nozzle. Finally, he outlined in rather vague terms the principle of employing stages such that, when the initial stage had "reached the upper limits of its flight," a second stage would be fired while the first stage dropped back to Earth.[29]

Goddard illustrated these points in a series of equations and tables, along with calculations of the reduced density and resistance to the rocket's motion as the altitude increased. In the process of calculating the minimum mass needed to raise the apparatus to an infinite altitude, he mentioned the possibility of sending a small amount of flash powder to the dark surface of a new Moon and igniting it upon impact, permitting observation of the light from Earth with a large-apertured telescope.[30]

Although the paper dealt mostly with solid-propellant rockets, in a note at the end Goddard discussed the greater efficiency of hydrogen and oxygen as propellants and mentioned their other advantages, including higher velocities. But he noted their "difficulty of application." In a summary, Goddard pointed out, among other things: "A theoretical treatment of the rocket principle shows that, if the velocity of expulsion of the gases were considerably increased and the ratio of propellant material to the entire rocket were also increased," an enormous increase in range would result. This stemmed "from the fact that these two quantities enter exponentially in the expression for the initial mass of the rocket necessary to raise a given mass to a given height."[31]

In a separate conclusion he stated that, while the paper did not describe a working model, he believed the theory and experiments he had discussed settled all serious questions about reaching high altitudes. All that remained was to perform "necessary preliminary experiments" so that "an apparatus" could be constructed to "carry recording instruments to any desired altitude."[32] As Frank Winter has written, this "publication established Goddard as the preeminent researcher in the field of rocketry" and "was unquestionably very influential in the space travel movement."[33]

Thereafter Goddard spent most of the interwar period performing the preliminary experiments he referred to in his conclusion and trying to construct a rocket that would achieve a higher altitude than sounding balloons could reach. With the moral support of his wife Esther after he married her in 1924—she was then a secretary in the office of the Clark president—he conducted rocket research in comparative secrecy, helped only by a small team of technical assistants. His method of proceeding seems to have been strongly influenced by the publicity following the release of his paper in January 1920. Spurred by a Smithsonian press release that commented on the ignition of flash powder on the dark part of a new Moon, newspaper headlines from Boston to San Francisco trumpeted a rocket that would reach the Moon.

An editorial in the *New York Times* called his paper "A Severe Strain on Credulity." But the writer failed to understand Goddard's proof that a rocket could function in a vacuum. He ignorantly pontificated, "That Professor Goddard, with his chair in Clark College and the countenancing of the Smithsonian Institution, does not know the relation of action to reaction, and of the need to have something better than a vacuum against which to react . . . would be absurd." Despite this ill-informed skepticism, the Roaring Twenties witnessed a temporary fascination with the notion of a Moon rocket. Goddard became "the moon rocket man" to part of the press. Songs and ballads appeared on the subject. And Goddard received a deluge of mail for and against the idea, including letters from volunteers anxious to take rocket flights. As a result, the retiring New Englander became even less prone to publicize his work than before.[34]

His reclusive tendency probably grew stronger when his projections to the Smithsonian of reaching high altitudes in a short time span proved hopelessly overoptimistic. After experiencing frustrating problems with solid propellants, Goddard had switched in 1921 to liquids.[35] But it was not until March 26, 1926, nine years after his initial proposal to the Smithsonian, that he was able to achieve what has generally been recognized as the world's first flight of a liquid-propulsion rocket. He launched it at the farm of Miss Effie Ward, a distant relative, on Pakachoag Hill in Auburn, Massachusetts.[36] Like the Wright brothers, Goddard had a very short first flight, 41 feet vertically and 184 horizontally.[37]

As with the Wrights' achievement in aviation, the distance traveled did not diminish Goddard's feat, but the nine years this had taken him did temper his optimism, if only slightly. When astrophysicist Charles Greeley Abbot of the Smithsonian congratulated him on this first flight and inquired what funding would be required to build a high-altitude rocket, Goddard replied, "I believe that the cost would be at least $2500 and the time required from eight months to a year."[38]

In fact, as table 1.1 has shown, the funds Goddard ultimately required were almost eighty-four times that figure, and when he finally turned from development of high-altitude rockets to wartime work in 1941, the highest altitude one of his rockets had reached (on March 26, 1937) was estimated at 8,000 to 9,000 feet—less than two miles, still far short of the twenty miles that high-altitude balloons reached.[39]

Why had his efforts over more than twenty years not been more success-

Figure 1. Robert H. Goddard observing his New Mexico launch site from his launch control shack. The launch control panel is next to his left hand. Courtesy of NASA.

ful on his own terms? One answer came from Theodore von Kármán, the brilliant aerodynamicist who was one of the founders of the Jet Propulsion Laboratory in Pasadena. Von Kármán had offered to cooperate with Goddard during the late 1930s; he believed Goddard "was an inventive man [who] had a good scientific foundation, but he was not a creator of science and he took himself too seriously." According to von Kármán, Goddard should have taken others into his confidence. Had he cooperated with others, von Kármán said, "I think he would have developed workable high-altitude rockets and his achievements would have been greater than they were."[40]

Frank J. Malina reached a similar conclusion. The coincidence of their views is not surprising, since Malina was a graduate student of von Kármán's and an early member of the Guggenheim Aeronautical Laboratory, California Institute of Technology (GALCIT), which von Kármán directed. Malina came to play an important role in American rocket development. In preparation for his efforts, he had arranged in 1936 to visit Goddard near Roswell, New Mexico, where the Clark physicist was pursuing his own research. Malina related Goddard's continuing bitterness toward the press for its reaction to his first paper in 1920. He also recalled his own impression that Goddard regarded rockets as his private preserve. Goddard's attitude, Malina wrote, "caused him to turn his back on the scientific tradition of communication of results through scientific journals, and instead he spent his time on patents."[41] In Malina's view, "Goddard had not succeeded in constructing a successful sounding rocket because he had underestimated the difficulties involved—the day of the isolated inventor of complex devices was over."[42]

Despite their confusion of science and engineering, these two complementary assessments are valid as far as they go. During the years of his support by the Daniel and Florence Guggenheim Foundation, Goddard had pursued rocket research in the isolation of New Mexico with a handful of technical assistants. He had conferred with Abbot, Daniel Guggenheim, and the noted aviation pioneer Charles Lindbergh, but for most of his innovations he relied on his own resourcefulness, his knowledge of physics, and the technical proficiency of his assistants in fabricating components for his various rocket models.[43]

Perhaps more important than Goddard's failure to create a high-altitude sounding rocket was the effect of his secretiveness upon other American rocket projects. Although his example influenced others in the United States,[44] and his 1919/1920 paper undoubtedly stimulated many to pursue research in rocketry, Goddard failed to inspire a cohesive group of followers to pursue his ideas and improve upon his techniques, as Oberth did in Germany during this period. As a result, on the whole Goddard's impressive technical and theoretical achievements did not contribute significantly to American rocketry.[45]

This was true despite the fact that, at the urging of Guggenheim and Lindbergh, he did publish a second paper in the *Smithsonian Miscellaneous Collections* in 1936, entitled "Liquid-propellant Rocket Development." Here Goddard addressed liquid propellants much more explicitly than in his longer and more theoretical "Method of Reaching Extreme Altitudes." He noted their higher energy, an important advantage over powder rockets. He discussed ways of feeding the liquids into the chamber and, if only very vaguely, the problem of cooling the chamber so the heat of combustion would not burn through it. He mentioned the use of gyroscopically controlled vanes to obtain stabilized vertical flight and the need for lightness in rocket construction. He also discussed some details of the rockets he had developed and included many illustrations. But in general the rather low level of detail and Goddard's failure to discuss many of the problems encountered at every step

and Frank Malina, and/or if he had employed a more systematic, step-by-step procedure in his tests, designing each test to perfect or at least improve a single, faulty component instead of making several "improvements" on the rocket as a whole all at once.

Given Goddard's nature, a systematic procedure probably was the more likely option. He was not totally averse to cooperation, but he seems to have been too much an independent inventor to work as part of a team. Lehman records Esther Goddard's comment after her husband's death that "perhaps we were wrong, perhaps we should have joined some team, but it didn't seem so to us at the time. I'm not sure Bob knew how to give up control of his rocket." Lehman also quotes Harry Guggenheim's characterization of Goddard as "one of those lone wolves who didn't want to hunt with the pack."[56]

Substantiating that position of Goddard's wife and Harry Guggenheim is a letter by Professor William F. Durand. Retired as head of the Department of Mechanical Engineering at Stanford, a recognized authority on aerodynamic theory, a longtime member and former chairman of the National Advisory Committee for Aeronautics, and during World War II the chairman of its Special Committee on Jet Propulsion, Durand commented in 1941: "Mr. Goddard does not work comfortably with other organizations, and . . . if a contract is arranged with him [to develop jet-assisted takeoff devices for

Figure 4. Goddard towing a rocket behind a Ford Model A to his launching tower northwest of Roswell, New Mexico, in the 1930s. Courtesy of NASA.

the Navy], it should, I believe, be with him alone and he should be left to work it out in his own way without expectation of cooperation with other organizations."[57]

Since this trait of Goddard's presumably was so well established by the 1930s as to be unalterable, why did he not compensate by being more systematic in his experimental practices? Perhaps the reason was the uncertain continuation of his grants and his own eagerness to bring about space travel.[58]

Or perhaps the reason was not simple but reflected the man's complex psychology. A clue in this regard may be his earlier mentioned Mars vision in the cherry tree. When his beloved grandmother lay dying, according to Lehman, Goddard felt compelled to tell her how his rocket would pose great dangers to mankind but would also offer tremendous possibilities. Two days after she died, he recorded in his diary that he had gone over to see her remains in the parlor of their farmhouse and had—by coincidence—noted it was Anniversary Day and that he "Saw cherry tree."[59]

We know that Goddard was optimistic about his chances for success, and Lehman has recorded his resiliency in the face of repeated problems and failures.[60] Perhaps, in line with his fixation on the cherry tree epiphany of his youth, Goddard had developed an almost mystical belief in the success of his venture that induced him to take scientific shortcuts.

The mixture of scientific and mystical elements in his character is confirmed by an essay he wrote in 1925 entitled "The Doctrine of Recurrence." In it, he began with scientific sorts of observations: "It is now generally regarded that the universe as a whole is infinite in time; that there is an interchange of matter and energy, even electromagnetic radiant energy, and that there is rebirth of elements as well as disintegration." From these, he concluded in a mystical vein that—however long the interval between these events—there would be times when the universe would repeat previous configurations and all events would reproduce themselves. Realization of this, he said, and the existence of "this kind of immortality for everyone" should lead people to avoid "sordid, narrow, and unidealistic" behavior because the prospect of reliving them would be unpleasant.[61]

Several other notations in his diaries point in this same direction. For example, in August of 1905 while he was on summer break from Worcester Polytechnic Institute, he read about reincarnation, theosophy, and Buddhism. On February 2 the following year he wrote, "Read the occult world in afternoon." On February 6, 1937, he made the entry "Morning in the desert—when the impossible not only seems possible, but easy." At the end of the diary for 1940 he quoted Rudyard Kipling, "When your Daemon is in charge, do not try to think consciously. Drift, wait, and obey." And on October 28, 1944, he copied another quotation, this one unattributed, "Logic is only the art of going wrong with confidence."[62] Individually, none of these entries would be particularly noteworthy, but taken together with the essay on recurrence and the influence of his epiphany in the cherry tree, they suggest a strong mystical aspect of his makeup that may have affected his decision making.

Another facet of Goddard's life, his tuberculosis, might paradoxically have reinforced an inclination toward a mystical belief in the ultimate success of his work. When the disease was first diagnosed in 1913, he later learned, the family doctor had given him only two weeks to live.[63] Yet he survived. Advised repeatedly thereafter to avoid overexertion or suffer the consequences, he persisted in ignoring this medical counsel. Referring to the suggestion that he enter a Swiss sanitarium, his wife described his fatigue in 1938 and characterized him as "A man who 'should be in bed in Switzerland' coming in at 7:30 at night after a day that began at 3 in the morning. And yet," she added, "when I remonstrate, he looks as though I had insulted him."[64]

If Goddard could thus defy the dictates of medical science with relative impunity, might he not have believed that he could take shortcuts with the methodology dictated by physical science? Clearly he was a complex and enigmatic man, and we probably will never know the answer to this question.

On June 14, 1945, following years of what he thought was laryngitis, Goddard experienced a fit of choking. Two days later, he and his wife traveled to Baltimore to see a surgeon. The doctor found a malignant growth and on June 19 removed it along with Goddard's larynx. Less than two months later, on the morning of August 10, 1945, the rocket pioneer died, just four days after the bombing of Hiroshima.[65] His wife recorded as the concluding entry in her "Excerpts" from his diary, "Darling Bob slipped away . . . end of a love story."[66]

While his truly remarkable achievements surely earned him a share of the title Father of Modern Rocketry,[67] Goddard failed to live up to his full potential. Nevertheless, his dedication, despite tuberculosis, yielded an impressive array of accomplishments. His 214 patents, many of them submitted by Esther Goddard after his death, led to a settlement in 1960 by NASA and the three armed services of $1,000,000 for use of more than two hundred of them covering innovations in the fields of rocketry, guided missiles, and space exploration. Half of the money went to the Guggenheim Foundation and the other half to Mrs. Goddard.[68] Throughout his adult life, Goddard kept notebooks that included whimsical reflections and comments, as well as details of his rocket research. In 1944–45 he condensed the more important of these experimental details, which his wife and G. Edward Pendray edited and published in 1948.[69] By then his achievements had already been surpassed by the Germans under von Braun and the Malina rocket team at the Jet Propulsion Laboratory in Pasadena. So his own observations were already mostly of historical interest. But they attest to a remarkable imagination. Perhaps if Goddard could have known at the beginning of the 1930s that Malina's group and the Germans under von Braun would eclipse his achievements, he would have shared details of his research earlier, when it would have given his country an important head start. As it was, according to noted rocket expert George Sutton, the large liquid-propellant rocket engines "developed later by General Electric, Rocketdyne, and Aerojet" came to be "designed and produced without the benefit of the work done by Goddard."[70]

Oberth

In early May of 1922, Goddard received a letter in awkward English from Hermann Oberth, who was then totally unknown to him. At the time, Goddard was in the midst of liquid-propellant rocket research and development and had recently announced his engagement to his future wife, Esther Kisk.[71] Oberth's letter highlights the differences in the approach to their work of these two rocket pioneers. Dated May 3 from Heidelberg, Oberth's missive read:

> Already many years I work at the problem to pass over the atmosphere of our earth by means of a rocket. When I was now publishing the result of my examinations and calculations, I learned by the newspaper, that I am not alone in my inquiries and that you, dear Sir, have already done much important work at this sphere. In spite of my efforts, I did not succeed in getting your books about this object. Therefore I beg you, dear Sir, to let me have them. At once after coming out of my work I will be honored to send it to you, for I think that only by common work of the scholars of all nations can be solved this great problem.[72]

Although Goddard's reply was reportedly lost to Allied bombings in World War II, he did send Oberth a copy of his 1919 paper.[73] Still, he hardly shared the view "that only by common work of the scholars of all nations" could the problems of rocketry be solved.

What had made Oberth so much more willing to cooperate with others and to expose even his most exotic ideas without fear of jeopardizing his reputation as a scientist? The answer is not entirely clear, because in some ways his background and early experiences resembled those of Goddard. He was born almost twelve years after Goddard on June 25, 1894, in the partly Saxon-German town of Hermannstadt, Transylvania, which was then part of the Austro-Hungarian Empire but experienced a name change to Sibiu after it became part of Romania in 1918.[74] At the crossroads of cultures in mountainous Transylvania, the Saxon Germans had settled in the middle of the twelfth century, and by the sixteenth century they constituted a solid group of usually Lutheran burghers in the towns of the region. Transylvania had been part of Hungary when the Saxons arrived but became an autonomous principality within the Ottoman Empire during most of the sixteenth and seventeenth centuries before returning to Hungary near the end of that period.[75]

Surrounded by Romanians and Hungarian Magyars, the Saxons retained a strong ethnic identity that one contemporary of Oberth described as "more thoroughly Teutonic" than that of Germans in the Fatherland. They "clung stubbornly, tenaciously, blindly to each peculiarity of language, dress, and custom." Their principal source of strength lay "in their schools, whose conservation they jealously guard, supporting them entirely from their own resources, and stubbornly refusing all help from Government."[76]

Such a background might have tended to make Oberth self-confident from the success of his ethnic Saxon forebears in preserving their identity among the ruling Magyars and the subject Romanians. If so, his family background appears to have strengthened his self-confidence and assertiveness. Grandfather Krasser had the audacity to predict such a seemingly unthinkable event as a Moon landing within a century; he was also a freethinker, socialist, and poet as well as a doctor. Hermann's mother told the young man about his grandfather and said her son resembled the old man in thinking, speech, and interests. The boy's father, Julius Gotthold Oberth, was a highly capable physician, which may also have contributed to the young Oberth's strong self-image.[77]

In any event, Oberth himself remembered one place he lived, Schässburg (later, Sighişoara), as "a somewhat patriarchal town of some eleven thousand inhabitants," about half of whom were German.[78] Like Goddard—and also the Russian rocket researcher Tsiolkovsky—Oberth was significantly influenced by science fiction. The Transylvanian related in 1959, "At the age of eleven, I received from my mother as a gift the famous books, 'From the Earth to the Moon' and 'Travel to the Moon' by Jules Verne, which I had read at least five or six times and, finally, knew by heart." In a separate reminiscence published in English in 1974 he added, "I was fascinated by the idea of space flight, and even more so, because I succeeded in verifying the magnitude of the escape velocity" Verne had used in his writing.[79]

Oberth was able to make these calculations despite a weakness in his education in physics and mathematics. He had already begun attending the Bishop Teutsch *Gymnasium* (humanistic secondary school) in Schässburg, which followed the German tradition for such schools and emphasized Latin and Greek, philology, and history instead of the sciences. As he put it colorfully, his school "resembled a car which has only small headlights in front, but which illuminates very brightly the way it has already traveled, thus helping light the way for others." In school, the young man said he had learned "infinitesimal" (integral) calculus and the laws of gravity but had to teach himself differential calculus.[80] Thus, like Goddard, Oberth was partly self-educated.

Throughout his life Oberth exhibited an independent streak. According to biographer Hans Barth, on Oberth's first day of primary school he was reprimanded by his teacher, "Hermann, in school one must obey!" Oberth's reply was characteristic of his response to authority throughout his life: "Why? I don't obey my mother either." In secondary school the young Oberth breezed through physics and mathematics but, partly because of a bout with scarlet fever, had to be tutored in history and Hungarian. As he later reported, "our backward school system also did me something good. I became quite immune against the sayings of so-called authorities who were of a different opinion from my own."[81] This attitude is far removed from Goddard's concern about jeopardizing his scientific reputation by releasing his findings before he had achieved experimental results. While Goddard was arguably as bold as Oberth in his conceptions, he certainly was not as intrepid in publishing them.

Once the self-assured young doctor's son had graduated from secondary school on his eighteenth birthday in 1912, he prepared to study medicine at the University of Munich. His mother was convinced he had no qualifications for medicine but a real gift for physics and mathematics. However, Dr. Oberth apparently insisted that his son follow in his own footsteps. Hermann at first did not seem to object, but before he went off to university in Germany, as was the practice of most Saxon Germans in Transylvania, his father sent him to Italy to recuperate from his scarlet fever.[82]

Like Goddard, Oberth therefore had his education interrupted by illness, though not to the same degree. A larger interruption came with World War

I. Oberth had begun his studies at the University of Munich in 1913. Besides the premedical courses, he attended the lectures of the noted professor of theoretical physics Arnold Sommerfeld at the university and those of the Swiss astrophysicist, meteorologist, and teacher of aeronautics Robert Emden at the Munich Technical Institute. But with the outbreak of war, he was inducted into an infantry regiment and sent to the Eastern Front, where he suffered an abdominal wound in February 1916 and went back to Schässburg to become an ambulance sergeant and orderly for the duration of the war.[83]

In 1917, according to Oberth, he submitted to the German War Department his design for a long-range missile using an alcohol-water mixture and liquid air as propellants. It featured steering by rudders controlled by a gyroscope and servomotors. The design used a potentiometer and a system for generating electricity proportional to acceleration to close the propellant valve via an ammeter and thus shut off the propulsion at the proper time. It also featured regenerative cooling (by circulation of a propellant around the outside of the combustion chamber) to keep the engine from burning through. In short, it supposedly exhibited many features later used successfully on rockets and missiles, although according to Oberth the design itself was lost along with all of his early notes. In a letter accompanying his design, Oberth said he explained why rockets, so far, had not performed better. The reply he said he received from Berlin in 1918 ignored his explanation and declared that experience had shown rockets would not travel farther than five kilometers (seven in another account). The authorities from Berlin stated that considerably longer distances could not be expected, given the Prussian thoroughness in such tests.[84]

On July 6, 1918, Oberth married Mathilde Hummel, an orphan living with her siblings and contributing to the family upkeep through work as a housekeeper and then in a boutique. She brought to the marriage a practical sense and social graces lacking in her husband, who required solitude for his reflections and had the "habit of suddenly breaking off a conversation in the middle and becoming quite oblivious to his surroundings—a sign that his restless mind was on the track of a new idea." His wife, variously called Tilly or Tilla, also had to take all responsibility for raising their children, whom he essentially neglected. When he was not away for education, rocket projects, and employment, he closeted himself in his study, seeing the children only at dinner.[85]

Meanwhile, after the war, Oberth resumed his education, this time in mathematics and physics rather than premedicine. He studied at the University of Cluj (formerly Klausenburg) in Transylvania in 1919, then in Munich in 1920–21 at both the university and the technical institute, at Göttingen University during 1920–21, and the University of Heidelberg in 1921–22. While attending classes, Oberth continued his rocket designs and his development of spaceflight theory. By the fall of 1921 he had completed a manuscript on his designs and theories. He submitted it to the University of Heidelberg as his doctoral dissertation, but the famous astronomer Maximilian F. J. C. Wolf, discoverer of more than a hundred asteroids and many nebulae plus a comet that bears his name, would not accept the study as a dissertation because it did not deal with astronomy. Wolf did pay Oberth the compliment of calling the manuscript an "ingenious and scientifically irreproachable study," according to Barth. Oberth next submitted his manuscript to Philipp Lenard, who in 1905 had won the Nobel Prize in Physics for work on cathode rays, who like Oberth was from the Austro-Hungarian Empire, and who soon was to become one of the earliest supporters of Nazism. As Barth quoted him, evidently from Oberth's recollections, Lenard pronounced the manuscript "an amazing achievement, but unfortunately not classical physics." Consequently, unlike Goddard, Oberth never received a Ph.D. (other than honorary ones), although he learned about physics and related disciplines from many other distinguished scientists including the "father of aerodynamics," Ludwig Prandtl, and the geophysicist and meteorologist Emil Wiechert.[86]

While undaunted by the rejection of his manuscript, Oberth "refrained from writing another" dissertation on a more acceptable and conventional topic. Goddard had done just that, producing a dissertation, "Conduction of Electricity at Contacts of Dissimilar Solids," on a valid subject in which he had no particular interest.[87] Instead, Oberth determined to prove to the world that he was a greater scientist than some of his teachers, even without a doctoral degree. He attempted to get his manuscript published, but in the middle of hyperinflation in Weimar Germany, he was unsuccessful at first. Returning to his native land, now part of Romania, he began a year of practice teaching at the teachers college in Sighişoara and underwent his teaching examination at the University of Cluj in mathematics, pedagogy, Romanian, and German. In 1923 he received his diploma from that institution, accompanied by the title of professor. In the fall of that year he began a teaching career that lasted, with leaves of absence, until 1938. He taught mathematics, physics, and chemistry at the secondary school he had attended, plus mathematics at the teachers college. Then in February 1925 he started teaching at the *Gymnasium* in Mediaş, also a small town in Transylvania not far from Sighişoara.[88]

Meanwhile, in October 1922 a friend informed Oberth that he had finally found a publisher for the manuscript. The Oldenbourg publishing firm in Munich agreed to print it if Oberth would pay the costs. He had no money, but because publication was so important to him, Tilla gave him what she had managed to save. *Die Rakete zu den Planetenräumen* (the rocket to interplanetary space) appeared in 1923. Although Goddard always suspected that Oberth had borrowed heavily from his 1919/1920 paper,[89] in fact Oberth's account bears little resemblance to Goddard's study. Not only is Oberth's book much more filled with equations, but it is also considerably longer than Goddard's paper—some 85 pages of smaller print compared with 69 in the version of Goddard's paper reprinted by the American Rocket Society in 1946.

In addition, Oberth devoted far more attention to liquid propellants and multiple-stage rockets than Goddard did. Oberth also set forth the basic principles of spaceflight to a greater extent than Goddard had in a work much more oriented to reporting on his experimental results than theoretical elaboration. Although Oberth conceded his lack of experimental investigation of the matters he described, he discussed such topics as liquid-propellant rocket construction using both alcohol and hydrogen as fuels, the use of staging to escape Earth's atmosphere, the use of pumps to inject propellants into the rocket's combustion chamber, employment of gyroscopes for control of the rocket's direction, physiology in space, chemical purification of the air in the rocket's cabin, spacewalks, microgravity experiments, space telescopes, and the concepts underlying a lunar orbit, space stations, and reconnaissance satellites.

Toward the end of his book, as he had promised, Oberth discussed Goddard's paper. While admitting that Goddard had done considerable experimentation, he said that he, Oberth, had tried to present primarily a theoretical treatment of the issue, and for that reason the two works complemented one another. He also accepted the necessity of experimental confirmation of theory in rocket work. Oberth noted Goddard's consideration of hydrogen and oxygen as propellants but pointed out the older scientist's failure to follow up on his own comment because of the difficulty of obtaining them. Oberth admitted, too, the greater understandability of the American's calculations and equations, as well as their general similarity to his own, but he added that the very difficulty of his own calculations made them perhaps more useful. Still, he generously recommended Goddard's paper to anyone able to read English because the American had conscientiously carried out his experiments and written about them in an easily understood and interesting fashion. Oberth concluded his commentary by emphasizing his own work's independence of Goddard's, stating that he had begun working on rocketry in 1907 and had completed a simple design for a rocket as early as 1909.[90]

Due to Germany's inflation and no doubt also to the book's complexity for the lay reader, the small initial edition did not sell out quickly, but it did gradually generate a spaceflight movement that necessitated a second printing in 1925. Perhaps because of its comparative difficulty and its open treatment of human spaceflight, it had a much more profound effect on the world than Goddard's more empirical descriptions. An early disciple of sorts was Max Valier, an Austrian who wrote for a popular audience and had obtained a copy of Oberth's book in January 1924. He immediately wrote to the spaceflight pioneer proposing that he, Valier, write propaganda for Oberth's cause and that they collaborate on a brochure with Valier writing the easier-to-understand parts. Oberth accepted, telling Oldenbourg that he had already thought of writing something popular but believed Valier could do that better than he. In 1924, with some technical help and criticism from Oberth, Valier composed a number of articles for popular journals and a book, published by Oldenbourg, entitled *Der Vorstoss in den Weltenraum, eine technische Möglichkeit* (The thrust into outer space, a technical possibility). The 4,000 copies of the book quickly sold out, and between 1925 and 1929 the work appeared in five more editions, the last two retitled *Raketenfahrt* (Rocket travel).[91]

According to several sources, Oberth's first book directly inspired Wernher von Braun to study mathematics and physics, so necessary for his later work. Von Braun had already been interested in rocketry but was failing math and physics. However, in 1925 he had seen an ad for Oberth's book and ordered a copy. When he confronted what were to him its complex formulas, he took it to his math teacher, who told him that the only way he could understand what Oberth was saying was to study his two worst subjects. He did,[92] and ultimately earned a Ph.D. in physics. Without the spur Oberth evidently provided, who knows whether von Braun would eventually have assumed positions of leadership in the German and American rocket programs?

Similarly, von Braun's boss in Germany, Walter Dornberger, wrote to Oberth in 1964 that reading his book in 1929 had opened a new world

Figure 5. Hermann Oberth, in foreground, with officials of the Army Ballistic Missile Agency in Huntsville, Alabama, in 1956. Behind Oberth are *(clockwise)*: Ernst Stuhlinger *(seated)*; Maj. Gen. Holger N. Toftoy, commanding officer; and Wernher von Braun, director, Development Operations Division. NASA Historical Reference Collection, Washington, D.C. Courtesy of NASA.

to him. And according to Konrad Dannenberg, who had worked at Peenemünde and come to this country in 1945 with the rest of the von Braun team, many members of the group in Germany had become interested in space through Oberth's books. Also in response to Oberth's book, in 1927 the German Society for Space Travel was founded to raise money for him to perform rocket experiments. Oberth served as president in 1929–30, and the organization provided considerable practical experience in rocketry to several of its members, including von Braun. Some later served under von Braun at Peenemünde, although they constituted a very small fraction of the huge staff there—some 6,000 by mid-1943.[93]

By contrast with this sort of influence, magnified by the promotional efforts of Valier and of such other publicists as spaceflight enthusiast Willy Ley and novelist Otto Willi Gail, Oberth faced very considerable skepticism in the academic community. Not only was his work rejected as a doctoral dissertation; once published, it was also refuted by some of Germany's scientific and engineering elite. Already in 1924 a Professor Riem, who held a Ph.D., attacked Oberth in the popular science magazine *Der Umschau* (Survey, or Review), arguing (as had the errant *New York Times* editorial against Goddard) that in the upper atmosphere his rocket would not work because the air was too thin for it to push against. Another professor named Spies, writing in the same publication and citing his experience as a ballistics expert, stated that even when fired out of a cannon, rockets would not climb higher than five kilometers (about three miles).[94]

Without mentioning names, Oberth subsequently referred to other respected figures who disputed his findings, but two more examples will suffice to depict the opposition he faced. In 1927 Professor Hans Lorenz, who was an expert on refrigeration, materials testing, and ballistics at the Technical Institute in Danzig, published a refutation of Oberth's calculations about rockets achieving the velocities necessary for space travel. Later the Wissenschaftliche Gesellschaft für Luftfahrt (Scientific Society for Aviation) invited both men to speak. According to Theodore von Kármán, after Oberth made an enthusiastic presentation on the possibility of escaping from Earth's gravitational pull, the "distinguished" Professor Lorenz argued that this was "beyond engineering capability." It would require so much of the best available fuel that the rocket would weigh "34 times as much fueled as empty." Von Kármán said he spoke out in defense of Oberth. So the Romanian secondary teacher was not without his eminent defenders. But the Association of German Engineers, which had published the comments of its second chairman, Lorenz, refused to publish Oberth's reply, claiming that its journal lacked space for his comments. Oberth was able to demonstrate in print that Lorenz's calculations were wrong, but he had to do so in the journal of the German Society for Space Travel, which was undoubtedly much less prestigious than the journal of the engineering association.[95]

That this was true is suggested by Oberth's comment in the late 1960s that his "very highly esteemed colleague, Professor Klaus Oswatisch, is now a member of the International Academy of Astronautics, although 15 years ago he maintained that it would be unworthy of a serious scientist to occupy himself with astronautics, especially manned space travel." If this was still the case in the early 1950s, it is easy to imagine how little prestige the German Society for Space Travel enjoyed among most scientists and engineers at the time of Lorenz's criticisms. In any event, Oberth also mentioned a Dr. von Dallwitz-Wegner, who along with a Professor Dr. Kirchberger argued essentially the same point as Lorenz—that the propellant for a rocket would not even contain sufficient "energy to lift its own weight beyond the earth's field of gravitation," let alone lift the rocket as well.[96]

All of this was remarkably analogous to the situation that faced spaceflight enthusiasts in the United States about the same time. Again, Riem's comments about Oberth were almost identical to the ones made about Goddard in the *New York Times*. And as late as the fall of 1938 at a meeting of a committee sponsored by the National Academy of Sciences in the United States there was a discussion of two research problems for the Army Air Corps. One had to do with visibility for bomber aircraft under icing conditions and the other involved rocket-assisted takeoff for heavy bombers. Jerome Hunsaker of the aeronautics department at the Massachusetts Institute of Technology said MIT would work on the problem of visibility while von Kármán and Caltech could take the "Buck Rogers job," meaning an effort in the realm of fantasy. While Caltech did undertake rocket research at what became the Jet Propulsion Laboratory in Pasadena, von Kármán wrote that "the word 'rocket' was in such bad repute that for practical reasons we decided to drop it from our early reports and even our vocabulary," hence the name Jet (rather than Rocket) Propulsion Laboratory.[97]

Even at Caltech and even in 1945, it was not respectable for physicists to discuss the prospect of a trip to the Moon. One professor reportedly proved to a seminar, very much in the spirit of Lorenz and von Dallwitz-Wegner in Germany, "that it would take a rocket as big as Mt. Everest to reach the moon." And in similar fashion to Oswatisch in Germany, Dr. Vannevar Bush, who headed the National Defense Research Committee in the United States during World War II, once reportedly said to Caltech physicist and Nobel

Prize winner Robert Millikan and von Kármán, "I don't understand how a serious scientist or engineer can play around with rockets."[98]

While the reactions of engineers to the idea of spaceflight were quite similar in the United States and Germany, the response of Oberth to the authorities' disdain for his ideas was diametrically different from Goddard's. Oberth continued to engage in popularizing spaceflight. In 1929 he produced a considerably revised and expanded edition of his book, now called *Wege zur Raumschiffahrt* (Ways to Spaceflight). It was more popularly written than *Die Rakete*, with the highly technical material indicated by a vertical line in the margin to show it was intended only for specialists. Oberth also served as a technical advisor to the silent movie *Die Frau im Mond* (called in English *The Girl in the Moon* and *By Rocket to the Moon*). So great had been the interest in spaceflight aroused by Ley, Valier, and the others that Universum Film Aktiengesellschaft (UFA) and the already celebrated director Fritz Lang decided to make this movie. For additional publicity, Lang and UFA even provided Oberth with funds to build a liquid-propellant rocket to be launched when the movie premiered. But Oberth possessed few engineering skills and failed to get the rocket ready in time. He was injured in an explosion during the effort and finally returned to Romania.[99]

The movie provided inspiration for other recruits to rocketry and spaceflight. Krafft Ehricke—who later worked at Peenemünde and then came to the United States, where he was generally credited as the father of the Centaur upper stage, the first vehicle to use liquid hydrogen as a fuel—saw the movie at the age of twelve and was so impressed that he saw it eleven more times and began to read materials he needed to understand Oberth's book. Another youngster the movie won over was Richard Gompertz, at age nine. Gompertz was born in Germany and educated at the Institute of Technology in Berlin but ended up serving in the United States Army Air Forces during World War II. He continued working for what became the Air Force in 1947 and eventually became chief of the Rocket Engine Test Laboratory at Edwards Air Force Base in California.[100]

In addition, Lt. Col. Karl Emil Becker, a doctor of engineering in the German Army Ordnance Office's Weapons Testing Division, was evidently influenced by Oberth's and others' publicity for rocketry. This led him to investigate the military potential of liquid-propellant rockets. His interest and that of Capt. Walter Dornberger led to the V-2 program.[101]

Meanwhile, Oberth returned to teaching in Romania. He participated in further rocket experiments there and was later employed by the German Air Force in Vienna and then by the German Army in Dresden and eventually at Peenemünde in rocket work, but his direct contributions were no longer seminal. In Vienna he set up an experimental station for rockets and carried out some combustion tests, but nothing significant came of the work. In Dresden, where he transferred in 1940, he worked on a liquid-propellant pump for the V-2, but meanwhile another one had already been developed for Peenemünde. When he finally obtained German citizenship and moved to Peenemünde, the V-2 had already been essentially designed; there were still details to be worked out, and even after the missile's first successful launch in October 1942 there were many problems to fix, but there was nothing significant for Oberth to do. Neither trained nor inclined to be an engineer, he regarded the enormous missile as too complicated. He began looking at all available patents to see if they could be used for rocket development. He made some recommendations but said they were never followed. He also conducted a study of the best methods of employing multistage rockets, but nothing came of that during the war.

In early 1943 he transferred to the supersonic wind tunnel, also at Peenemünde. At the end of 1943 he moved to the Westfälisch-Anhaltische Sprengstoff AG in Reinsdorf near Wittenberg, where he worked until the end of the war on a solid-propellant antiaircraft rocket employing ammonium nitrate. However, owing to the short supply of this chemical because of Allied bombing, his project never reached fruition. Even the admiring biographer Hans Barth stated that the spaceflight pioneer's presence at Peenemünde had not contributed to the rocket program there, though he added justly that Oberth's earlier theoretical and technical writings had made that program possible.[102]

Younger than Goddard, and blessed with better health, Oberth still had a long life ahead as the war ended in Germany. But nothing he did in his subsequent career made a significant impact on rocketry. The eccentric spaceflight pioneer lived to be ninety-five before he died on December 28, 1989.[103]

Comparisons

Both Goddard and Oberth were remarkable individuals. They shared an interest in rocketry and spaceflight that few of their contemporaries recognized as respectable or possible, and both were unusually gifted in many ways. Both men made important, if quite different, contributions to space and missile developments—particularly, as it turned out, those in the United States, although their influence was much more widespread than that. Goddard's launching of the first liquid-propellant rocket was something of a turning point in history, and when the launch became widely known, it assumed a symbolic and inspirational importance. More concretely, Goddard's variable-thrust rocket engine was also a noteworthy, if limited, contribution to rocket aircraft research and development. But Goddard's other remarkable innovations remained mostly without direct influence upon further rocket developments because of his failure to publish the details of his research. He might well have published them, of course, if he had achieved the success he sought and expected in launching one of his rockets to the sort of altitude reached by the WAC Corporal (see chapter 3) a few months after his death. But Goddard was prevented from reaching such altitudes not only by his general unwillingness to cooperate with other physicists and engineers but also by his failure to follow carefully controlled, step-by-step engineering procedures.

If Goddard was somewhat lacking as an engineer, Oberth was even more so. While Oberth did some rocket development work of his own—the publicity project for the movie *Die Frau im Mond* and various efforts on the side in his native Transylvania during the 1930s and later at La Spezia, Italy—clearly he had no gift for engineering, and these efforts were insignificant. Goddard did have many of the technical skills necessary for engineering but failed to follow the correct principles.

Far more important, Goddard's interest in patents (understandable, given his background, and fruitful in supporting his wife after he died) as well as his caution and concern for respectability kept him from having greater influence. Oberth had none of Goddard's inhibitions, and his writings led to a robust spaceflight movement in Germany. Without that, there might have been no V-2 and its successor rockets and missiles down to the Saturn V in the United States. Unlike Goddard, Oberth was not deterred by criticism from publishing his important theoretical findings. As a result, he arguably had a more profound impact upon future developments than Goddard did. Through the spaceflight movement he was instrumental in launching, he influenced a number of important rocket engineers such as von Braun, Dannenberg, Ehricke, and Gompertz who later developed rockets in the United States. He thus helped lay the foundations for the later achievements of the National Aeronautics and Space Administration, in particular. Both Goddard and Oberth exemplified the pronouncement of Goddard in a high school graduation speech "that the dream of yesterday is the hope of today and the reality of tomorrow."[104] But ironically it was Oberth who apparently made the more important, if indirect, contribution to the realization of both men's dreams.

Alternatively, one could argue that Goddard and Oberth had a complementary effect on American rocketry. Although Goddard did not achieve the altitudes he sought, he did pioneer many technologies later developed independently by others. While the details of his use of these technologies were not available at the time when they could have been useful to other rocket developers, his general achievements were known to many engineers and apparently helped to inspire them. Oberth had few such engineering accomplishments to pass on, but he wrote about rocket theory in greater detail than Goddard did. Thus, at least for those who could read German, he provided a basis in theory for further development of actual rocket designs. According to this view, both men in different ways influenced the implementation of their common dreams.[105]

2

Peenemünde and the A-4 (V-2), 1932–1945

A German army rocket-development program got started at the Kummersdorf proving grounds near Berlin in 1932 under the leadership of Capt. Walter Dornberger and an aristocrat named Wernher von Braun, who had begun developing rockets as an amateur under Hermann Oberth and others. After going to work for the army, von Braun and a growing number of engineers started with simple rockets, proceeding through models designated A-1, A-2, A-3, and A-5 before developing the A-4 (V-2) missile that was destined to be a starting point for a number of American missiles and rockets. Not only the technology of the V-2 but many of its designers moved to the United States after World War II and contributed to the development of missiles and launch vehicles ranging from the Redstone to the Saturn V. This important series of developments stemmed from the movement that Oberth's writings started.

The Beginnings of German Rocketry

Following the publication of Oberth's *Die Rakete* and Max Valier's book and articles, plus speaking tours, a spaceflight movement of significant proportions emerged in Weimar Germany. There were science fiction novels, rocket car demonstrations, and the founding by Johannes Winkler of the spaceflight journal *Die Rakete* (The rocket) in 1927. Also in 1927, Valier, Winkler, and a handful of others founded the Verein für Raumschiffahrt (Society for Space Travel), often known as the VfR for short. The journal *Die Rakete* became its organ and Winkler its first president, with Oberth succeeding him. All of this plus the movie *Die Frau im Mond* generated a popular enthusiasm for rocketry and space travel probably exceeding that of any other country except Russia. The publicity and enthusiasm led to a number of amateur experiments with rocketry that, in turn, paved the way for the more professional rocketry program undertaken by the German army.[1]

Among the amateurs, Winkler apparently got started first, in activities independent of the VfR. He launched some biplane models with skyrockets in late 1927 and then measured solid-propellant thrust curves at the Technische Hochschule (Technical Institute) of Breslau's machine laboratory. Thereafter, with the financial support of hatmaker Hugo A. Hückel, he began work with liquid-propellant rockets. He also worked with the Junkers Aircraft Company in Dessau, trying out such propellants as ethane and nitrogen monoxide before deciding to use liquid oxygen and methane. Employing these latter propellants, he launched the first verifiable liquid-propellant rocket flight in Europe in early 1931.[2]

Meanwhile, Oberth had hired two technically incompetent assistants in his effort to build the rocket that would help publicize *Frau im Mond*. Although the trio never succeeded in launching the rocket, by July 1930 they and their growing team of assistants had tested a redesigned version of Oberth's cone-shaped motor, burning gasoline and liquid oxygen, before the head of the Chemische-Technische Reichsanstalt (State Chemical-Technical Institute), who affirmed in an affidavit that it performed without mishap. Oberth returned to Romania shortly after this static test, and Rudolf Nebel took over the leadership of the rocket efforts for the VfR. By this time Lt. Col. Karl Emil Becker of the Army Ordnance Office had become interested in military uses for rocketry.[3]

Born in 1879 when Otto von Bismarck was chancellor of the German Empire, Becker had become a lieutenant in the army in 1900 and earned his doctorate of engineering in 1922 at the Technical Institute of Berlin, where Professor Carl Julius Cranz was his mentor. He contributed to the third edition of Cranz's *Lehrbuch der Ballistik* (Textbook on ballistics, 1926), which contained a seventeen-page section on "reaction, or rocket projectiles" that has been attributed to Becker. Whether he wrote it or not, it offered analyses of both Goddard's 1919/1920 paper and Oberth's writing, and it showed that Becker was aware of rockets as possible weapons. His goal seemed to be their use as a way of delivering poison gas to a considerable distance from the launching point.[4]

Accounts differ about Becker's initial interest in liquid-propellant rockets. According to Heinz Dieter Hölsken, Becker thought only solid-propellant rockets were viable, but after World War II, Walter Dornberger stated that Becker had directed him in 1930 to develop solid rockets to deliver missiles from five to six miles *and* liquid rockets to carry a payload heavier than any artillery shell over a distance greater than the maximum range of any existing gun.[5]

Dornberger, the second son of a pharmacist in the university town of Giessen, was born in 1895 and entered the army in 1914 when World War I

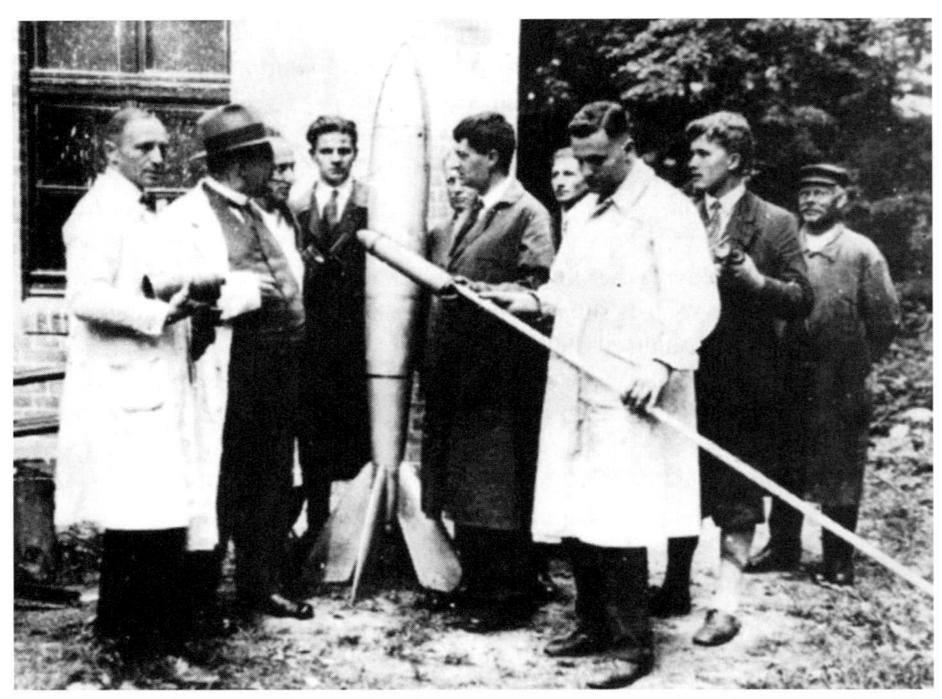

Figure 6. Scene from 1930, showing key figures in early German rocketry. Hermann Oberth is at center (*facing left*), and a young Wernher von Braun is second from the right, behind Claus Riedel *(in white coat)*. Courtesy of NASA.

started. As an artillery lieutenant in 1918, he was captured by the U.S. Second Marine Division and was held as a prisoner of war for two years in southern France. He remained in the army after the war, when the Versailles Treaty had restricted its size to 100,000 men. From 1925 through 1930 he studied engineering at the Technical Institute of Berlin under a program Becker had helped to establish. After his assignment to Becker's office, Dornberger received a "diploma engineer" degree, roughly equivalent to a U.S. master's degree.[6]

By the time Dornberger, now a captain, joined Becker's organization, Nebel, who made up for a lack of technical skills with salesmanship, had begun looking for funding and a place to test and launch the VfR's rockets. He and a new VfR member, engineer Klaus Riedel, found an abandoned storage depot in a suburb named Reinickendorf north of Berlin. It contained concrete bunkers in which the army had stored ammunition during the war. Becker's ballistics and munitions section of Army Ordnance's testing division provided limited funding and helped Nebel obtain an inexpensive lease on the property, which Nebel dubbed the Raketenflugplatz (Rocket Aerodrome, or Rocketport). He begged, traded for, and scrounged money and materials to conduct liquid-rocket tests at the site, which opened in September 1930.[7]

This is not the place to catalog the various rockets tested at the Raketenflugplatz from 1930 until it closed in 1934. Although Nebel and Riedel had been working on regenerative cooling, already described by Oberth and Tsiolkovsky, the rockets suffered explosions, lack of adequate guidance and control, burn-throughs of the combustion chambers, leaks, and related problems. It was probably experience with these problems by participants who later worked for Army Ordnance that was the real contribution of the Raketenflugplatz to the technical history of rocketry, rather than any specific designs developed there. There is, in any event, only spotty information about those designs, because Nebel and the others kept few records.[8]

Unquestionably the most important of the individuals who joined Nebel at the Raketenflugplatz was the young Wernher von Braun. He was an unlikely participant in that ragtag environment, since he was born into the old nobility of eastern Germany—those whose families had belonged to the noble estate before 1400. His full name was Freiherr (Baron) Wernher Magnus Maximilian von Braun. Wernher's father was the *Landrat,* or principal magistrate, of the governmental district around the eastern German town of Wirsitz (later Wyrzysk, Poland). It was there that Wernher was born on March 23, 1912. Even though Germany became a parliamentary democracy after the war, being born and growing up in an aristocratic family undoubtedly contributed to the remarkable self-possession that von Braun later exhibited.[9]

His mother, née Emmy von Quistorp, was a well-educated woman with a strong interest in biology and astronomy. She helped to inspire her son's later interest in spaceflight by giving him the science fiction works of Jules Verne and H. G. Wells as well as a telescope. The last was a gift when he was confirmed into the Lutheran Church in his early teens, in place of the customary watch or camera. Despite this background, he was not a good student. He was especially weak in mathematics and physics when he attended the French *Gymnasium* in Berlin. This prompted his father to enroll him in a less humanistically oriented *Oberrealschule* founded by the pedagogue Hermann Lietz. At two different Lietz schools, he responded positively to the progressive teaching methods Lietz had encouraged, but it was not until he encountered the difficult formulas of Oberth's *Die Rakete* that he began to study mathematics and physics seriously. When he completed secondary school, he enrolled in the Technical Institute of Berlin-Charlottenburg, where he earned a prediploma (*Vordiplom*) in mechanical engineering in 1932, followed by a Ph.D. in physics at the University of Berlin in 1934.[10]

By 1930 the eighteen-year-old von Braun had joined Oberth and Nebel along with Klaus Riedel. He was working as an apprentice at the Borsig factory as part of his training at the Technical Institute of Berlin, but he devoted his spare time to rocketry.[11] He later commented about the amateur methods and equipment at the Raketenflugplatz. For example, when the VfR rocket enthusiasts took one of their rockets to Army Ordnance's proving grounds at Kummersdorf in 1932 to demonstrate its capabilities in return for what they hoped (in vain) would be further funding, von Braun wrote, "Dornberger guided us to an isolated spot on the artillery firing range where were set up a formidable array of phototheodolites, ballistic cameras and chronographs—instruments of whose very existence we had theretofore been unaware." The rocket reached an altitude of only 200 feet before its trajectory became almost horizontal and it crashed. Becker then offered a "degree of financial support" if the group would work at Kummersdorf. Nebel rejected the offer, as did Klaus Riedel at the time. Only von Braun accepted and went to work at Kummersdorf in late 1932.[12]

Dornberger, writing after the war, said of von Braun: "I had been struck during my casual visits to Reinickendorf by the energy and shrewdness with which this tall, fair, young student with the broad massive chin went to work, and by his astonishing theoretical knowledge. It seemed to me that he grasped the problems and that his chief concern was to lay bare the difficulties. In this respect he had been a refreshing change from most of the leading men of the place."[13] Von Braun initially had a single assistant, an "enthusiastic and able" mechanic named Heinrich Grünow. His "laboratory" consisted of "one half a concrete pit with sliding roof; the other half being devoted to powder rockets."[14]

Von Braun's and Dornberger's recollections differ with respect to the first engine von Braun constructed. Von Braun remembered that his first "small, water-cooled motor" was ready for static firing only in January 1933. "To the amazement of the authorities, it developed a thrust of 140 kilograms [308 pounds] for 60 seconds at its first test," he claimed.[15] Dornberger put the test at December 21, 1932, and said the engine had exploded, destroying the test stand.[16] Probably the actual date was March 4, 1933.[17]

Von Braun and Dornberger agreed that troubles ensued. Von Braun characterized them as "ignition explosions, frozen valves, fires in cable ducts and numerous other malfunctions," to which Dornberger added a description of a burn-through of a combustion chamber. And they agreed that these failures, which von Braun said "taught me the hard way," led to what the younger man identified as the A-1 engine (also known as the 1B for *brennstoffgekühlten* or cooled by fuel, to distinguish it from the 1W or water-cooled engine). Von Braun said he had learned from Nebel "that modern inventing consists largely in combining known and manufactured elements," so he called in "welding experts, valve manufacturers, instrument makers and pyrotechnicists . . . and with their assistance a regeneratively cooled motor of 300 kilograms [660 pounds] thrust and propelled by liquid oxygen and alcohol was static tested and ready for flight in the A-1 rocket which had been six months a-building." Dornberger added that the "650-pound-thrust chamber . . . gave consistent performance" but an exhaust velocity that was slower than they needed even after they "measured the flame temperature, took samples of the gas jet, analyzed the gases, [and] changed the mixture ratio."[18]

Another feature of the A-1 that resulted from going to outside experts was a process of anodization (*Eloxierung*, electrochemical oxidation) of the inner and outer surfaces of the aluminum combustion chamber that both hardened the chamber and increased the heat emitted from it by a significant percentage. This process had been developed by the Vereinigte Aluminiumwerke AG (United Aluminum Company) in the town of Lauta in eastern Germany.[19] This firm apparently provided von Braun and the Army Ordnance operation with information about another company named Zarges in Stuttgart, leading to initial contracts for combustion chambers and alcohol tanks.[20]

The researchers at Kummersdorf had thought to stabilize the A-1 rocket (the A standing for *Aggregat*, German for "assembly") by having it spin about its longitudinal axis. But if they rotated the rocket, the propellants would rise up the walls of their tanks from centrifugal force and cause problems in feeding into the combustion chamber. Dornberger therefore suggested rotating only the nose, which would form a sort of gyroscope rotating on an axis. Since a gyroscope, like a spinning top, is resistant to disturbing forces, it would tend to keep the rocket from deviating off its intended course by what von Braun called "brute force." The A-1 launched vertically from a launching rack. Pressurized nitrogen fed the propellants, liquid oxygen and alcohol at 75 percent strength, into the combustion chamber. The rocket was four and a half feet long and a foot in diameter. It weighed 330 pounds. Unfortunately, half a second after ignition on its first launch, the vehicle exploded. A buildup of propellants before a delayed ignition was the cause.[21]

"A second A-1 with improved ignition might have flown well," von Braun said, "but certain other factors called for a complete redesign named A-2." He does not characterize the "other factors," but one apparently was the location of the gyro-flywheel, which they moved from the nose to near the center of gravity of the rocket between the propellant tanks.[22] Another factor evidently was the performance of the engine. While the A-1 engine was still in the test stand, they built a larger test stand for higher-performance engines and designed a new propulsion unit. Here, however, they encountered the problems of scaling up that were to plague many subsequent engines. They experienced repeated burn-throughs as a result and essentially got nowhere with developing a new engine before the A-2 was ready to be launched.[23]

In the midst of all this development, von Braun was still working on his doctoral degree. He delivered the dissertation on his work at Kummersdorf to his Ph.D. committee at the University of Berlin on April 16, 1934, while the A-2 was still very much a work in progress. In the course of developing the A-2, the liquid-propellant rocket group gained two key new researchers. Both came from a project begun by Valier at Paul Heylandt's Aktiengesellschaft für Industriegaswertung (Industry Gas Utilization Company) at Britz, near Berlin. Heylandt allowed Valier to work on an engine to propel his rocket car with the assistance of several of his people, including two engineers named Walter Riedel (no relation to Klaus) and Arthur Rudolph. The three of them were testing an engine burning liquid oxygen and an emulsion of diesel oil and water on a Saturday, May 17, 1930. The engine exploded, killing Valier. Rudolph tried to figure out what had caused the explosion. The injector was egg-shaped and the holes drilled in it were irregular. The result was poor atomization and resultant combustion instability, as the phenomenon is called today. Rudolph studied flow patterns and came up with an injector design in which the fuel sprayed out regularly from the top and flowed to the wall of the combustion chamber. Liquid oxygen flowed from the wall inward to mix with kerosene. He, Riedel, and their supervisor, Alfons Pietsch, also built a cooling jacket around the engine so that it was regeneratively cooled. It produced smooth combustion.[24]

Following these events and a contract with Heylandt for a small engine, the Army Ordnance operation at Kummersdorf hired Walter Riedel in January 1934. Von Braun said Riedel's "skill and experience at design problems" complemented his own "preference . . . for formulating specific tasks and outlining problems." Dornberger called Riedel a "versatile practical engineer" whose "deep knowledge" and "practical experience" guided "the bubbling stream of von Braun's ideas into steadier channels." Older than von Braun by a decade, the "short, sedate man with a permanently dignified and serious expression" became "Papa Riedel" to the young group of researchers at Kummersdorf and later Peenemünde, where he headed the design office at the ripe old age of thirty-five in 1937 when the operation moved to the Baltic coast.[25]

Meanwhile, both Rudolph and Pietsch had lost their jobs at the Heylandt works in 1932, and Pietsch had obtained a contract to build an engine for the army. He and Rudolph started working on it according to Rudolph's design, but Pietsch used some of the contract money for his personal needs. Seeking funding to complete the work, Rudolph went to Dornberger, who provided enough money to complete the job. Rudolph built an engine made of copper with a bullet-shaped liquid oxygen tank and a combustion chamber built into the alcohol tank to reduce costs. It used the injector design he had developed at Heylandt plus a cooling jacket. Rudolph said it did not quite meet the specifications of 100 kilograms (220 pounds) of thrust but it burned longer than the stipulated ten minutes. Von Braun decided it met the specs because the product of the thrust and time did satisfy the requirements. Rudolph hoped for another contract, but Dornberger told the "lean, starved-looking engineer" he could either join the group at Kummersdorf or not work at all. Hired on the spot in August, he went to work in December after going through security checks.[26]

The A-2, like the A-1, employed a 660-pound-thrust engine. Two A-2s named Max and Moritz completed development in December 1934. (The names came from the German comic strip called *The Katzenjammer Kids* in the United States, although there the kids were named Hans and Fritz.) The A-2s had the same dimensions as the A-1. A small group including von Braun, Riedel, and Rudolph took them to the island of Borkum in the North Sea where they launched to altitudes of more than a mile on December 19 and 20, 1934.[27]

The A-3 and the Move from Kummersdorf to Peenemünde

Near the beginning of von Braun's work at Kummersdorf, the Weimar Republic collapsed and National Socialist (Nazi) Party leader Adolf Hitler became chancellor of Germany on January 30, 1933. Appointed by President

Paul von Hindenburg, Hitler headed a coalition cabinet with former chancellor Franz von Papen as his vice-chancellor. There were only two other Nazis in the cabinet, Wilhelm Frick as minister of the interior and Hermann Göring as minister without portfolio. Despite this seemingly weak position, Hitler began almost immediately to consolidate his power and to rearm in violation of the Versailles Treaty.[28]

Hitler's rearmament was much more extensive than anything that had preceded it during the Weimar years but was not unprecedented. Germany had never abided by the restrictions imposed by the victorious powers in World War I. For its effect on the rocket research project of the army, the most important violation by Hitler was the establishment of a German air force, the Luftwaffe, which came into existence secretly in 1933. But Germany had violated Versailles' banning of an air force as early as 1922 when, in the Rapallo Treaty, it and the Soviets had agreed that the Germans could set up army and air force training facilities in the USSR as a quid pro quo for German technological and training assistance to Soviet forces. Germany sent up-to-date aircraft to the Soviet Union, and its crews trained using them. This laid a firm foundation for the air force that emerged in 1933, with Göring, a fighter ace in World War I, becoming air minister and commander in chief of the Luftwaffe.[29] Even though the project at Kummersdorf remained with the army, this was an important step for von Braun and Dornberger because Göring's influence with Hitler ensured that the Luftwaffe was favorably situated to obtain funding and resources.[30] The army researchers soon benefited from them.

The cramped conditions at Kummersdorf and plans for a much larger rocket than the A-2, which already had necessitated a trip to the Baltic to be launched, required facilities with more space than was available near Berlin. As Dornberger related in testimony before a committee in 1943, after the launching of Max and Moritz in 1934 he and his group had already planned for the much larger A-4. By 1935 hopes for a larger facility were much on their minds. In January of that year, von Braun recalled, the rocket researchers at Kummersdorf received a visit from Maj. Wolfram von Richthofen, a cousin of Manfred von Richthofen, the Red Baron of World War I fame. Major von Richthofen was chief of development within the Luftwaffe's Technical Division, and he was interested in using liquid-propellant rocket engines for powering aircraft. The two services cooperated on a jet-assisted takeoff project for the Heinkel He 112, and von Richthofen urged further collaboration on a rocket-powered fighter and a jet-assisted takeoff unit for heavy bombers.[31]

As von Braun wrote, "Our facilities at that time consisted of . . . a small experimental station sandwiched in between two of the Kummersdorf artillery ranges. Von Richthofen's demands could not be accommodated in such close quarters." The Luftwaffe offered some five million reichsmarks for building more ample facilities at some other location. Becker, now a general and soon to become chief of Army Ordnance, was indignant at the offer by the junior service and announced—in a fit of interservice jealousy—that he would provide six million to top the air force's five. Although it took some convincing of Becker's superiors in the army, funding for the new facility was forthcoming from both services.[32]

Meanwhile, von Braun had located the site they would use for the expanded facilities. "It was during the Christmas season of 1935," he recalled in one version of his reminiscences, and he was visiting what he called his "father's farm in Silesia." He explained his need for a site along the coast that would enable his group to "fire rockets for two hundred miles or more." His mother suggested looking at Peenemünde where her father used to shoot ducks.[33] Dornberger said that the place was "far away from any large towns or traffic" and quite sylvan, with deer as well as ducks and other birds. Seaside resorts along the Baltic were far enough away not to affect the "lonely inlet of Peenemünde," which offered a "range of over 250 miles eastward along the Pomeranian coast." A Luftwaffe civil engineer was soon in residence, and many of the rocket researchers, elevated in the summer of 1936 to an independent group under Dornberger, moved to the new center in April 1937, with an enlarged staff of 123 white-collar and 226 blue-collar workers as of the opening in May. Because test stands were not yet ready, a propulsion group stayed behind at Kummersdorf until 1940.[34]

The Luftwaffe's facilities were located at Peenemünde-West on the northwest corner of the island with Peenemünde village located south of it. There, an oxygen plant and a new harbor complemented the army's research establishment extending from the northern tip of the island along the Baltic coast to the small resort town of Karlshagen, which the combined army–air force experimental center took over. Furthest north were the test stands of Peenemünde-East, as the army facilities were called. Below them were the development works—headquarters building, administrative offices, laboratory, and workshops. Further down the coast was a production plant completed after 1939, then a settlement where people from both the army and the air force lived, with barracks to the west of Karlshagen.

Test Stand 1 was not ready until the spring of 1939. It was a vertical test facility used for static firings of the A-4. At the time it was being built, the researchers already knew that the A-4 might have a thrust of 25 to 30 tons, but they built the stand with a capacity of 100 tons. Known as P-1 after the German word *Prüfstand*, the test stand had a flame deflector consisting of tubes through which water flowed under high pressure to cool them. This was an engine test stand only, with no facilities for attitude testing of the guidance and control system, although whole vehicles could be accommodated.[35]

It is extremely difficult to determine when most of the other test stands were ready. Information about them is scattered among many sources. It appears, however, that some of them, even for the A-4, were not ready until near the end of 1941.[36] There were at least eleven different test facilities, some of which were not used for the A-5 and A-4 but for the Wasserfall (Waterfall) antiaircraft missile, which never went into production.[37]

P-6 was a copy of the larger test stand at Kummersdorf, used initially for testing components of the guidance systems for the A-5 while the rocket engine burned. After tests there, Peenemünde personnel sent suggestions to the manufacturers for improvements. Then they carried out and tested modifications, which were continuous. P-7 was the main A-4 test facility, completed sometime after April 1941. It accommodated both launches and static firings, with thrust capacity up to 200 tons. And to mention just one other stand, P-11 was a production-missile test stand for the A-4. The researchers used it for calibration and quality control of missiles produced at the Mittelwerk underground production facility that was ultimately constructed and operated with prisoners of war and concentration-camp inmates. Not activated until June 1943 and destroyed in an air raid on July 18, 1944, the stand was used in only forty firings.[38]

Even before planning and construction of Peenemünde began, the rocket group proceeded from the A-2 to design and development of the A-3. Rather than relying on the brute force of a single flywheel, the A-3 would have a gyroscopically operated control system. This meant that instead of simply using a single gyro-flywheel's resistance to disturbing forces (gyroscopic inertia), the control system would also use more sophisticated properties of a gyroscope. As a spinning body, a gyroscope responds to a disturbing angular force by rotating slowly (precessing) in a direction at right angles to that of the force. The precession is predictable, so by mounting gyroscopes in such a way that they respond in only one or two directions, they can be used to indicate acceleration or angular velocity. These indications together with electrical pickoff devices can then direct servomechanisms to vanes, gimbals, or other devices to adjust the attitude of a rocket and thus, in effect, steer it. The arrangements to use this property of gyros can be complicated.

The initial expertise for designing such a system came to the rocket researchers from a firm named Kreiselgeräte GmbH (Gyro Instruments, Limited). This firm resulted from the German navy's need for gyroscopic devices to provide accurate fire control, among other needs. In 1926 the navy bought an aerial survey firm in Amsterdam named Aerogeodetic and made it into a secret gyro firm with a branch in Berlin. After Hitler's rise to power, the firm changed its name to Kreiselgeräte. Knowing nothing about the gyro field, von Braun got the navy to recommend this source of expertise. The firm's technical director was Johannes Maria Boykow, a visionary who had served as an officer in the Austrian navy before World War I. Then, also before the war, he became a professional actor. Called back to active duty once the war started, he became in turn a destroyer captain and a naval aviator before getting involved with torpedo development and becoming obsessed with gyroscopes for the rest of his life (1879–1935).[39] Dornberger described him as "a tall, robust man with bright eyes in a shrewd face dominated by a tremendous nose." Fritz Mueller, who had graduated from the Thüringische Technische Staatslehranstalt (Thuringian Technical State Teachers College) in electrical engineering and gone to work for the firm in 1930, described Boykow further as an impressive figure with a deep voice whose ideas were sometimes fantastic or ahead of their time but whom no one contradicted. Mueller said that von Braun had contacted Boykow in 1934 and told him his plans for the A-3 control system. Boykow then sent Mueller to Kummersdorf to work on guidance and control for rockets.[40]

The group at Kummersdorf had debated using aerodynamic surfaces (rudders) on the outside of the rocket to control its attitude, but for some time after launch, the speed of the rocket would be so slow that the control forces imparted by the rudders would be negligible. It occurred to the engineers, however, that the speed of the exhaust gases from the combustion chamber through the nozzle would be almost unchanged during the entire period the propellants were burning. They could use what came to be called jet vanes in the exhaust stream to control the rocket. But according to Mueller, no one had any knowledge of the forces the vanes would generate. He designed test devices to measure those forces.

More importantly, Boykow had by then drawn up plans for the guidance and control system to be used on the A-3. Mueller actually designed the system based on Boykow's ideas. It used a platform stabilized by two gyros, one for pitch and the other for yaw (the two directions in which the rocket could deviate from the vertical; if the rocket were to pitch over toward horizontal flight, yaw then became the left-right axis and pitch the up-down one). There was no roll gyro in the system. It required servos to keep the platform horizontal in accordance with signals from the gyros. A motor to operate these servos caused unwanted oscillations in the gyros, so the researchers used a pneumatic servo. An electric contact picked up the precession of each gyro and sent a signal to a magnetic valve device. This opened a valve in the pneumatic pressure line, providing the impetus to move the platform, thereby stabilizing it in the pitch and yaw axes.

In addition to the two gyros for the stable platform, the system, which came to be called the SG-33, featured three rate gyros mounted below the platform and fixed to the rocket, not the stable platform. These were comparatively large gyros that precessed when affected by the slightest angular motion. One was for roll, one for pitch, and one for yaw deviations. When one of them precessed, it hit a contact that activated control motors. These, then, moved the jet vanes, made of molybdenum and tungsten, in such a way as to counteract the deviation in pitch, roll, or yaw. However, there were also little carriages on the stable platform that moved as the platform adjusted in the pitch or yaw. These also had electrical contacts that fed into a torque generator on the precession axis of the particular rate gyro that corresponded to the motion being picked up. These carriages served as integrating accelerometers that resulted in the torque generator's receiving two signals that were mixed, with the sum of the two going to a servomotor that moved the vanes.

Two of the four vanes rotated in the same direction to control pitch or yaw, in opposite directions to control roll. It took a couple of years to develop the system and test it under static conditions in the Kreiselgeräte labs by moving heavy weights with the vanes, there being no system at the firm to test them on engines. Then the firm delivered the platforms to Kummersdorf, where they were put into the assembled A-3s and tested in a gimbal system so that they tilted the rocket when the combustion chamber was operating. The system worked in this static testing.[41]

Meanwhile, because the rocket researchers initially expected the A-3 to travel faster than the speed of sound, there was a need to design the fuselage and tail fins so that it could operate through the difficult transonic speed range just before the rocket reached supersonic speed. At this time, very little was known about bodies flying at or near supersonic speeds. All the experimental data then available dealt with spin-stabilized projectiles, but the rocket group needed to use fin stabilization even though some people believed this was not possible in supersonic flow, a view in which Cranz's textbook on ballistics concurred. Dornberger referred to fin stabilization as "arrow stability," and Cranz's book maintained that "experience had proved . . . it was impossible for bodies with arrow stability to accomplish perfect flight at supersonic speeds." Nevertheless, since it had low drag, the 8-millimeter infantry bullet provided the first shape for the A-3 with the addition of fins, which were anchored by an antenna ring to give them structural strength.

It was no doubt the anticipated problems with stability in supersonic flow that prompted Dornberger and von Braun to contact the Technical Institute of Aachen for help. In this, they made a fortunate choice. Germany was home to the man who generally was regarded as the father of aerodynamics, Ludwig Prandtl, who taught at the University of Göttingen. The renowned mathematician Felix Klein had brought him there in 1904 as part of an effort to bring together the disciplines of science and engineering, which in Germany tended to be taught in separate institutions. Klein also wanted to promote the study of aerodynamics at Göttingen. Prandtl set up aerodynamics and hydrodynamics institutes at the university that rivaled any others in the world, and he wrote foundational papers in the discipline of fluid dynamics, of which aerodynamics was a subdiscipline. By 1930 his student and rival Theodore von Kármán had transformed the Aerodynamics Institute at the Technical Institute of Aachen to compete with the one at Göttingen, then moved on to an appointment in the United States at the California Institute of Technology. After 1934 Professor Carl Wieselsberger, another student of Prandtl's, and Dr. Rudolf Hermann, both at the Technical Institute of Aachen, had constructed a supersonic wind tunnel 10 centimeters (4 inches) square in cross section (10x10), funded by the Luftwaffe. Before building it, Hermann had consulted with Dr. Adolf Busemann, who had built an even smaller 6-centimeter-square (6x6) supersonic tunnel at Göttingen. The Aachen tunnel was still very small, and it was a blowdown device, meaning that it would not operate for long. The operators had to pump for many minutes to create a vacuum in the tunnel. Then they opened a quick-acting valve to start the test. Blowers from a 200-horsepower motor provided pressure for only 20 to 25 seconds. The square shape of the tunnel made it less accurate in the data that it provided than a circular tunnel, and because of its small size, models tested in the tunnel had to be very small indeed.

In an interview in 1988, Hermann recalled that von Braun had visited him at the Aerodynamic Institute at Aachen in January 1936, beginning the work of the wind tunnel for the German army. Hermann performed tests on various A-3 shapes, showing that the tail fins on the third shape for the

A-3 could stabilize the rocket when it was flying supersonically. However, as design proceeded, the weight of the rocket became too great for it to reach supersonic speeds anyway.[42]

One innovation in the A-3 that apparently reduced rather than increased its weight was a liquid nitrogen pressurization system that used a vaporizer to provide the pressure needed to feed the propellants into the combustion chamber. This did away with the thick walls in the nitrogen tank of the A-2, needed to hold the nitrogen under pressure. The A-3 also featured alcohol and oxygen valves that operated pneumatically by means of magnetic servo-valves. These permitted the propellants to flow in two stages, largely eliminating the danger of explosion on ignition. The new valves subsequently carried over into the design of the A-4.[43]

Design of the engine for the A-3 partly involved a scale-up of the A-2 engine, with a long, cylindrical combustion chamber still welded inside the alcohol tank, as had been true of the A-2. The A-2 and A-3 engines were both regeneratively cooled, but the A-3 engine changed, as a result of Walter Riedel's influence, from the A-2 injector system that had featured fuel and oxidizer jets simply pointing at one another. Following similar principles to the design Rudolph had come up with at Heylandt, the A-3 engine featured a mushroom-shaped injector near the top of the combustion chamber that sprayed alcohol upward to mix with oxygen sprayed downward from jets at the top of the chamber. This evidently increased the efficiency of the mixing, resulting in an improvement of the engine's exhaust velocity from 1,600 m/sec to 1,700 m/sec (about 5,300 ft/sec to 5,600 ft/sec). The thrust was 3,300 pounds. However, the improved performance was bought at the cost of higher burning temperatures, necessitating a series of tests with a variety of different aluminum alloys as well as modifications to the basic design of the engine.[44]

Finally, in early December 1937 four A-3s were ready for launch. They were rather large rockets for the time, 22 feet long, 2.3 feet in diameter, and they weighed about 1,650 pounds when loaded with propellants. This time the site of the launch was the Greifswalder Oie, an island only 300 feet wide by 1,000 feet long in the Baltic Sea not far from Peenemünde. After a variety of delays, on December 4 the first A-3 rose vertically for a few seconds, turned into the wind, and crashed onto the island. On December 6 the second A-3 turned into the wind, rose several hundred yards, and fell into the sea. In both cases the rocket had deployed its parachute prematurely. Unable to determine a cause of the failures from fragments salvaged from the first two launches, the group launched the third A-3 without a parachute. It went up on December 8, followed by the fourth rocket on December 11. Both rose, turned into the wind, and fell into the sea from altitudes of between 2,500 and 3,000 feet.[45]

Discussions, examination of films, and tests on the ground finally revealed the problem. Its source was the control system. But as Mueller said much later, no one at the time had the experience with control systems for rockets to say that Boykow's basic design was impractical before it actually flew. Simulations on the ground had not turned up any flaws. But after the four flights, it became obvious that they should have introduced a simpler system. One problem was that the stable platform was capable of operating as long as the rocket did not turn far before it could be stabilized, but if the rocket did turn abruptly, as happened at Greifswalder Oie, the gyros could not follow it. They would hit the end of their range of motion (about 30 degrees) and the whole platform would tumble. Since the contacts for the parachute were connected to the gimbal system on the platform, when the platform tumbled, the parachute deployed. (The idea was that this would happen only when the rocket reached its peak altitude, at which point it was desirable for the parachute to unfurl.) At the same time, the roll rate gyro sent signals to the controls to compensate for roll, but the movements of the jet vanes were not powerful enough to stop the rolling. They needed to move more quickly and have their control forces increased.

In hindsight, it also became clear that there had not been sufficient wind-tunnel analysis. The fins with the antenna ring were not well designed. For one thing, if the rockets had risen high enough, the jet plume would have expanded outward with the decreasing pressure at higher altitudes, badly burning the fins and destroying the antenna ring and the stabilization.

Reserving the designation A-4 for the more ambitious vehicle the group had planned to follow the A-3, the researchers proceeded to the development of the A-5. This vehicle would have a new control system but would be about the same length as the A-3. It would also have differently designed fins. If the flight tests of the A-3s had clearly been failures, the group learned a great deal from them.[46] Many other developmental rockets would experience failures before their designers could correct problems that they had not anticipated in design and ground testing. Such surprises were common in the development of missiles and rockets. The A-3s were thus harbingers of the future.

The A-5

Whereas the A-3 had been intended to carry a payload of recording instruments, the A-5 became essentially a test bed for a new guidance system in preparation for the A-4, including graphite jet vanes and a new stabilization system.[47] For the aerodynamic work on the A-5—and later on the A-4 and other projects—Dr. Rudolf Hermann, the man who had built and directed the small supersonic wind tunnel at Aachen, joined the staff at Peenemünde to construct a new supersonic wind tunnel. Experience with the A-3 tests in the tiny wind tunnel at Aachen, which was in any case distant and under the control of the Luftwaffe, plus discussions with Hermann had convinced both von Braun and Dornberger that they needed their own supersonic tunnel. Dornberger in turn convinced Becker of the necessity. Hermann started to work at Peenemünde on April 1, 1937. Interestingly, as director of the supersonic wind tunnel, he reported directly to Dornberger, not to von Braun as did the other directors.

Hermann was born in 1904 in Leipzig. He attended the University of Leipzig from 1924 to 1929, taking courses in math, physics, astronomy, and physical chemistry. He became a doctoral candidate in 1927, working under Professor Ludwig Schiller, a physicist described by Peenemünder Peter P. Wegener as "a well-known innovator in fluid mechanics." Hermann worked on a topic in this discipline, the resistance to water flowing in pipes at high speeds and temperatures, earning his Ph.D. on July 12, 1929.

After working as an assistant to Schiller, in 1933 he joined the Department of Aeronautics at the Technical Institute of Aachen, completing a thesis for his habilitation as a lecturer in 1935. The topic of his habilitation thesis was heat transfer. Besides being the head of the supersonic wind tunnel division at the institute and building the 10-by-10-centimeter tunnel, he lectured in the fields of hydrodynamics, aerodynamics, and gas dynamics. The day he started at Peenemünde, Hermann said, he received the offer of a full professorship in mechanical engineering at the University of Braunschweig, but he did not accept because he "saw all the possibilities at Peenemünde, big problems to solve."[48]

Already in 1937 Rudolf Hermann hired Hermann Kurzweg, another Ph.D. trained by Schiller at the University of Leipzig, as his deputy and head of the research division in the aerodynamics, or wind tunnel, division at Peenemünde. Wegener, a later subordinate of Kurzweg's, described Hermann's deputy as "an exceptionally pleasant person." Kurzweg had the job of overseeing the aerodynamic testing of the A-5 and designing the fins. The new test vehicle turned out to be slightly longer and thicker than the A-3—24.2 feet long to the A-3's 22 feet, and 2.5 feet in diameter to the A-3's 2.2.[49]

As Kurzweg and his colleagues began the task of evaluating the A-5 shape, similar to that of the A-3, they had little theoretical guidance to draw upon.

They relied heavily on the publications of Busemann, especially an article on gas dynamics he had published in 1931 in a handbook on hydro- and aerodynamics. They also had the talks that aerodynamicists like Prandtl and von Kármán had given at the important Volta Congress on high-speed aerodynamics in Rome in 1935, subsequently published. Kurzweg made three fin designs for the A-5 based on rough estimates of the pressure distribution over the fuselage "and a few available subsonic normal force and centre of pressure data for flat plates" with aspect ratios less than one (that is, if the plates were envisioned as fins, they would extend further in a longitudinal direction along the rocket's fuselage than they would outward into the airstream). All three were swept 60 degrees in front to a point about a quarter of the distance behind the leading edge of the fin, with the rest of the fin going straight back on a line roughly parallel to the overall length of the body. The inside trailing edge was swept outward 45 degrees in back to leave ample space for the expanding jet of rocket exhaust at high altitudes, fixing one of the problems with the A-3 fins. There also was no antenna ring holding the fins in place.

Since he needed information right away and the supersonic tunnel at Peenemünde had not been built, Kurzweg carved the first model of the A-5 from a 1-inch pine branch, with the three fin configurations made from hard scrap rubber. He established the estimated center of gravity by placing brass plugs in holes in the model, much as children do with metal clips on toy gliders to make them fly straight. He then suspended the model on a wire placed through the center of gravity and pulled it behind an automobile traveling at 100 kilometers per hour (62 mph). The crude test suggested that two of the fins provided stability, whereas the third failed to do so. This reliance on primitive methodology to meet time constraints is one illustration that Kurzweg and the entire rocket group were not doing "rocket science" but were engaged in engineering. Although Peenemünde ultimately would go to great lengths to obtain the best information it could, the goal was never knowledge for its own sake but rather the ability to make its vehicles work as effectively as possible.

More systematic tests followed on much more precise models in various subsonic wind tunnels, including one at Aachen, with the Zeppelin tunnel at Friedrichshafen on Lake Constance making detailed pressure measurements over the whole fuselage. Three-dimensional data from the Friedrichshafen wind tunnel about pressure distribution over the body and fins were very useful in getting a sense of aerodynamic loads on the shape. From such data, Kurzweg and his fellow aerodynamicists could visualize the forces operating all over the A-5, so they could determine the centers of pressure at various angles with respect to the prevailing wind (angles of attack). They could thereby get a better idea about the size and position of the fins necessary to stabilize the rocket in flight but still leave it without excessive stability. That way, the jet vanes could correct deviations from the intended flight path more easily than with a highly stable design.[50]

The air resistance of the new fins evidently would be significantly lower than for the A-3, giving promise that the A-5, with the same engine as the A-3, could fly supersonically. As yet there were no wind tunnels anywhere, even supersonic ones, that could give accurate data about flights at the speed of sound,[51] so there were concerns that flight at that speed (about 760 mph at sea level, lower at higher altitudes) would result in oscillations so powerful they could shatter the rocket. The most satisfactory available recourse was to drop A-5 models from airplanes at high altitudes and see what happened to them. The researchers built several solid-iron models and began dropping them from a Heinkel He 111 at 20,000 feet in September 1938, recording the trajectory with cine- and phototheodolites. They determined that the models reached a maximum speed of 800 mph at 3,000 feet, well in excess of the speed of sound. There had been no oscillations exceeding 5 degrees. Later the aerodynamics group launched some forty small A-5 models, powered by hydrogen-peroxide engines developed by the firm of Hellmuth Walter, to test fin shapes. Eventually this work resulted in a fin with a rudder assembly for the A-5, the A-4 then intentionally keeping the same basic shape as the A-5 to avoid having to gather new aerodynamic data.[52]

The research team at Peenemünde also went to work on a parachute to deploy at the peak of the trajectory. It would reduce the velocity of the A-5 from as much as 250 mph to 45 mph. The group engaged the Graf Zeppelin Flight Research Institute at Stuttgart to develop a ribbon parachute for braking. Additionally, there would be a large parachute to slow the descent speed of the A-5 sufficiently for the group to retrieve the vehicle and troubleshoot any failures. Like drop models of the A-5, the parachutes were tested with dummy rockets released from airplanes. One further innovation for the A-5 came from an unnamed technical draftsman recently hired at Kummersdorf. He suggested using graphite for the jet vanes instead of the more expensive tungsten-molybdenum ones used in the A-3. Successful tests by the propulsion group led to the adoption of the graphite, reducing the price of a set of vanes by two orders of magnitude, from 150 to 1.5 marks.[53]

The development of the guidance and attitude control system for the A-5 was a complicated process for a variety of reasons, ranging from the number of different outside firms involved and the overlapping of work on the A-5 and the A-4 to the sheer complexity of the technologies involved. Von Braun provided an overview but no details. He said that during launchings in the period 1939 to 1941, the A-5 tested three different control systems, "all of which worked perfectly. The vertical firings, reaching altitudes of eight miles, were followed by tilted trajectories. Radio guide beams were also tried." In the summary of developments that Dornberger presented in 1943, he added that the A-5 had also demonstrated a mechanism for cutting off combustion at the appropriate place in the trajectory.[54]

The first control system for the A-5 was the Kreiselgeräte system, referred to as SG-52. Fritz Mueller, who was involved in its development, described it as simpler than the system used for the A-3. It was "a 3-gyro, 3-axis stabilized platform that provided attitude control and a tilt programme." Rate gyros "located above the stabilized platform" sensed angular deviations. A device he called a torquer, mounted on each of the rate gyros and "influenced by the attitude pickup of the stabilized platform," mixed the signals and fed them into a "control system also mounted above the platform." This "complete guidance and control system" was in a single package. Through aluminum rods, it made connections to the jet vanes. The tilt program built into the stabilized platform caused the A-5 to turn in the direction it was intended to go. According to Mueller, the first A-5 with this guidance system was launched in late 1939.[55]

H. A. Schulze gave a more precise date for the first launchings of the A-5 with the "new guidance system," stating that these occurred in October 1939 with two vertical launchings before use of the tilt program to turn the rocket into a 45-degree trajectory. All three launches were fully successful. Dornberger added that the rockets had "not yet reached the speed of sound, but . . . we had shown that the liquid-propellant rocket was equal to the tasks set for it." An undated document in the Peenemünde records identifies a series of four flights of the A-5 from October 24, 1938, to December 12, 1939, as including three with the SG-52. The first flight was without a "steering machine," while the last three used the Kreiselgeräte control system. This document reported the system as working without problems.[56]

Meanwhile, even before the instructive failures of the A-3s, Dornberger and von Braun had met on November 9, 1937, with a number of representatives of Siemens Apparate- und Maschinenbau GmbH (Siemens Instrument and Machine Manufacturing Corporation) in the Berlin suburb of Marienfelde to discuss possibilities for cooperation on what became a second guidance and control system for the A-5. In the 1930s the Siemens electrical firm had done considerable business in ship navigation and airplane autopilot design and construction. To this end, it had purchased rights to Boykow's patents covering his contributions to gyroscopic control mechanisms. Among those attending the meeting were Austrian engineer Karl W. Fieber, who had earned a Ph.D. at the Technical Institute of Vienna and had begun to work with gyro and control problems.

Von Braun and Dornberger proposed that the Siemens group develop a control mechanism within a year and a half (that is, by spring 1939) for a rocket that could fly up to 4.5 times the speed of sound and be controlled up to about 30 kilometers (some 19 miles). These requirements greatly exceeded anything Siemens had dealt with in the control of airplane flight. Fieber also pointed out, among other problems, that the flight characteristics and stability of rockets were virtually unknown. However, after the failures of the A-3s, Siemens accepted a contract to build a control system for the A-5. At first, a group of technical people around Fieber proposed using just two gyros to provide control signals in the roll, yaw, and pitch axes. One gyro, vertical with respect to the earth and called the *Vertikant*, had two degrees of freedom (ability to move in two axes), and the other, a horizontal gyro, moved in one direction. The *Vertikant* would measure movements of the rocket in roll and yaw, and its name came to designate the entire system. The *Horizont* would measure movements in pitch and would also initiate signals to the servomechanisms controlling the jet vanes to tilt the rocket after liftoff. This proposal would have implications for the A-4, but to simplify the A-5 control system, Siemens shifted from the two-gyro to a three-gyro system.

The three gyros, one to precess in each axis, were arranged in such a manner as to send electrical signals to hydraulic servomotors that would move the jet vanes in the appropriate fashion for pitch or yaw corrections and roll control. It did this by means of a potentiometer pickoff device. Three additional rate gyros sensed angular deviations for stabilization. A clockwork mechanism with a timing program rotated an electrical pickoff to connect with the pitch gyro and send signals to the servomechanisms, causing the jet vanes to force the rocket to pitch over and fly in the intended direction. The A-5 with the Siemens control system first flew successfully on April 24, 1940.[57]

The third control system resulted from an autopilot being developed by the Luftwaffe at the Rechlin test center in northern Germany by diploma engineer Waldemar Möller, who transferred to a precision-instrument firm named Askania for the production of the system. Called variously the Möller,

Askania, or Rechlin system, it involved position gyros, a mixing system, and a servo system. It had to be developed further at Peenemünde. In its first flight on April 30, 1940, the A-5 turned sideways, crashed into the water off of Greifswalder Oie, and exploded. The second flight with the Askania control system on August 1, 1940, was successful.[58]

Two final elements in the A-5 testing were the guide-plane system for cross-range (lateral) control and the radio system for cutoff of propulsion at a prescribed point in the trajectory so that the later A-4 (V-2) missile would hit its designated target area. Ernst Steinhoff became the leader of the guidance effort before this system was developed. A friend of Dr. Hermann Steuding—another important recruit at Peenemünde, whose recently established flight mechanics computation office was working on guidance—Steinhoff met von Braun in Darmstadt during March 1939 and joined the staff at Peenemünde in July of that year, becoming the chief of guidance, control, ballistics, flight mechanics, and instrumentation within a year of his arrival. He was born in Treysa in western Germany in 1908 and had earned a bachelor's degree in mathematics and natural sciences at the Technical Institute of Darmstadt in 1931, followed by a master's in aeronautics and meteorology there in 1933. He then received a Ph.D. in applied physics from the same school the year after arriving at Peenemünde, with a dissertation on aviation instruments. From 1933 to 1936, Steinhoff had taught aeronautical engineering at the Polytechnic Institute of Bad Frankenhausen in east central Germany. Then in 1936 he became the research scientist and group leader of flight mechanics and automatic controls at the German Aeronautical Research Institute at Darmstadt. A pilot, he held the world distance record in a glider, flying from Frankfurt to Brno in Czechoslovakia, a distance of 500 kilometers (310 miles).[59]

Mueller had some dealings with Steinhoff, since he continued to work on guidance and control systems for Kreiselgeräte, and the contract engineer emphasized Steinhoff's energy and ability to push developments forward. However, Mueller had reservations about Steinhoff's grasp of reality. He tended to think things could be done in shorter periods of time than were realistically possible, Mueller thought. Thus, besides his qualities as a pusher, perhaps Steinhoff's major contribution was his recruitment of skilled staff, making his department a "particularly effective one."[60]

One such recruit was Helmut Hoelzer, born the same year as von Braun (1912) in Bad Liebenstein in central Germany. He attended an *Oberrealschule* in nearby Bad Salzungen, benefiting from that type of school's emphasis on science. After an apprenticeship in a repair shop for the national railroad, he attended the Technical Institute of Darmstadt, working on the side to support his education. For a time he taught mathematics, electronics, and other subjects at an aviation college and an Institute for Motorless Flight in Thuringia. Finally finishing his master's degree in electronics at Darmstadt, he joined the electronics firm of Telefunken in Berlin, working on wave propagation and wireless aircraft guidance for control systems. He had known Steinhoff and Steuding at the Technical Institute. One night those two engineers and another man Hoelzer didn't know (von Braun) whistled under his window and asked him to join them for a beer. After some discussions that night about controlling unspecified flying bodies, the electronics engineer received a civilian draft notice to Peenemünde, where he went to work about the end of October 1939.[61]

At Peenemünde, Hoelzer and two other control personnel named Otto Hirschler and Otto Hoberg, worked with others to develop a radio guidance system to provide information for cross-range guidance. The cross-range system, which they developed themselves, was based on a blind landing system of an electronics firm named Lorenz but involved a more accurate way of measuring deviations from a desired trajectory along a radio guide plane.[62] A transmitted signal less than 10 miles behind the missile at the point where it reached the desired cutoff of the engine provided cross-range guidance through corrections for deviations from the vertical plane of the signal.

There was one problem with the cross-range system in its basic form. With a control correction to return the A-5 to the radio guide plane in the yaw axis, the rocket would rotate to the correct trajectory. But without a device to measure or calculate the angular velocity of the rocket as it swung back to the center of the guide plane and then compensate for that velocity so as to stop the correction short of the center, the vehicle would overshoot. That would result in another correction in the opposite direction and so forth, with each oscillation getting larger than the last because of the time it took for servomotors and other components to operate. The deviations would continue until propulsion cutoff. Then, with no rocket plume to make the jet vanes effective and with low atmospheric pressure at altitude precluding significant aerodynamic controls, the rocket would be likely to fly in an incorrect direction. At the time, there was no available device to measure the lateral velocity, so Hoelzer devised an analog condenser that used its charging current to calculate the correction to be sent to the jet vanes. This would eliminate the oscillations. Following its initial development, the guidance team used aircraft to test the system, with Steinhoff serving as pilot on many of the flights. There were numerous problems to be worked out, measuring devices to perfect, trajectory calculations to make, and bugs to be eliminated. Finally, however, the system was successfully tested on the A-5.

For propulsion cutoff at a preselected speed and point in the trajectory, a radio device developed separately by Professor Walter Wolman, an electronics engineer from the Technical Institute of Dresden, used the Doppler effect whereby a receiver on the missile took a frequency from the transmitter and doubled it. The doubled frequency, transmitted back to the ground station, went to a device called a Wien bridge preset to the doubled frequency calculated to correspond with the intended cutoff velocity. When the frequency coming back from the A-5 corresponded with the preset frequency in the Wien bridge, a command to the rocket's control system initiated the cutoff of propulsion, following which the A-5 (later the A-4, but with a two-stage cutoff) would follow a ballistic trajectory to the intended target area. Hoelzer and the others at Peenemünde, of course, had to integrate Wolman's contribution into the overall control system.[63]

Accounts differ as to the total number of A-5 launchings. Von Braun said that after March 1939, "some twenty-five A-5s were launched during the next two years, some of them several times." It is not clear if each of the multiple launchings of single vehicles was counted in his total and, if not, how many total launches that "some twenty-five" represented. Schulze said that "approximately 70 to 80 launchings took place until late 1942." Dornberger put the number at about seventy. And in a lecture in January 1946, Walther Riedel (known as Riedel III to distinguish him from Walter and Klaus Riedel), who had become chief of the mechanical design section at Peenemünde in 1942 and soon became head of the development and design section for liquid-propellant rockets, said that there had been fifty A-5 rockets launched through October 1943. Whatever the total number, the important point is that the multiple launches gave the increasingly large group at Peenemünde lots of experience and data. The A-5 served as a test vehicle for a variety of modifications. It proved that the different control systems were effective at least on a vehicle the size of the A-5, that the graphite (or carbon) jet vanes worked, and that parachute recovery was effective.[64]

The A-4

During the course of the A-5's development, most of the work dealt with aerodynamics and control, which were, of course, related. Although the control work was far from over in 1940, the annual report of the division under

Steinhoff for that year gives some insights into one of the processes involved in developing the flight controls. The report states that numerous tasks were entrusted to institutions of higher education. Already many noteworthy suggestions and exemplary proposals to solve problems had resulted. A few schools had made contributions that led directly to improvements.[65]

The work of the institutions of higher education, mostly technical institutes, began in September 1939 with the so-called Wisdom Day, which von Braun described at length. The group at Peenemünde had invited thirty-six professors of "engineering, physics, and chemistry" to Peenemünde to "enlist their interest and co-operation." Since the universities and technical institutes were subject to conscription, the professors, von Braun says, were "eager to participate" (and thereby avoid conscription because they were working on a defense project). Each returned to his school with one or more problems to investigate, selected to match available facilities. The problems ranged from "integrating accelerometers, improvement of pump impellers, trajectory tracking by Doppler radio, gyroscope bearings, [and] research on wave propagation" to "new measuring methods for [the] supersonic wind-tunnel" and "computing machines for flight mechanics." Von Braun added that discussions, visits, and symposia followed, stimulating "many creative contributions."[66] The specific contributions that resulted can better be discussed in conjunction with the development of the A-4's individual systems, but the list of tasks suggests that the work was all in engineering disciplines.

Meanwhile, the staff at Peenemünde continued to grow. It had stood at 349 in May 1937. This number was to expand to some 12,000 at the peak of activities in 1943. It shrank to 7,278 in November 1943 and 4,863 in August 1944. It dropped still more to 4,325 at the beginning of 1945. Only the last number is broken down by project: 135 assigned to the Taifun antiaircraft missile, 1,220 to the Wasserfall antiaircraft missile, and 1,940 to the A-4. A further 270 worked on a winged A-4b and 760 as supply or administrative personnel. These figures undoubtedly are not strictly comparable, with many of the 12,000 probably consisting of construction workers. The last breakdown and the fact that only about half of the peak figure was engaged in development efforts suggest that some rough figures compiled by former Peenemünde managers Eberhard Rees and Arthur Rudolph after the war (see table 2.1) may give a roughly accurate if incomplete and conservative picture of those employed exclusively in the research and development of the A-5 test missile and then the A-4 itself.[67]

Table 2.1. Peenemünde Employees Working on A-5 and A-4

1937	250	1941	1,500
1938	400	1942	2,000
1939	700	1943	3,000
1940	1,000	1944	3,000

Source: Eberhard Rees and Arthur Rudolph, "Short Report about the Time for Development and Manufacturing of the A4 Rocket in Germany," folder "A4 Missile," von Braun Collection, USSARC.
Note: These figures, which are estimates, exclude military personnel, the general maintenance crew, power plant personnel, railroad and ship crews, motor pool personnel, and so forth.

A-4 Propulsion

During the period when the A-5 was being developed and then tested, the rocket research group gained one further key player. This was Dr. Walter Thiel. "A pale-complexioned man of average height, with dark eyes behind spectacles with black horn rims," this fair-haired man with "a strong chin" joined the experimental station at Kummersdorf in the fall of 1936. He was born in Breslau in 1910, the son of an assistant in the post office. After attending an *Oberrealschule*, he matriculated at the Technical Institute of Breslau both as an undergraduate and graduate student in chemistry, earning his doctorate in 1935 for a dissertation on the addition of compounds with strong carbon-halogen bonding to unsaturated hydrocarbons. He had served as a chemist at another army lab before coming to Kummersdorf.[68]

Dornberger said he was put "in complete charge of propulsion, with the aim of creating a 25-ton motor" (the one used for the A-4), although since he remained at Kummersdorf from 1936 to 1940 instead of moving to Peenemünde with the rest of the von Braun group, testing facilities limited him to engines of no more than 8,000 pounds of thrust. While Thiel was "extremely hard-working, conscientious, and systematic," Dornberger said he was also difficult to work with. Ambitious and aware of his abilities, he "took a superior attitude and demanded . . . devotion to duty from his colleagues" equal to his own. This caused friction that Dornberger claimed he had to mollify. Martin Schilling—who became chief of the testing laboratory at Peenemünde for Thiel's propulsion development office and later became Thiel's successor in some sense after Thiel died in a bombing raid in 1943—noted that his boss was "high strung." He said that "Thiel was a good manager of such a great and risky development program. He was a competent and dynamic leader, and a pusher. At the same time, he could not match von Braun's or Steinhoff's vision and optimism."[69]

A memorandum Thiel wrote on March 13, 1937, after he had been on the job about six months, gives some idea of where he was in developing a viable large engine close to the beginning of his efforts as well as what approach he brought to his task. Although he certainly lacked optimism at some points in his career at Kummersdorf and Peenemünde,[70] he did not betray this failing in the memo in question. He referred to "a certain completion of the development of the liquid rocket" that had been achieved by the Weapons Test Organization "during the past years," surely an overstatement in view of the major development effort that still was needed. "Combustion chambers, injection systems, valves, auxiliary pressure systems, pumps, tanks, guidance systems, etc. were completely developed from the point of view of design and manufacturing techniques, for various nominal sizes. Thus, the problem of an actually usable liquid rocket can be termed as having been solved."

Despite this assessment, Thiel listed a number of "important items" that required further development. One was performance of the rocket engine, which needed to improve. He noted that the existing engines at Kummersdorf, burning alcohol and liquid oxygen, were producing a thermal efficiency of only 22 percent and that combustion chamber losses were on the order of 50 percent—that is, about half of the practically usable energy was going to waste through incomplete combustion. The use of gasoline, butane, and diesel oil theoretically yielded an exhaust velocity some 10 percent higher, but measurements made with these hydrocarbon fuels showed actual exhaust velocities no higher than those with alcohol. Thiel felt that "for long range rockets, alcohol will always remain the best fuel" because hydrocarbons increased the danger of explosion, produced coking in the injection system, and presented problems with cooling.

Thus, he said, the way to improved performance lay in exploiting the potential 50 percent energy gain available with alcohol and liquid oxygen through more complete combustion. This could come from improving the injection process, relocating the mixture of oxygen and fuel into a premixing chamber, increasing the speed of ignition and combustion, and increasing chamber pressure by the use of pumps, among other suggestions. He knew, of course, about the tremendous increases in performance available through the use of liquid hydrogen. But he cited the low temperature of this propellant (-259°C), its high boil-off rate, the danger of explosion, and the huge tank volume resulting from its low specific weight (it being the lightest element of all), plus a requirement to insulate its tanks, as "strong obstacles" to its use.

Thiel made repeated reference to existing rocket literature, including a mention of Goddard, but noted that "the development of practically usable models in the field of liquid rockets . . . has far outdistanced research." Nevertheless, he stressed the need for cooperation between research and development, a practice he would follow. He concluded by noting the need for further research in materials, injection, heat transfer, "the combustion process in the chamber," and "exhaust processes,"[71] in line with what was done after Wisdom Day (as described above by von Braun)—which lay some two and a half years in the future at the time Thiel wrote all of this.

Despite the optimism Thiel evinced here, Martin Schilling referred in a postwar discussion of the development of the V-2 engine to the "mysteries of the combustion process."[72] Thiel, indeed, listed the combustion process as a topic for further research but did not discuss it in such an interesting way. Dornberger, like Thiel, failed to use such a term, but his discussion of the development of the 25-ton engine suggested that there really were mysteries to be dealt with. He pointed out that to achieve complete combustion of the alcohol before it got to the nozzle end of the combustion chamber, the rocket researchers before Thiel had elongated the chamber. This gave the alcohol droplets more time to burn than a shorter chamber would, they thought. And their analysis of gases in the engine exhaust seemed to prove the idea correct. "Yet performance did not improve." The researchers realized that combustion of the droplets was not "homogeneous," and they experienced frequent burn-throughs of the chamber walls.

Dornberger said he suggested finer atomization of both oxygen and alcohol by using centrifugal injection nozzles and igniting them after they mixed "to accelerate combustion, reduce length of the chamber, and improve performance." Thiel, he said, developed this idea, then submitted it to engineering schools for research while he used the system for the 1.5-ton engine then under development. It took him a year, but he shortened the chamber from almost six feet to about a foot. This increased exhaust speed to 6,600 and then 6,900 ft/sec, from the roughly 5,300 to 5,600 produced heretofore. It was a significant achievement, but it brought with it a rise in temperature and a decrease in the chamber's cooling surface. Thiel "removed the injection head from the combustion chamber" by creating a "sort of mixing compartment," which separated the flames from the brass injectors. This kept the injectors, at least, from burning.[73]

In conjunction with the shortening of the combustion chamber, Thiel had gone to a spherical shape to create the most efficient geometry by increasing volume. This also served to reduce pressure fluctuations and increase the mixing of the propellants. Until he could use Test Stand 1 at Peenemünde, however, Thiel was restricted in scaling these innovations in the 1.5-ton engine up to 25 tons. He thus went to an intermediate size of 4.2 tons that could be tested at Kummersdorf, and he moved from one injector in the smaller engine to three in the larger one. Each had its own "mixing compartment" or "pot," and the clustering of three pots actually increased the efficiency of combustion further. But to go from that arrangement to the eighteen "pots" necessary for the 25-ton A-4 combustion chamber created considerable problems of scaling up and of how to arrange the pots. At first Thiel and his associates favored a "star" arrangement of six or eight bigger injectors around the sides of the chamber, but von Braun suggested eighteen pots of the size used for the 1.5-ton engine, arranged in concentric circles on the top of the chamber. As Schilling said, this was a "plumber's nightmare" with the many oxygen and alcohol feed lines that it required. But it avoided the problems of combustion instability—as we now call it—that other arrangements had created.[74]

Cooling the engine remained a problem. Regenerative cooling used on earlier, less-efficient engines did not suffice for the larger engine. Oberth had already suggested the solution, which was film cooling[75]—introducing an alcohol flow not only around the outside of the combustion chamber (regenerative cooling) but down the insides of the wall and the exhaust nozzle to insulate them from the heat of combustion by means of a film of fuel. Apparently, others in the propulsion group had forgotten Oberth's suggestion. And it is not clear that the idea came from Oberth as it was applied to the 25-ton engine. Several sources agree, however, that diploma engineer Moritz Pöhlmann, who headed the propulsion design office at Kummersdorf after August 1939, came up with the suggestion. Tested on smaller engines, the idea proved its validity. So on the 25-ton engine, there were four rings of small holes drilled into the chamber wall that seeped alcohol along the inside of the motor and nozzle. This film cooling took care of 70 percent of the heat from the burning propellants, the remainder being absorbed into the alcohol flowing in the regenerative cooling jacket on the outside of the chamber. Initially, 10 percent of the fuel flow was used for film cooling, but Pöhlmann refined this by "oozing" rather than injecting the alcohol without loss of cooling efficiency.[76]

In conjunction with the combustion chamber, Thiel's group had to come up with a pumping mechanism to transfer the propellants from their tanks to the injectors. The large quantities of propellant that the A-4 would use made it impractical to feed the propellants by nitrogen-gas pressure from a tank, as had been done on the A-2, A-3, and A-5. Such a tank would have had to be too large and heavy to provide sufficient pressure over the 65-second burning time of the engine, creating unnecessary weight for the A-4 to lift. This in

Figure 7. V-2 engine, with nearly spherical shape and cumbersome 18-*Topf* (pot or burner cup) ignition system. National Museum of the United States Air Force, Wright-Patterson AFB, Ohio. Courtesy of the museum.

turn would have reduced its performance. In 1937 Thiel had mentioned the need for pumps to increase the chamber pressure and that some pumps had already been developed. Indeed, von Braun had already begun working in the middle of 1935 with the firm of Klein, Schanzlin & Becker, with factories in southwestern and central Germany, on the development of turbopumps. In 1936 he began discussions with Hellmuth Walter's engineering office in Kiel about a "steam turbine" to drive the pumps through use of hydrogen peroxide and a catalyst to produce the steam.[77]

In the final design, a turbopump assembly contained separate centrifugal pumps for alcohol and oxygen on a common shaft, driven by the steam turbine. As planned, the turbine was powered by hydrogen peroxide that was converted to steam by a sodium permanganate catalyst. It operated at a rate of more than 3,000 rpm and delivered roughly 120 pounds of alcohol and 150 pounds of liquid oxygen per second to create a pressure in the combustion chamber of about 210 psi (pounds per square inch). This placed extreme demands on the pump technology of the day, especially given the huge differential between the heat of the steam (+385°C) and the boiling point of the liquid oxygen (-183°C).[78]

Moreover, the pumps and turbine had to weigh as little as possible so as to reduce the load the engine had to lift. As a consequence, there were problems with the development and manufacture of both devices. Krafft Ehricke, who worked under Thiel after 1942, said in 1950 that the first pumps "worked unsatisfactorily" so the development was "transferred to Peenemünde." He claimed the steam generator was also developed at Peenemünde. This is suggested as well by Schilling, who wrote that for the steam turbine "we borrowed heavily from" the firm of Hellmuth Walter at Kiel. He said a "first attempt to adapt and improve a torpedo steam generator [from Walter's works] failed because of numerous details (valves, combustion control)." Later a successful version of the steam turbine was developed, and mass production was done by Heinkel in Bavaria. As for the pumps, there are references in Peenemünde documents as late as January 1943 to problems with them but also to orders for large quantities.[79]

The problems with the pumps included warping of the pump housing because of the temperature difference between the steam and the liquid oxygen, cavitation because of bubbles in the propellants, difficulties with lubrication of the bearings, and problems with seals, gaskets, and choice of alloys. The cavitation problem was especially severe, since it could lead to vibrations in the combustion and possible explosions. It was solved by the interior redesign of the pumps and carefully regulating the internal pressure in the propellant tanks to preclude the formation of the bubbles.[80]

The manufacturers of the turbines and pumps were among many contributors to the A-4 who were not a part of Peenemünde proper. As of mid-March 1942, a Peenemünde document (see table 2.2) showed the percentages of development work on different portions of the A-4 (called the *Rauchspurgerät II* or smoke-emitting apparatus II).

This table gets ahead of the story in some categories but shows that most of the work on the pumps occurred at the contract firm and that, while the same was true for the steam turbine, a large percentage of the work was done in-house by von Braun's group. It thus qualifies but does not contradict what Ehricke stated about the work on pumps transferring to Peenemünde. Eh-

Table 2.2. Percentage Contributions to A-4 Development

	Kummersdorf and Peenemünde Staff	Collaborating Institutes	Contract Firms
Combustion chamber[a]	85	13	0
Pumps	20	10	70
Steam turbine	40	2	53
Gyro controls	35	5	60
Guide plane	40	25	35
Engine cutoff mechanism	5	85	10

Source: "Prozentualer Anteil der mitarbeitenden Firmen, Institute und Dienststellen an der Entwicklung des Rauchspurgerätes II," as of mid-March 1942, dated March 27, 1942, microfilm roll 59, FE 692f, NASM.
Note: It is not clear who compiled these numbers, which must be only estimates, but they seem plausible.
a. For the combustion chamber, only 98 percent of the contribution is accounted for here.

ricke related as well that development of what he called "control devices for the propulsion system, i.e. valves, valve controls, gages, etc." presented "especially thorny" problems. The ones available from commercial firms either weighed too much or could not handle the propellants and pressure differentials. A special laboratory at Peenemünde had to develop them during the period 1937 to 1941, with a pressure-reducing valve having its development period extended until 1942 before it worked satisfactorily.[81]

As table 2.2 shows, technical institutes contributed a small but significant share of development for both the combustion chamber and the pumps. A Professor Wewerka of the Technical Institute of Stuttgart provided valuable suggestions for solving design problems in the turbopump. He had written at least two reports on the centrifugal turbopumps, in July 1941 and February 1942. In the first, he investigated discharge capacity, cavitation, speed relationships, and discharge and inlet pressures on the alcohol pump, using water instead of alcohol as a liquid to pass through the pump. Because the oxygen pump had dimensions almost identical to those of the alcohol pump, he merely calculated corrections to give values for the oxygen pump with liquid oxygen flowing through it instead of water through the alcohol pump. In the second report, he studied both units' efficiencies, effects of variations in the pump inlet heads upon pump performance, turbine steam rates, discharge capacities of the pump, and the pump's impeller design. These tests also used water, with pump speeds going up to 12,000 rpm.[82]

Schilling pointed to important work Professor Wewerka's institute did in nozzle design, which was critical to achieving the highest possible performance from the engine by establishing an optimal expansion ratio. This problem was complicated by the fact that an ideal expansion ratio at sea level, where the missile was launched, quickly became less than ideal as atmospheric pressure decreased with altitude.[83] Wewerka wrote at least four reports during 1940 studying such things as the divergence of a Laval nozzle and the thrust of the jet discharged by the nozzle. In one report in February, he found that a nozzle divergence of 15 degrees produced maximum thrust. Gerhard Reisig agreed with Schilling that this was the optimal exit-cone half angle for the A-4. Reisig, chief of the measurement group under Steinhoff until 1943, also gives Wewerka credit along with Thiel for shortening the nozzle substantially. In another report, Wewerka found that the nozzle should be designed for a discharge pressure of 0.7 to 0.75 atmospheres, and Reisig says the final A-4 nozzle was designed for 0.8 atmospheres, which corresponded to the atmospheric pressure at 2,000 meters (about a mile and a quarter).[84]

Schilling pointed to Professor Hase of the Technical Institute of Hannover and Professor Richard Vieweg of the Technical Institute of Darmstadt for their contributions to the "field of powerplant instrumentation." Other "essential contributions" that Schilling listed included those of Professor Schiller of the University of Leipzig for his investigations of regenerative cooling, and of Professors Pauer and Beck of the Technical Institute of Dresden "for clarification of atomization processes and the experimental investigation of exhaust gases and combustion efficiency, respectively."[85] In an immediate postwar interview at Garmisch-Partenkirchen, an engineer named Hans Lindenberg—who had been doing research on fuel injectors for diesel engines at the Technical Institute of Dresden from 1930 onwards and, from 1940 on, partly at Dresden and partly at Peenemünde, on the combustion chamber of the A-4—even claimed that the design of the A-4's fuel injection nozzles "was settled at Dresden." This was undoubtedly an exaggeration, but he added that Dresden had a laboratory for "measuring the output and photographing the spray of alcohol jets." And surely it and other technical institutes contributed ideas and data that were important in the design of the propulsion system.[86] Along similar lines, Konrad Dannenberg, who worked on the combustion chamber and ignition system at Peenemünde from mid-1940 on, described their development in general terms and then added, "Not only army employees of many departments participated, but much of the work was supported by universities and contractors, who all participated in the tests and their evaluation. They were always given a strong voice in final decisions."[87]

One final innovation, of undetermined origin, involved the ignition process. This used a pyrotechnic igniter. In the first step of the process, the oxygen valve was opened by means of an electrically activated servo system, followed by the alcohol valve. Both opened to about 20 percent of capacity, but since the propellants flowed only through gravity (and slight pressure in the oxygen tank), the flow was about 10 percent of normal. When lit by the igniter, the burning propellants produced a thrust of some 2.5 tons. The launch team then started the turbopump by opening a valve, permitting air pressure to flow to the hydrogen peroxide and sodium permanganate tanks. The permanganate solution flowed into a mixing chamber, and as soon as pressure was sufficient, a switch opened the peroxide valve, allowing it too to enter the mixing chamber. After pressure rose to 33 atmospheres as a result of the decomposition of the hydrogen peroxide, the oxygen and alcohol valves opened fully, and the pressure on the turbines in the pumps caused them to operate, feeding the propellants into the combustion chamber. It required only about three-quarters of a second from the time the valve in the peroxide system was electrically triggered until the missile left the ground.[88]

Even after the propulsion system was operational, the propulsion team had by no means solved Schilling's "mysteries of the combustion process" in a scientific way. The engine ultimately developed an exhaust velocity of 2,050 m/sec (6,725 ft/sec), which translated into a specific impulse of 210 lbf-sec/lbm, or pounds of thrust per pound of propellant burned per second, the more usual measure of performance today.[89] This was quite low by later standards. But it was sufficient to meet the requirements set for the A-4 and quite a remarkable achievement for the time. As von Braun said after the war, however, "the injector for the A-4 [was] unnecessarily complicated and difficult to manufacture."[90] Certainly the eighteen-pot design of the combustion chamber was inelegant. And despite all the help from an excellent staff at Peenemünde and from the technical institutes, Thiel had to rely on a vast amount of testing. Von Braun said, "Thiel's investigations showed that it required hundreds of test runs to tune a rocket motor to maximize performance," and Dannenberg reported "many burn-throughs and chamber failures," presumably even after he arrived in 1940.[91]

Still, through a process of trial and error, use of theory where it was available, further research, and testing, the team under von Braun and Thiel had achieved a workable engine that was sufficient to do the job. As of 1958 in the United States, a knowledgeable practitioner in the field of rocketry wrote, "The development of almost every liquid-propellant rocket has been plagued at one time or another by the occurrence of unpredictable high-frequency pressure oscillations in the combustion chamber"—Schilling's mysteries still at work—and "after some fifteen years of concentrated effort in the United States on liquid-propellant rocket development, there is still no adequate theoretical explanation for combustion instability in liquid-propellant rockets."[92] That the propulsion team at Kummersdorf and Peenemünde was able to design a rocket engine that did the job despite the team's (and later researchers') lack of fundamental understanding of the combustion processes at work is testimony not only to their skill and perseverence but to the fundamental engineering nature of their endeavor. Their task was not necessarily to understand all the "mysteries"—although they tried—but to make the rocket work. This they did. And when they did it, it was rocket engineering, not rocket science, because they still did not fully understand *why* what they had done was effective, only that it was.

Even without a full understanding of the combustion process, the propulsion group went on to design engines with better injectors. They did so for both the Wasserfall and the A-4, although neither engine went into production. Both engines featured an injector plate with orifices so arranged that small streams of propellants impinged upon one another. In the process, the streams produced oscillations in the engine (combustion instability, characterized by chugging and screeching), but the developers found an angle of impingement that, while not completely eliminating the oscillations, at least minimized them. They also designed a cylindrical rather than a spherical combustion chamber for the A-4, but it had a slightly lower exhaust velocity than the spherical engine.[93]

A-4 Aerodynamics

While Thiel's division was producing engineering solutions to propulsion problems, Rudolf Hermann's was designing and developing Peenemünde's supersonic wind tunnel. Starting with a staff of twenty people in April 1937, the aerodynamic group expanded roughly threefold by the summer of 1939 when the 40-centimeter-square (40×40) supersonic tunnel was first put into operation, according to a postwar account by Hermann. However, a much more nearly contemporary Peenemünde document stated that the aerodynamic group did not begin using the wind tunnel until November 1939, and then only with damp air and uncorrected nozzles. Since the nozzles in the tunnel were extremely important in producing the intended supersonic velocity, in keeping the air flow at a constant Mach number (speed in relation to that of sound), and in keeping it free of the shocks produced as the flow accelerated through the speed of sound, the first tests must have been somewhat crude.

This was especially true as Hermann had pioneered in demonstrating experimentally that condensed water particles were the cause of the "shocklike phenomena" Ludwig Prandtl had reported at the Volta Congress as occurring in his supersonic nozzle. Hermann's mentor, Carl Wieselsberger, had suggested the cause to his subordinate at Aachen, and the results of Hermann's experiments relating the "existence and strength of the disturbances in the flow to the relative humidity in the air" resulted in the installation of a drying system in the Peenemünde tunnels that was reportedly not used elsewhere. This system made Hermann's tunnel superior to contemporary facilities. The drying system began operating in the wind tunnel on April 18, 1940.[94]

Correcting the nozzles was as difficult as it was important. Nozzle theory is complex. Mathematical design of nozzles is "cumbersome," to use Hermann's word, and their mechanical design and construction requires highly qualified specialists. The nozzles at Peenemünde were noteworthy for having been built according to Busemann's method involving the shortest length for each nozzle compatible with a "given throat size and shape." Over the course of the years until near the end of the war—first at Peenemünde and then at Kochel in the Bavarian Alps after an air raid of August 1943 at Peenemünde led Army Ordnance to move the wind tunnels to Kochel in stages—the aerodynamics staff completed seven nozzles ranging in Mach number from 1.22 to 5.18. The Mach 5.18 nozzle was completed only in January 1945 and never fully tested, but the other six saw service over a number of years.

The first corrected nozzle was ready for a Mach number of 1.86 on August 8, 1940. Reliable measurements at Mach numbers of 1.56 and 2.50 could be made with nozzles for those velocities after September 24, 1940, and through the end of that year the aerodynamicists did work for the A-5 and A-4 as well as other projects including some for artillery. For example, already in October 1940 there was a report on aerodynamic studies using an A-4 model in the wind tunnel under conditions ranging from high subsonic velocities to Mach 2.5. The report sought to determine the aerodynamic characteristics of the vehicle and the location of the center of pressure. By April 1942 a report by staff member Siegfried Erdmann, trained at the Technical Institute of Berlin, was reporting on pressure distributions for the A-4 models at Mach numbers from 0.56 to 3.31 and 4.48, showing that the last two high-speed nozzles had been brought on line by then. Erdmann also reported on the distribution of axial forces by evaluating pressure distribution at Mach 2.5 and then calculating the pressure distribution for the missile's reentry speed of Mach 5.1.[95]

The supersonic tunnel was a blowdown variety like Hermann's smaller tunnel at Aachen. It had a vacuum reservoir of 1,000 cubic meters at Peenemünde but only 750 at Kochel—where the tunnel and associated workshops took on a cover designation as a *Wasserbau Versuchsanstalt* (hydraulic experimental station), suggestive of research concerned with dams, canals, and the like. Peenemünde had six vacuum pumps with 800 kilowatts of total power, whereas there were only four pumps at Kochel in the beginning. At the Baltic site there were two 40×40-centimeter test sections, but only one in the Bavarian Alps for some time, the other remaining behind. In addition, at both sites there apparently was an 18×18-centimeter section capable of continuous operation at Mach 3.3. Finally, during blowdown operations, the running time for the larger test sections at Peenemünde was 20–25 seconds; at Kochel, it was only 15.

By the time the initial move of the aerodynamic staff and tunnel began in the fall of 1943, the A-4 fortunately was close to mass production; several test flights had apparently shown that the missile's external aerodynamics and graphite control surfaces, as then configured, worked well. It took roughly a full year and three hundred freight cars to transport the initial equipment to Bavaria. However, the first test section at Kochel began operations about half a year after it was freighted from the north. Operations at Peenemünde continued until May of 1944, with work continuing on Wasserfall. By November 1944 the test section that had remained at Peenemünde was reassembled at Kochel and could participate again in full-fledged testing there.[96]

Even before the supersonic wind tunnel was first functioning at Peenemünde, the aerodynamic staff had begun to make contributions not only to the A-5 but to the A-4 as well. At the end of March 1938, for example, Gerhard Eber and Rudolf Hermann completed a report on the expected increases in skin temperature of the A-4 during atmospheric flight. As might have been expected, temperatures rose during ascent through the atmosphere, cooled during flight in the stratosphere, and increased following reentry. By November 1938, Eber had found that the metal for the skin had to withstand temperatures estimated at more than 1,000°C at a Mach number of 5.0, higher than expected. What was not known was how much of this heat would actually "enter the skin."[97] No doubt the resulting data contributed to the ultimate decision to shift to sheet steel for the missile skin in place of lighter materials, although orders to avoid aluminum and magnesium alloys because they were scarce were also a factor.[98]

Structural designers needed pressure measurements for the body, fins, and air rudders to determine loading. Studies such as the ones mentioned above fulfilled these demands. The center of pressure moved with changes in Mach number, as did lift and drag, and "the pressure distribution over the fins." The entire missile had to be designed with compromises made between the ideal arrangements for subsonic and supersonic speeds, and for this, data provided by the aerodynamicists were surely critical.[99]

Additionally, during the early stages of the A-4's development, inaccuracies in the construction of the vehicle produced an angular velocity about the longitudinal axis of something like 30 to 60 rpm. This made control difficult. Wind tunnel tests revealed these inducements to roll ("rolling moments") caused by the asymmetries in construction. Aerodynamicists informed construction engineers how much error could be tolerated in the lineup of the fins, and a machine was developed to measure these tolerances. Then residual spin caused by asymmetries in the construction could be compensated by air rudders on the bottoms of the fins.[100] (This problem would exist only in the atmosphere, where air pressure acting on the asymmetries would cause the roll; the air rudders could correct for it as soon as the rocket reached a speed where they were effective, which was probably no greater than the speed that would cause the rolling moments.)

One final contribution of the wind tunnel came when the specifications for the jet-vane servos failed to account for some "extreme strains" that could be measured only after a number of wind tunnel tests. This occurred late in the A-4 development. The tests showed that the control systems for the roll and yaw axes could be combined in a way that solved the problem. For example, in the case of a deviation in roll, the control engineers made the jet vane next to one tail fin move in one direction in concert with the air vane on that fin, while their opposite numbers deflected in the other direction. How this worked was complicated. Suffice it to say that wind tunnel data allowed control engineers to fix the problem.[101]

Despite the comparatively large staff and excellent facilities at their disposal, Hermann's aerodynamicists did not do all the work in-house. They relied heavily on the writings of leading aerodynamicists like Prandtl, Busemann, and von Kármán. For example, Kurzweg noted that an article von Kármán coauthored in 1932, "Resistance of Slender Bodies," became the basis of calculations on the pressure distribution for the A-5 and A-4. Wegener noted that "the underlying knowledge of fluid dynamics—including some aspects of supersonic flows—accumulated at Prandtl's laboratories and other institutions made possible the rapid developments at Peenemünde." But beyond simply drawing upon this body of engineering knowledge, Hermann, Kurzweg, and others farmed out research to be done at these outside institutions, including Prandtl's Göttingen University but also places like the Technical Institutes of Darmstadt and Aachen. At these schools Professors W. Tollmien of Darmstadt and R. Sauer of Aachen "developed their methods of characteristics for axisymmetric bodies and calculated for us the pressure distribution over the A-4 bare bodies."[102]

In short, by combining their in-house expertise and equipment with outside knowledge and contributions, the aerodynamic group around Hermann managed to make the A-4 stable enough to fly in its proper trajectory but not so stable as to make the requirements on movable rudders and jet vanes excessive.[103]

The aerodynamic problems of the A-4 seemed to be solved until some mysterious problems developed starting in November 1943. A-4 rockets began to be launched over land in Poland at a place called Heidelager. When they had been launched over water along the Baltic coast, dye marked where they impacted and there had been no suspicion of problems. But when they were launched over land, it soon became apparent that 70 percent or more of the missiles were breaking up in the air.

There was much speculation as to the causes, with suspicion falling on ignition of the propellant remaining in the tanks after cutoff, overpressurization of the tanks, and aerodynamic heating. The problems occurred despite verification of wind tunnel data showing that the rockets could stand the stresses they faced. Flight tests of vehicles in which the alcohol was designed to be used up entirely showed that the A-4s still experienced airbursts, ruling out the fuel tank as the problem. The engineers tried a further expedient of insulating the alcohol and oxygen tanks with glass wool to protect them from aerodynamic heating. All six of the vehicles modified in this way flew without problems, and the modification was adopted for missiles being shipped west for use against the Allies. However, it proved to be only a partial answer. Some 30 percent of the modified missiles still broke up before reaching their targets. With no clear indication where the problem lay, engineers next tried reinforcing the center section of the vehicle. This too produced fewer airbursts, making it appear that aerodynamic heating and structural loads from air pressure were the culprits, despite wind tunnel data to the contrary.[104]

If this interpretation was valid, rocket development was clearly not a science at this point, and the best available engineering data had proved flawed. Cut-and-try engineering yielded the best solution, although hardly a perfect one.

A-4 Guidance and Control Systems

The control systems on the A-5 were successfully tested, but they were far from the final solution for the A-4. Much research and development remained to be done, and several different systems flew on the A-4. Thus it cannot be said that there was a single control system for the missile. Moreover, the technologies employed were forbiddingly complex. For those reasons, only a general discussion can be provided here. Since, however, the systems tested on the A-4 provided some of the bases for later developments in the United States—as was true also of other A-4 components—it is important to cover the major points.

One new development for the A-4 that apparently had not been used for the A-5 was the so-called *Mischgerät* or mixing device developed by Hoelzer. When he first came to Peenemünde, he was told that the A-5 flew satisfactorily with gyros unless there was a stiff wind. His job was to solve that problem for the A-4. While still a student at the Technical Institute of Darmstadt, he flew gliders and had conceived of a device to measure velocity independent of wind, which was the germ of the condenser used on the A-5 to calculate corrections for lateral velocity. Now he extended that solution into a larger application. It applied more mathematical functions to obtain the angular rate and acceleration of the vehicle by differentiating signals from position gyros indicating attitude. His device in its final version could even perform double differentiation, doing away with the necessity of feedback from the jet-vane servo system. It did this by using the rate of change of the gyros' signals over time. Then it mixed the roll and yaw signals. The process required both an amplifier and a rate circuit, plus a ring modulator and a phase bridge rectifier. This is highly technical, but essentially the device eliminated the need for rate gyros. It also saved a great deal of expense, since rate gyros cost about 2,800 times as much as the very inexpensive mixing device, which was in effect an analog computer.[105]

The principal gyroscopic system used on the A-4 was a variant of the Siemens *Vertikant* system, sometimes referred to as the LEV-3. In some cases it was coupled with the Wolman radio control system for cutoff of the engine. In others, a gyroscopic integrating accelerometer (discussed below) provided cutoff. In some missiles, both cutoff systems operated. As in Siemens' original conception, one gyro "controlled" deviations in yaw and roll, while the other did so for pitch and also caused the vehicle to modulate its pitch according to a time-plan until it had tilted into the predetermined angle with the horizon that would carry it to its target. Each gyroscope was coupled with a control potentiometer pickoff that sent electrical current to the mixing device. Based on the signals it received, the *Mischgerät* in turn controlled the electrical-hydraulic servomotors that moved eight control surfaces: four jet vanes and four aerodynamic control surfaces, one at the bottom of each large fin. Two of the jet vanes moved together to control yaw and in opposite directions to control roll, and they were linked to their corresponding aerodynamic surfaces so that both sets would move together for control in either roll or yaw. The pitch vanes and aerodynamic surfaces operated separately from each other. The pitch gyroscope was matched with an electrical "controller" that was "driven by a constant-speed motor." It also had a "precessing coil" that caused the axis of the gyroscope to rotate and change the missile's pitch axis so that it would tilt toward a horizontal direction with respect to the earth and fly toward its intended target.[106]

The other principal gyroscopic system used on the A-4 was Kreiselgeräte's stabilized platform, initially the SG-66 and later an SG-70. Basically similar, the two platforms were suspended in gimbals with the outermost gimbal pivoting on the pitch axis. The next gimbal rotated on the yaw axis, followed by one rotating on the roll axis. Three gyros provided three-axis stabilization in space (without relation to the vehicle itself). This characteristic made the design a forerunner of several later guidance systems, although its components were much less accurate than those that followed. The stabilizing provided by the gyros made the system an inertial one. Torques acting on the individual single-degree-of-freedom gyros (free to move in only one axis) caused them to precess about their sensitive axes. The orientation of the missile relative to the platform was "then determined by potentiometer pickoffs at the various gimbal pivots." The output from the potentiometers went to the mixing device, which controlled the missile through servomotors acting upon the jet vanes and aerodynamic control surfaces.

Fritz Mueller was responsible for developing this platform. He also designed a double-integrating gyro accelerometer, used with the platform and coupled with a ball-and-disk integrator, which controlled propulsion cutoff in this system. Engineers and technicians mounted the gyro as a pendulum, oriented along the missile's major axis. The gyro used an on-off control loop that produced torques about the gyro's precession axis. When the gyro's precession axis rotated at an angle that represented the intended cutoff speed, it closed a contact to initiate the cutoff of propulsion. A clock provided a time reference for integration. This combination in effect, said Mueller, "calculated velocity and distance and determined propulsion cut-off time."

There were some fifteen to twenty successful test firings of A-4s equipped with the stabilized platforms and double-integrating accelerometers, but only late in the war when it was past the time for the system to be used operationally. According to one source, the Germans claimed an experimental accuracy of 500 yards over a range of 200 miles with this system, with the impact points distributed roughly within a circle. But Mueller himself stated that the test firings only "demonstrated a 50 percent probability that range error would not exceed 4.2 km (2½ miles) and lateral error would not exceed 2.4 km (1½ miles) for a range of 200 km (125 miles)." To reduce errors arising from friction in the gyros on this system, near the end of the war the Germans developed air bearings in which high-pressure air jets supported the shaft. They had completed experimental models and incorporated them in a modified SG-66 called the SG-X, but they never were able to test this system.[107]

The research and development group at Peenemünde used Mueller's and other accelerometers to cut off propulsion in about half of their tests. In the other half, they used radio control. While the radio control was generally more accurate (if less dependable), German troops firing the missile against England used radio control only about 10 percent of the time because of a shortage of ground radio equipment and the increased danger of air attack while the launch crew operated the radios. Accelerometers also were not subject to jamming and required less equipment than the radio cutoff, which used more than a hundred tubes in its component units. With either system, cutoff occurred in two stages for greater accuracy. Once the steam valves received a signal to close, it took some 2.7 seconds for them to stop the propellant flow. But until the flow ceased, the missile continued to accelerate. In the two-step process, the steam supply to the fuel pumps was reduced at a cutoff presignal given when the velocity reached a point about 5 percent lower than the intended cutoff speed. This reduced the thrust from about 25 tons to 8. Then at the cutoff point, steam and propellant supply ceased altogether.[108]

Like radio cutoff, the guide-plane system for cross-range control (used previously on the A-5) was subject to jamming, although multiple frequencies, coding, and use for only short periods reduced the likelihood that the enemy could do this. Nevertheless, although it also improved accuracy, the guide-plane system was used only sparingly during operational employment of the missile toward the end of the war, although it was used against Antwerp once the city was back in Allied hands.[109]

Another device used with the control system on the A-4 was a cross-range accelerometer that Dr. Helmut Schlitt developed. Schlitt initially worked at the Technical Institute of Darmstadt, but later joined Hoelzer's staff at Peenemünde. Although his accelerometer was tested on the A-4, apparently it was not flown in operational launches against the Allies. Schlitt's device used an electric coil that moved and thus acted like a pendulum. It was positioned in a gap of the sort of magnet used in a loudspeaker. A contact on the moving coil (pendulum) initiated an electrical current that was proportional to the acceleration the moving rocket imparted to the coil. This current was then integrated into the overall control system.[110]

For all the elements of the A-4 control system to function properly, much analysis, testing, simulation, integration, and actual flying of the entire A-4 vehicle was required. As table 2.2 suggests, for the guide plane and engine cutoff mechanism, a good deal of the contribution was made by outside firms and by institutes associated with Peenemünde, with the large percentage on the cutoff device evidently reflecting Wolman's contribution. It appears, however, that the major contributions to the overall control system and its integration into a functioning whole came from in-house efforts, with the institutes offering mostly studies and data.

As late as 1940, to be sure, Steinhoff's group was limited in the sophistication of the equipment it could use. In that year, testing consisted of airplanes flying individual components, flights of the A-5 with different control systems installed, and firings of a rocket with its control system installed in a test stand so that it could rotate in one axis about its center of mass as the appropriate controls operated. All three of these methods were expensive. In the test-stand method, engineers and technicians subjected the vehicle to specific torques supplied by heavy springs to simulate the aerodynamic forces on the vehicle. But the torques could not be adjusted to reflect the changing aerodynamic forces operating on the missile as it rose in the atmosphere, accelerated through the transonic region, tilted toward the horizontal, and the like. Moreover, the center of mass changed with fuel consumption, and the measurements were already inaccurate as a result of friction in the bearings used for suspension.

Hermann Steuding suggested a method of laboratory simulation with a control system mounted on a platform balanced on a knife edge instead of ball bearings. Called a mechanical-motion simulator, it had a servodrive on the platform to effect motion in proportion to the actual movement of the vanes on the A-4. This was still somewhat crude, so Walter Haeussermann developed an attitude control-loop simulation with a pendulum. Haeussermann had received his diploma degree at the Technical Institute of Darmstadt in electrical engineering in 1938 and joined Steinhoff's staff at the end of 1939. This device he developed tilted to simulate the dynamics of rocket motion in different flight situations. Besides the pendulum, it used an eddy current brake, potentiometers, a servoactuator, and a torquer to investigate various aspects of control system operation. These and other electromechanical simulators marked a step forward, but "time varying coefficients" still could not be investigated satisfactorily in the laboratory or on the test stand.[111]

To investigate these coefficients, the researchers needed another invention by Hoelzer, a fully electronic analog computer. As he wrote to historian James E. Tomayko in 1983, once he got accustomed "to the use of electronic circuits for the realization (the transformation of equations into hardware) of the guidance and control equations, it was easy" for him to see that he could do the same for the equations of motion of the vehicle itself. Interestingly, both Steuding—despite his own suggestion of the earlier simulator—and von Braun opposed this development. Von Braun even told him to stop "playing with electronic toys." Undeterred, Hoelzer and Otto Hirschler simply moved to a back room and proceeded to develop the computer, with Hirschler doing most of the wiring. Steinhoff agreed to the subterfuge, and in 1941 when the computer was built, the two builders demonstrated it to Steinhoff. He then arranged a demonstration for von Braun and Dornberger. The device simulated the operation of the A-4 in two degrees of freedom, lateral motion (yaw) and roll.

Hoelzer began by developing mathematical descriptions that mostly consisted of nonlinear differential equations to represent the motions he needed to simulate. Then he designed electronic circuits to represent the mathematical expressions in the equations. The functions the computer could perform included addition and subtraction, multiplication, division, "exponentiation," roots (such as square root), integration, differentiation, and functions of functions. Reflecting its engineering nature, the device simulated the dynamic behavior of the rocket "as exactly as possible" for the steady-state guidance and control equations, but less accuracy was needed for the transient functions. Thus, for simplicity, there were no feedback circuits in the differentiators and integrators. Once Hoelzer tested a guidance and control system in the simulator and it performed well, Steinhoff's division installed it in airplanes and tested it there. After it was successful there, the division changed the resistors in the system and installed it in the rockets

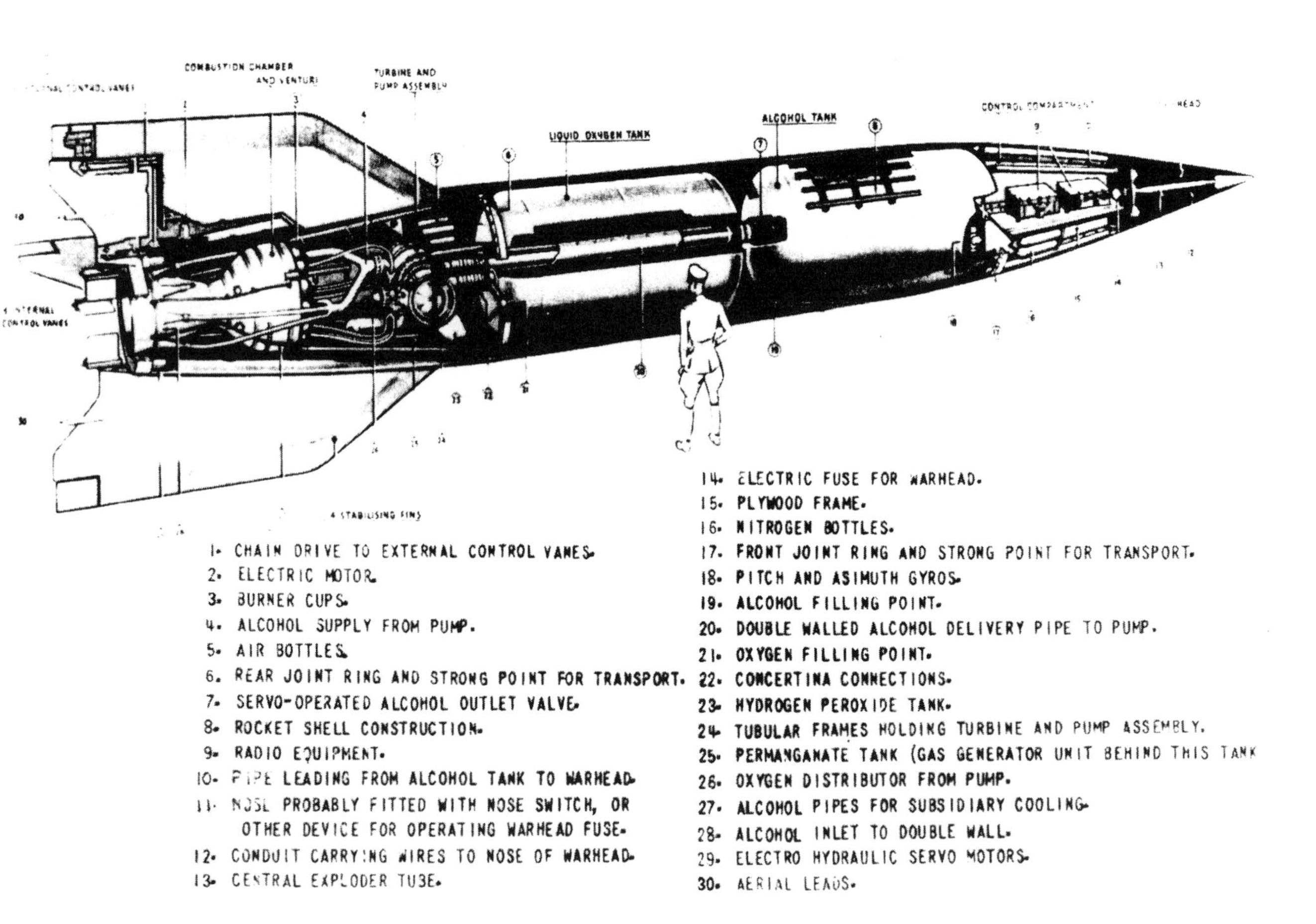

Figure 8. Drawing of the A-4 (V-2) missile, showing its component systems. NASA Historical Reference Collection, Washington, D.C. Courtesy of NASA.

for its final tests. Because of his inventions, Hoelzer claims at least partial credit for the first successful flight of the A-4 on October 3, 1942, after two partially successful launches on June 13, 1942, and August 16, 1942.[112]

Conclusions

The launch on October 3, 1942, was by no means the end of the development effort. There would be plenty of failed launches after that,[113] problems with a variety of systems, with manufacturing, and even with the operational employment of the A-4, renamed the V-2. This is not the place to recount that part of the missile's history in any detail, but some indication of the final part of the story will shed light on the success of the development effort. The A-4 that emerged was about 46 feet long (almost twice as long as the A-5) and 5 feet 5 inches in diameter, with an empty weight of 8,818 pounds. Loaded with propellants, it weighed 28,440 pounds. Engine thrust at takeoff was 55,100 pounds. The rocket stayed vertical for 4 seconds after takeoff and then began its tilt over to 49 degrees from the vertical, which it attained at 54 seconds. It reached a velocity of approximately a mile a second and a range of close to 200 miles, with an effective range of 162 miles.[114]

These were impressive numbers compared with any other rocket of the day, but the A-4 was, by all accounts, not very accurate, despite the best efforts of Steinhoff's division. MacKenzie gives the accuracy at the range of 200 miles as about a half a mile in either direction,[115] but that accuracy varied with the control system used. An Army Air Forces report in 1947 gives the following accuracy, scattered about the mean point of impact, with various methods of control for V-2s used against Antwerp during 1944–45 (see table 2.3). The report does not indicate how these statistics were gathered, but it does give one other indication of the effectiveness of the missile. Some

Table 2.3. Accuracy of V-2 Firings on Antwerp, 1944–45

Type of Control		Mean Deviation (miles)[a]	
Range	Cross-Range	Range	Cross-Range
Mechanical	Mechanical	2.1	2.9
Mechanical	Radio	1.8	0.25
Radio	Radio	4.2	0.25

Source: H. Friedman, "A-4 Control," 40, NHRC.

a. Deviations of "range" cause the missile to fall short of or go beyond its target, while "cross-range" deviations are lateral ones. In converting the report's kilometers to miles, I have rounded off the figures and omitted the plus-or-minus indications for the deviation, which were small. In kilometers, the first line showed 3.4 km ± 0.2 and 4.6 ± 0.3, line 2 showed 2.9 ± 0.3 and 0.4 ± 0.1, and line 3 showed 6.7 ± 0.8 and 0.4 ± 0.1 respectively for range and cross-range, which the report calls "line."

17 percent of the rounds fired against Antwerp failed to launch, and another 18 percent did not follow a normal trajectory, while some 65 percent fell in the vicinity of the target. The report does not reveal the number of rounds, but they are tallied in table 2.3 only for the 65 percent that fell in the vicinity of the target.[116]

According to another source, a total of somewhat more than 3,174 V-2s were shot at Allied cities, of which 1,610 went toward Antwerp and 1,359 against London, the remainder going in much smaller numbers against other Belgian, French, English, and Dutch cities. (The V-2 was too inaccurate to use against any target but a city.) Roughly 5,000 people were killed by these weapons, a tragedy for their families but a very small number compared to the deaths by Allied bombing raids on cities like Hamburg and Dresden. Thus, the V-2 was fortunately a comparatively ineffective weapon.[117]

It was, nevertheless, quite a technological achievement, and it would become an important influence on the development of U.S. missile and launch-vehicle technology, especially as many of its developers came to the United States and participated in developing missiles and launch vehicles. It is therefore relevant to the subject of this book to ask how the Germans achieved such an important technological breakthrough.

Reflections

One important ingredient in the relative technological success of the group formed by Dornberger and von Braun was the heterogeneous engineering of those two managers and Karl Becker. As defined by MacKenzie, heterogeneous engineering is "the engineering of the social as well as the physical world."[118] It was Becker who had gotten the rocket development program started, and he had played a key role in sustaining the program over the years until he committed suicide in 1940 over unspecified family scandals, hints to authorities about them, and verbal assaults against Army Ordinance. He had supported Dornberger and played a role in ensuring that Peenemünde was supplied with materials needed to sustain development there.[119] Dornberger himself piloted "the liquid rocket programme around all the reefs and shoals" it encountered, as von Braun said, and managed to keep support for the program going despite fluctuations in Hitler's support. Both he and von Braun met with key figures in the government and party, including Armaments Ministers Fritz Todt and Albert Speer, on up to Hitler himself. They wended their way skillfully through the "'polycracy' of competing power groups" that characterized the Nazi regime.[120]

It had been Dornberger who perceptively noted the qualities of knowledge and shrewdness in the young von Braun and recruited him to the program. Von Braun *was* young and, like all humans, he made mistakes. One was his discouragement of Hoelzer's "playing with electronic toys," when those "toys" became an important ingredient in the limited but real success of the A-4 control system. That system was undoubtedly ahead of its time, but it and research for it helped to lay the groundwork for future guidance and control systems.[121] Dornberger also recognized that von Braun had failings, being "erratic at first and not completely persistent. He would go from one thing to another, but only until he had a clear idea of what he wanted to achieve." Then "with infinite shrewdness, and full steam ahead, he would pursue the course he considered to be the right one." Self-servingly, Dornberger went on to say "it was a never-ending joy to me to contribute to the development of this great rocket expert by training him from his youth."[122]

It is unquestionably true that von Braun played a key role in integrating the various systems for the A-4 so that they worked effectively together. To this end, he fostered communication between different departments as well as within individual groups. He met individually with engineers and perceptively led meetings of technical personnel to resolve particular issues. According to Dieter Huzel, who held a variety of positions at Peenemünde in the last two years of the war, von Braun "knew most problems at first hand. . . . He repeatedly demonstrated his ability to go coherently and directly to the core of a problem or situation, and usually when he got there and it was clarified to all present, he had the solution already in mind—a solution that almost invariably received the wholehearted support of those present."[123] This was not only technical management of the first order but also heterogeneous engineering, meaning in this context the ability not only to envision a solution but to win its acceptance by a group.

Relatedly, Dornberger described how as "a good systems engineer, he [von Braun] knew how things should and must fit together[;] he not only coordinated but also directed the widespread effort of the many branches of research, engineering, manufacturing and testing in Peenemünde."[124] These lines, of course, were written much later and include the term "systems engineering," which belonged to the postwar vocabulary. But they square with other portraits of von Braun's role.[125] There is no doubt that the young manager did what later came to be called systems engineering. Whereas Goddard had been wonderfully inventive and had foreseen many of the kinds of innovations that the group at Kummersdorf and Peenemünde implemented, he had never succeeded in putting them all together in such a way that they worked effectively to reach the altitudes he sought. He also had not published his detailed findings and results of his research and development. Thus von Braun and the others had to reinvent the rocket and improve the way it operated. The resultant A-4 was not a very effective weapon, but within the limitations of existing theory, technology, and wartime shortages, the Germans had done a remarkable job. Von Braun was an important ingredient in this partial but significant success.

As another Peenemünder, Ernst Stuhlinger, and several others wrote in 1962, "Predecessors and contemporaries of Dr. von Braun may have had a visionary genius equal or superior to his, but none of them had his gift of awakening in others such strong enthusiasm, faith and devotion, those indispensable ingredients of a successful project team." They added, "It is his innate capability, as a great engineer, to make the transition from an idea, a dream, a daring thought to a sound engineering plan and to carry this plan most forcefully through to its final accomplishment." (Since Stuhlinger was principally a scientist, the use of the term "engineer" in reference to von Braun is significant.)

It would be tedious to continue quoting adulatory comments by von Braun's subordinates and admirers, but they did know him well, and one further quotation about his handling of meetings will help to round out the picture. Stuhlinger and Frederick Ordway, who knew von Braun in the United States, wrote in a memoir about him, "Regardless of what the subject was—combustion instability, pump failures, design problems, control theory, supersonic aerodynamics, gyroscopes, accelerometers, ballistic trajectories, thermal problems—von Braun was always fully knowledgeable of the basic subject and of the status of the work. He quickly grasped the problem and he formulated it so that everyone understood it clearly."[126]

Of course, it took many more people than von Braun to develop the A-4. He and Dornberger were successful in recruiting such able managers as Thiel, Hermann, and Steinhoff. They in turn, together with von Braun and others, recruited many capable engineers who provided key innovations. Pöhlmann and Hoelzer are two prime examples. Too little seems to be known about Pöhlmann to discuss how he was recruited and what lay behind his contribution of film cooling for the A-4 engine, but since that was a rediscovery of a process already described basically by Oberth and others, the gap is not critical. We do know a lot about Hoelzer, and his contributions to guidance appear to have stemmed from a combination of the proper technical education, teaching experience that may have sharpened his understanding of technical matters, practical experience in electronics, and being a pilot, which undoubtedly gave him a graphic understanding of the problems of control that he could relate to rocketry in much the way that the Wrights used their knowledge of bicycles to good advantage in understanding how to control their first airplane. As with the Wrights, however, one further ingredient would seem to have been some sort of innate genius that allowed him to come up with the idea of an analog computer in the first place.

Peenemünde was a very large organization, and while some organizational charts have survived, undoubtedly more important than lines of organization was its informal structure. Its functioning was not frictionless.[127] And some of this friction constituted a management failure. But people did apparently communicate and work together to get the job done. Managers sometimes pushed their subordinates pretty hard, but although people worked long hours, they seem to have found the work stimulating, as was true of many later rocket engineers in the United States. And there were sufficient resources at the center so that they could make the necessary progress, coordinated by von Braun and the other leaders under him.

Another essential ingredient in the mix consisted of the contributions of outside firms and technical institutes. Without the initial contributions of Siemens and the ongoing work of Fritz Mueller and Kreiselgeräte, it seems unlikely that the A-4 control systems would have worked as well as they did. The same is true of Wolman's contribution of the fuel cutoff device from the Technical Institute of Dresden. It was not the only system used for that purpose, but it was used to the end of the war and was a critical element. Technical institutes also made important contributions to propulsion and aerodynamics, as already seen. Less tangible were the many discussions that took place among representatives of firms, technical institutes, and engineers from Army Ordnance at Kummersdorf and Peenemünde. The pooling of their expertise probably contributed in innumerable ways to technological development, a practice that was also characteristic of American missile and launch-vehicle development.

Certainly, too, technical reports written by both staff at Peenemünde and people at the technical institutes contributed to the fund of engineering knowledge that Peenemünde passed on to the United States. Not only did Germans from Peenemünde itself immigrate to the United States after the war, carrying their knowledge and expertise with them; some engineers from supporting firms, like Mueller, and even some professors from technical institutes, such as Carl Wagner, joined their former colleagues in the United States. In addition, much of the documentation of the engineering work done in Germany was captured by U.S. forces after the war, moved to Fort Eustis, Virginia, and often translated. How much these documents contributed to postwar rocketry is impossible to know, but the information was available to those engineers who wanted to avail themselves of it and knew where to look.[128] Finally, many actual V-2 missiles were captured and taken to the United States. They too contributed to the fund of information available to American rocket engineers and played a vital role, along with a huge number of separate U.S. contributions to rocketry, in the story that will unfold in the following chapters.

3

JPL

From JATO to the Corporal, 1936–1957

At the time that Dornberger and von Braun in Germany were planning to move their research and development group to Peenemünde, a quite different group of researchers in the United States was beginning a project that bears comparison with the development of Germany's V-2. It was doing so under quite dissimilar circumstances, however. For one thing, the Jet Propulsion Laboratory, as it became in 1944, had far fewer resources available to it than did the German group.[1] The complexity and multidisciplinarity of work on ballistic missiles, however, complicate any comparison of the two efforts. In May of 1945, Dr. Homer E. Newell, who was then working as a theoretical physicist and mathematician at the U.S. Naval Research Laboratory and later became associate administrator for space science and applications in the National Aeronautics and Space Administration, wrote that the "design, construction, and operational use of guided missiles requires intimate knowledge of a vast number of subjects. Among these . . . are aerodynamics, kinematics, mechanics, elasticity, radio, electronics, jet propulsion, and the chemistry of fuels."[2] Newell could easily have added thermodynamics, combustion processes, and materials science, among other topics. It would require more space than is available here to do a thorough comparison, across so many disciplines, of the resources available to the managers at Peenemünde, on the one hand, and JPL, on the other. To the extent that theoretical physics was a factor, it seems clear that the United States in general lagged behind Germany early in the century but had caught up and perhaps surpassed it by the late 1920s or early 1930s.[3]

In the application of theory to engineering, however, it appears that the United States lagged significantly behind Germany until after World War II, although there were exceptions in some fields. In the 1920s and 1930s, some educational institutions and European engineers had begun to demonstrate the importance of theory to American engineers. The United States was slow in changing, though, and remained largely committed to a much more practical, nontheoretical approach than prevailed in Germany.[4] This was notably true in the vital discipline of aerodynamics.[5] In Germany around the turn of the century, as we have seen, the mathematician Felix Klein had established a tradition at the University of Göttingen of applying theory to engineering, and strong centers of theoretical aerodynamics had developed especially at Göttingen and the Technical Institute of Aachen by the beginning of World War I. Two decades later, when Rudolf Hermann moved from Aachen to Peenemünde, he had close ties with the leading aerodynamicists in Germany and could get reports done for Peenemünde by aerodynamicists at Göttingen, among other places.[6] By contrast, the United States was somewhat backward.

GALCIT

Fortunately for the development at rocketry at JPL, by 1930 Nobel laureate Robert A. Millikan and his associates at what had been Throop College of Technology had succeeded in converting that institution into the elite research university known after 1920 as the California Institute of Technology (Caltech, CIT); they had also lured a major theoretical aerodynamicist, Theodore von Kármán, from Aachen to become the director of the Guggenheim Aeronautical Laboratory at Caltech (GALCIT). There he trained a generation of engineers in theoretical aerodynamics and fluid dynamics.[7] Von Kármán's importance is suggested by a remark of his former assistant and student Allen J. Puckett: "There had been a school of thought—and I'm afraid MIT was part of this—that engineering was a sort of trade. You learned how to calculate things by reading equations out of a handbook. It was von Kármán who changed all that by introducing the whole concept of using fundamental theory and precise analysis, by relying on basic principles, to arrive at your result."[8] More generally, with its eminence in physics, physical chemistry, and astrophysics as well as aeronautics,[9] Caltech proved to be a favorable site for the early development of U.S. ballistic rocketry. This allowed it to rival in some degree the achievements of the much larger center at Peenemünde, even with the latter's more lavish funding and earlier start.

Despite von Kármán's presence and the work being done at Caltech in aerodynamics, but for a chance confluence of circumstances, the university probably would not have gotten involved in research on surface-to-surface ballistic missiles. In Germany, Army Ordnance's interest in the subject dated back to the early 1930s. At Caltech, according to von Kármán graduate stu-

Figure 9. Theodore von Kármán, a founder and the first director of the Jet Propulsion Laboratory (JPL), Pasadena, California, in front of a blackboard. Courtesy of NASA/JPL-Caltech.

dent Frank J. Malina, the start of rocket development occurred in 1936 when William Bollay, another of von Kármán's graduate students, gave a presentation on the prospect of rocket-powered aircraft, based primarily on the work of Eugen Sänger, a Viennese engineer who had studied liquid-propellant combustion chambers for rockets while employed at the Technical Institute of Vienna from 1933 to 1935.[10] A newspaper report about Bollay's lecture attracted two rocket enthusiasts to GALCIT—Edward S. Forman and John W. Parsons, described respectively by Malina as "a skilled mechanic" and "a self-trained chemist" without formal schooling but with "an uninhibited fruitful imagination." Together Forman and Parsons teamed up with Malina, a graduate student of Czech descent who, like many other rocket enthusiasts, had become interested in rocketry in childhood when he read Jules Verne—except that in his case he read it in Czech.[11] After an undergraduate degree in mechanical engineering from Texas A&M, Malina had earned an M.S. in mechanical engineering at Caltech in 1935 and was then completing an M.S. in aeronautical engineering there as well.[12] Parsons, for his part, was not entirely self-taught but did actually take some chemistry courses from the University of Southern California in 1935–36, although he never graduated. He worked as a chemist for Hercules Powder Company in Los Angeles from 1932 to 1934 and then was chief chemist from 1934 to 1938 for Halifax Explosives Company in Saugas, California.[13] Besides rocketry, Parsons was deeply interested in magic, a fascination that Malina, as a rationalist, said he found suspect.[14]

Although they constituted a rather surprising partnership, these three young men agreed to work together at the suggestion of Bollay, who was occupied with other efforts, and in March 1936 von Kármán allowed Malina to do his doctoral dissertation on rocket propulsion and rocket flight with the assistance of Parsons and Forman, "even though they were neither students nor on the staff at Caltech."[15] The trio, who were quickly joined by three other individuals, did their work initially without much apparent likelihood of outside support. The U.S. military, unlike the German Army, displayed no early interest in large, long-range surface-to-surface missiles,[16] though it did support an extensive program at Caltech and elsewhere to develop guided bombs and short-range rockets[17] (see chapter 5).

In some respects, the efforts of the small group of six individuals around Malina, Parsons, and Forman—with its usual experiences, for early rocket developers, of mishaps and explosions[18]—can be considered during its first few years as comparable to pre-1932 rocket development in Germany. From three important perspectives, however, the period from 1936 to roughly 1939 at the GALCIT rocket research project was more comparable to the years von Braun spent at Kummersdorf from 1932 to about 1934. First, it is evident that Malina and his fellow rocket researchers had more sophisticated equipment for regulating and measuring propellant pressures and temperatures, chamber pressure, and thrust than did von Braun and his colleagues before 1932.[19] Second, both von Braun and Malina were working on dissertations about similar topics at prestigious universities. And third, the group around Malina made extensive use of theory in its research and development efforts, like the group at Kummersdorf and later Peenemünde.

This point about theory revealed itself in a number of ways. In Malina's memoirs, he repeatedly refers to this use of theory. For example, he states that the initial program of his group included "theoretical studies of the thermodynamical problems of the reaction principle and of the flight performance requirements of a sounding rocket"; it also involved experiments to determine problems to be faced "in making accurate static tests of liquid- and solid-propellant rocket engines." This approach, he added, "was in the spirit of von Kármán's teaching."[20] Besides Malina's own comments, written much after the fact but based upon reports contemporary with the efforts of the late 1930s, there is also much evidence in the reports themselves of the theoretical approach being used in conjunction with experiment and testing. This is true even of a report written by Parsons in 1937. According to Malina, Parsons was often at odds with Caltech people like himself who were dissatisfied with their grasp of fundamental information about rocketry, when the two nonacademics simply wanted to launch rockets. Malina said further that Parsons had only a general understanding of the theory of chemistry, since he had no talent for mathematics.[21] Yet Parsons's 1937 paper, part of a collection of such early papers that the Malina group called its "bible,"[22] relies upon a number of disparate sources including the American Rocket Society, Goddard, and Sänger. It includes numerous calculations about energy, temperature, velocity, volume, pressure, and theoretical impulse. It provides tabular data for a variety of different chemicals that could serve as propellants for rockets, including hydrogen, alcohol, and trimethyl aluminum. And it compares theoretical values with those obtained experimentally, then speculates about reasons for discrepancies.[23] Other reports and articles from this period also attest to a wide-ranging examination of theoretical and experimental literature from all over the world and to the use of that literature to inform the tests and experiments the Malina group performed.[24]

JATO Development and the ORDCIT Contract

The character of the project changed in some respects in January 1939 when the National Academy of Sciences (NAS) Committee on Army Air Corps Research agreed to von Kármán's offer, in return for $1,000, to study jet-assisted takeoff (JATO) of aircraft and prepare a proposal for research on the subject. This led to an NAS contract for $10,000 to this end effective July 1, 1939, and to subsequent contracts from the Army Air Corps and the Navy for JATO units, both liquid- and solid-propellant. But it was not until the fall of 1943, in the wake of vague intelligence reports about German missiles, that the U.S. military evinced any interest in long-range surface-to-surface missiles. There resulted a contract with Army Ordnance beginning in 1944 to develop such a missile under what was called the ORDCIT (Ordnance–California Institute of Technology) contract. Meanwhile, the organization that was gradually assuming the identity of JPL had also agreed to contracts with the Army Air Forces (AAF) Materiel Command to perform hydrobomb (torpedo) and ramjet engine research.[25]

Thus, like Peenemünde, GALCIT/JPL was involved with considerably more than one project and with both army and air forces. At first, the number of personnel and the available facilities were limited, although throughout the history of the projects, Malina and his coworkers were able to call upon the considerable equipment and expertise of Caltech, which itself grew in the course of World War II. The annual prewar budget at Caltech had been $1.25 million. In the course of the war the budget for a single project, the short-range rocket project for the National Defense Research Committee (NDRC) of physics professor Charles Lauritsen, had grown from $200,000 a year in 1941 to $2 million a *month* in 1944, and Caltech's overall personnel had multiplied more than tenfold to almost 5,000.[26] The much more modestly funded and staffed effort under Malina and von Kármán, which was completely separate from Lauritsen's project, grew slowly from the initial half-dozen individuals plus von Kármán to only 85 in May of 1943, 264 in June of 1945 (55 professional, 51 administrative, and 158 skilled and unskilled workers), and 385 in June of 1946.[27] This was a far cry from the 12,000 people working at Peenemünde at its high point, although it's unclear to what extent the statistics for JPL and Peenemünde are comparable.

A couple of important early crises in the development of liquid- and solid-propellant JATO engines will illustrate further the ways in which the GALCIT/JPL team operated. In 1939–40, Parsons sought a solution to the problem of controlled burning for many seconds in a solid-propellant rocket motor. This was critical to the development of a JATO unit. It was Parsons, apparently, who conceived the concept of "cigarette burning" whereby the propellant would burn at only one end. But when powder was compressed into a chamber and coated with a variety of substances to form a seal with the wall of the chamber, tests repeatedly resulted in explosions. Authorities von Kármán consulted advised him that a powder rocket could not burn for more than two or three seconds. Not satisfied with this expert opinion alone, von Kármán characteristically turned to theory for a solution. After discussing the matter with Malina, he devised four differential equations that described the operation of the rocket motor and handed them to Malina for solution. In solving them, Malina discovered that, theoretically, if the combustion chamber were completely filled by the propellant charge and if the ratio of the burning propellant to the throat area of the chamber's nozzle (as well as the physical properties of the propellant) remained constant, the thrust also would remain constant. In other words, there would be no explosions. Encouraged by these findings, Parsons and others came up with a compressed powder design that worked effectively (after one initial explosion) for 152 successive motors used in successful flight tests of JATO units on an Ercoupe aircraft in August of 1941, convincing the Navy to contract for a variety of assisted-takeoff motors.[28] After storage at varying temperatures, however, the motors usually exploded. It was then that Parsons's fertile imagination supplied a solution. Apparently watching a roofing operation in the spring of 1942, he decided that asphalt as a binder and fuel mixed with potassium perchlorate as an oxidizer would provide a stable propellant. Thus the theory of von Kármán and Malina combined with the practical knowledge and imagination of Parsons yielded what came to be called a castable, composite solid propellant that, with later improvements, made large solid-propellant rockets possible.[29] This was a fundamental technological breakthrough.

In developing a liquid-propellant JATO, Malina and his coworkers had to accede to the Air Corps' objections to liquid oxygen as an oxidizer, though

Figure 10. Flight test crew for the Guggenheim Aeronautical Laboratory, California Institute of Technology's solid-propellant JATO (jet-assisted takeoff) experimental booster flown on August 12, 1941 *(left to right)*: F. S. Miller, J. W. Parsons, E. S. Forman, F. J. Malina, Capt. H. A. Boushey Jr. (pilot of the Ercoupe on which the JATO was tested), Pvt. Kobe, and Cpl. R. Hamilton. Courtesy of NASA/JPL-Caltech.

Figure 11. Ercoupe aircraft, piloted by Capt. Homer A. Boushey Jr., taking off from March Field, California, with the aid of a solid jet-assisted takeoff device developed by the Guggenheim Aeronautical Laboratory, California Institute of Technology (GALCIT). On August 12, 1941, this flight constituted America's first known rocket-assisted takeoff, an early milestone in the development of U.S. rocketry. Courtesy of NASA/JPL-Caltech.

it was used by the Germans on the V-2 and by Goddard on both his rockets and the JATO units he worked on separately for the Navy. The Air Corps insisted on an oxidizer that presented fewer problems in production, storage, and transport than the cryogenic (extremely cold) oxygen in its more compact liquid form, which boiled at -297°F and caused even steel to become brittle and organic materials including fabrics and lubricants to ignite or explode.[30] So Parsons suggested red fuming nitric acid (RFNA), a solution of nitrogen dioxide and nitric acid, as an oxidizer. Although it was poisonous and corrosive, requiring aluminum or stainless steel for its propellant tank, it was more acceptable to the Air Corps than liquid oxygen. In late December 1939 the GALCIT team performed tests with the RFNA and determined it would burn in an open crucible with gasoline and benzene. Meanwhile, in 1936 Malina and Hsue-shen Tsien had begun a theoretical study of the "characteristics of an ideal rocket motor consisting of a chamber of fixed volume and an exhaust nozzle." Tsien was a Chinese student of von Kármán whom the latter described as "an undisputed genius." He had earned a B.S. in mechanical engineering in Shanghai, an M.S. in aeronautical engineering at MIT in 1936, and his Ph.D. in aeronautics and mathematics at Caltech in 1939.[31]

To obtain better experimental data than the existing literature offered, the group tested gaseous oxygen and ethylene, burned in a large combustion chamber with high heat capacity and nozzle dimensions selected to permit low rates of consumption for the propellants, supplied under high pressure. Researchers injected propellants into the chamber separately with injection nozzles at the opposite end of the chamber from the exhaust nozzle and ignited them with an ordinary automotive spark plug. Then they measured the thrust, chamber pressure, and weights of propellants consumed, keeping photographic records of the instruments.[32] It would appear, however, that there had not been great progress on liquid propulsion as of July 1940 when Martin Summerfield joined the group. Summerfield was a roommate of Malina who had completed coursework on his Ph.D. (awarded in 1941) in the physics department at Caltech in x-rays and infrared radiation. At the Caltech library he consulted the literature on combustion-chamber physics and found a text with information on the speed of combustion. Using this, he calculated—much in the fashion of Thiel at Kummersdorf—that the combustion chamber being used by the GALCIT team was too large, resulting in heat transfer that degraded performance. So he constructed a smaller, cylindrical chamber that yielded a 20 percent increase in performance.

Summerfield analyzed the heat transfer and heat loss through the combustion chamber. He recalled that von Kármán believed roughly 25 to 30 percent of the heat would be lost, based on information about combustion chambers in reciprocating engines. The eminent aerodynamicist had therefore concluded that it would be impossible for the engines to be self-cooling, restricting both their lightness and their length of operation. Summerfield's calculations indicated that these assumptions about heat transfer were far too high and that a self-cooling engine could operate for a sustained period. Subsequent tests measuring the heat transfer confirmed Summerfield's calculations, and Malina learned about the technique of regenerative cooling from James H. Wyld of Reaction Motors during one of his trips to the East Coast.[33]

For the moment, the group worked with uncooled engines propelled by RFNA and gasoline. Successive engines of 200, 500, and 1,000 pounds of thrust with various numbers of injectors provided some successes but presented problems with throbbing or incomplete initial ignition, leading in either case to explosions. After four months of efforts to improve combustion and ignition as well as to stop throbbing, Malina paid a visit to the Naval Engineering Experiment Station in Annapolis in February 1942. There he learned that Ensign Ray C. Stiff, the chemical engineer in a group headed by Lt. Robert C. Truax, had discovered in the literature of chemistry that aniline ignited spontaneously (or hypergolically) with nitric acid. Malina telegrammed Summerfield to replace the gasoline with aniline. He did so, but it took three different injector designs to make the 1,000-pound engine work. The third involved eight sets of injectors each for the two propellants, with the stream of propellants washing against the chamber walls. This must have provided some film cooling, but Summerfield recalled that after 25 seconds of operation, the heavy JATO units glowed cherry red. Nevertheless, the units successfully operated on a Douglas A-20A bomber for forty-four successive firings in April 1942, the first successful operation of liquid JATO in the United States, leading to orders by the AAF with the newly formed Aerojet Engineering Corporation.[34]

Aerojet was established in March 1942 by Malina, von Kármán, Summerfield, Parsons, Forman, and von Kármán's lawyer Andrew G. Haley for the purpose of producing the rocket engines developed by the GALCIT group. It did considerable business with the AAF and Navy for JATO units during the war, and became by 1950 the largest rocket engine manufacturer in the world and a leader in research and development of rocket technology. Until its acquisition by General Tire and Rubber Company in 1944–45, when it became known as Aerojet General Corporation, Aerojet and the GALCIT project maintained close technical relations.[35] This was quite different from the relations of Peenemünde and Mittelwerk in many respects, although people from Peenemünde had to perform quality control and oversee production of V-2s at the underground facility.[36]

Although GALCIT/JPL was involved essentially with JATO work from 1939 to 1944, already in the summer of 1942 the project began designing pumps to deliver liquid propellants to a combustion chamber instead of feeding the propellants by using a gas under pressure. By the fall of that year, project engineers were also working on using the propellants to cool the combustion chamber of a 200-pound-thrust engine.[37] Thus, even before the group resumed formal work on larger rockets in 1944, it was already addressing technical issues that would contribute to the design of those rockets—both liquid and solid. While JATO remained the focus of the project, it was divided into liquid- and solid-propellant sections, headed in May 1943 by Summerfield and Parsons respectively, supported by design, data, and materials sections, plus one for production and maintenance. Where needed, these were all supplemented by other specialized facilities and equipment at Caltech itself.[38] Once the project began working on larger rockets, however, it reorganized into a larger number of sections, some devoted to specific technologies such as the liquid- and solid-propellant rockets, underwater propulsion, and the ramjet, while others fulfilled general functions like research analysis, materials research, propellant research, engineering design, and field testing.[39]

The dynamic von Kármán remained director of the project until the end of 1944, when he left to establish the Scientific Advisory Board for the AAF. Malina was chief engineer of the project until he succeeded von Kármán as (acting) director. But, according to Summerfield, there was no counterpart at GALCIT/JPL for von Braun at Peenemünde. Instead, the way the professionals in the project integrated the various components of the rockets and the various developments in fields as disparate as aerodynamics and metallurgy was by the simple practice of discussing them as colleagues.[40] Summerfield seemed to suggest that much of this was done informally, but JPL, like Peenemünde, also had many formal meetings where such issues were discussed. In addition, a good deal of what later was called systems engineering was done by JPL's Research Analysis Section, headed first by Tsien and then by Homer J. Stewart.

Stewart had earned a B.S. in aeronautical engineering in 1936 at one of the larger and more advanced aeronautical engineering schools in the country, the University of Minnesota. There he took lots of math courses. He then studied aeronautics and meteorology at Caltech, where he took a number of physics courses as well as one in compressible fluid theory, earning his Ph.D. in aeronautics in 1940. Stewart recalled that he had joined the project as soon as it had money available to hire him in 1939, and he had performed some early systems engineering on the JATO units, making calculations about performance on takeoffs with the JATOs and the results of a given amount of rocket assistance, for example. Later in Research Analysis, which he says was another term for systems engineering, he and his staff did applied mathematics, solving messy problems and then turning them over to regular engineering groups once they had been resolved enough to be considered routine. They performed trajectory analysis and studied heat transfer in nozzles, integration problems, and external aerodynamics, including testing in wind tunnels.[41]

Another important factor in the process of innovation in rocketry at JPL was the amount of information, data, and cooperation received not only from Caltech but from other sources outside the JPL complex. Beginning in 1942, a group under Bruce H. Sage of Caltech's Department of Chemical Engineering undertook analysis of chemical problems in propellants for JPL. Sage also made enormous contributions to Lauritsen's shorter-range rocket project in Eaton Canyon, but he used his expertise to assist the propellant engineers at JPL as well.[42] Besides Sage and Tsien, others serving as consultants to the project in 1943 alone included the noted aerodynamicist Clark Millikan (son of R. A. Millikan) and the Nobel Prize–winning chemist Linus Pauling, who also did work during the war analyzing double-base solid propellants for a separate NDRC project.[43] By 1945 when JPL began firing its rockets at various ranges in California and New Mexico, the Ballistics Research Laboratory of Aberdeen Proving Ground in Maryland was gathering radar tracking and ballistic data for JPL. Stewart and his staff could obtain wind tunnel data not only from the 10-foot subsonic wind tunnel at GALCIT but also from a supersonic tunnel at Aberdeen designed in part by Caltech's Allen Puckett. In addition, the Sperry Gyroscope Company was developing "long-range missile auto-pilots and servo mechanisms" for JPL.[44] In August 1944, Stewart discussed aerodynamic forces with Dr. Wolfgang Benjamin Klemperer from Douglas Aircraft Company, which had built a number of winged missiles and had valuable information to share.[45] These are only some of the examples of outside information made available to the members of the GALCIT/JPL program. It is impossible to measure their precise influence upon rocket development, but it is highly likely that the synergy resulting from the mixture of backgrounds, talents, and knowledge brought by the participants contributed significantly to the work of JPL in rocketry, just as it was doing at Peenemünde.

Private A

Apart from the earlier successes with JATOs, the JPL contribution to rocket development in the United States was essentially threefold in the period of the war and the immediate postwar era, and all three were interrelated. The first achievement was the successful launching in December 1944 of the Private A test rocket in the California desert north of Barstow. This was a fin-stabilized, 92-inch rocket propelled by a castable asphalt-perchlorate substance known as GALCIT 61-C and four 4.5-inch T22 standard Army solid-propellant rockets developed by the NDRC and used as boosters. Aerojet had manufactured the main rocket motor, which provided 1,000 pounds of thrust for 30 seconds. The sixteen rounds of Private A that were fired for record, after initial tests and adjustments, achieved ranges averaging about 18,000 yards—just over ten miles—with a maximum of 20,000 yards. Besides providing data on trajectories and the use of boosters, the rocket served as a precursor for later solid-propellant rocket developments.[46]

It also provided data on launching and external ballistics that were useful in the Corporal E missile development under a 1944 contract with Army Ordnance as part of the ORDCIT project. The Private A had been conceived from the beginning as a test vehicle to provide aerodynamic and structural information for designing and developing the Corporal E. The results of the testing were also intended to help solve stabilization problems and to provide familiarity with control issues for the Corporal E.[47]

WAC Corporal

The second major success of the JPL rocket development effort in this period was the launching of the WAC Corporal sounding rocket (the term WAC standing for Women's Army Corps or Without Attitude Control, depending upon the source consulted). Army Ordnance had requested that the project investigate the feasibility of a rocket carrying 25 pounds of meteorological equipment to a minimum altitude of 100,000 feet to satisfy a requirement of the Army Signal Corps for a high-altitude sounding rocket. The JPL team redesigned an Aerojet motor that used monoethylene as a fuel and nitric acid mixed with oleum as an oxidizer. The original motor was regeneratively cooled by the monoethylene. JPL adapted the motor to use RFNA containing 6.5 percent nitrogen dioxide as oxidizer and aniline containing 20 percent of furfuryl alcohol as a fuel, thereby increasing the exhaust velocity from 5,600 to 6,200 ft/sec. According to one source, the specific impulse was 200 lbf-sec/lbm (slightly lower than the V-2's). The rocket was 194 inches (16.17 feet) long and 12.2 inches in diameter (thus quite a bit smaller than the 22-foot A-3). It had an empty weight of 296.7 pounds and a weight with propellants loaded of 665 pounds. The propellants were fed by air at an initial pressure of 2,100 psi. Boosted by a modified Tiny Tim aircraft rocket developed by the Lauritsen group of the NDRC and guided by a launching tower with three guide rails, the WAC Corporal reportedly reached a maximum altitude on October 11, 1945, of between 230,000 and 240,000 feet—more than forty miles, and more than double the specifications set by the Signal Corps. Besides achieving this altitude, the WAC Corporal led directly to the successful Aerobee sounding rocket built by Aerojet, which was used by the Applied Physics Laboratory of Johns Hopkins University for research in the upper atmosphere. Also benefiting from the data provided by the WAC Corporal were the Lark and Nike missiles.[48]

A total of ten firings of the WAC Corporal took place in September–

October 1945. In addition to meeting the requirements of the Signal Corps, the small liquid-propellant rocket functioned as a test version of the Corporal E, providing valuable experience in the development of that larger vehicle. In the midst of the testing, the program decided to redesign the WAC Corporal to reach higher altitudes. To this end, there was a substantial modification of the engine, reducing its weight from 50 to 12 pounds. The initial WAC A version of the rocket had a comparatively thin, cylindrical inner shell of steel for the combustion chamber, with an outer shell that fit tightly around it but was equipped with a joint to permit expansion. Helical coils, spiraling around the outside of the combustion chamber like a screw thread, provided regenerative cooling, while a shower-type injector sent eight fuel streams to impinge on eight oxidizer streams, thus producing atomization and mixing.

For the modified WAC B engine, designers reduced the combustion chamber from 73 to 61 inches in length. It had an inner shell spot-welded to the outer shell, with helical cooling passages retained as a design feature. There were minor modifications to the injector, which remained a showerhead with eight pairs of impinging jets. The B-model rocket was 4 inches longer than the WAC Corporal A, and the diameter of the fins also increased by 4 inches. The diameter of the overall rocket remained unchanged. The A-model had used one long steel tank divided into three compartments for the pressurized air, fuel, and oxidizer. In the WAC Corporal B, the single tank gave way to three separate tanks, allowing the oxidizer to be stored in an aluminum structure for reduced weight. The other tanks now were made of chrome-molybdenum steel, with the air tank in front of the other two. The gauge of the skin for the outside of the rocket declined to reduce weight, with most of the weight carried by inner trusses rather than the skin. The researchers reduced the propellant weight by 40 pounds.[49]

Flight testing of the WAC Corporal B began on December 3, 1946, with an intermediate round having the characteristics of the A-model except for the larger fins and a nose system with a Signal Corps remitter inside. Four actual WAC Corporal B vehicles followed in the series by December 13, 1946. Despite the modifications to achieve a higher altitude, none of them rose more than 175,000 feet. Presumably the test team suspected that the cause of the less-than-stellar performance was cavitation (gas bubbles) in the injector system, since team members constructed three more B-model vehicles with orifice inserts that were screwed in, rather than drilled as before, to achieve cavitation-free injection of the propellants. They also used a lighter-weight steel for the air tanks in two of the modified vehicles, reducing their weight from 91 to 75 pounds. Douglas Aircraft of Santa Monica, California, modified the vehicles and sent them to White Sands Proving Ground for flight testing. Because only two modified Tiny Tim motors were available to serve as boosters, the third modified WAC Corporal B had to use a similar Navy motor with a longer body that weighed 150 pounds more, reducing booster velocity. From February 17 to March 3, 1947, the test team launched the three vehicles. After the first round reached an altitude of only 144,000 feet, the researchers concluded that its comparatively poor performance resulted from a malfunction of the disconnect coupling on the air line. The addition of a check valve to the air-fill line on the next two vehicles led to higher altitudes. The vehicle launched on February 24 reached 240,000 feet, but a parachute failed to open. The final round with the heavier booster reached 206,000 feet. The parachute functioned properly, and the team recovered the rocket nearly intact.

As far as the Corporal E missile was concerned, aside from experience with a variety of problems, the WAC Corporal firings engendered at least one concrete recommendation. The stressed-skin construction limited access to the propulsion system, so the test team suggested that for future applications, truss-type structures be substituted, with the skin carrying little or none of the load on the rocket. Overall, the WAC Corporal demonstrated that the propulsion system was sound and the nitric acid–aniline–furfuryl alcohol combination was viable as its propellant. Corporal E took over WAC Corporal B's placement of the air tank in front of the propellant tanks. "New materials, improved hardware, and more efficient manufacturing techniques in fabricating the propulsion system of WAC B contributed to Corporal E's development." Because the WAC Corporal was "a scaled-down version of the larger missile, WAC contributed considerable knowledge of aerodynamic forces and trajectories" to the Corporal E program.[50]

The Corporal

Both the Private A and WAC Corporal projects led to the third area of success for JPL in the wartime and immediate postwar periods. This was the development of the Corporal missile with an air-pressure system for supplying the propellants to the combustion chamber. Corporal was the culmination of the work on Private A, the WAC Corporal, and some other efforts such as general propellant, material, and control-system research and development. As first conceived, Corporal E (as it came initially to be called) was a research vehicle for the study of guidance, aerodynamic, and propulsion problems of long-range rockets. An initial schedule called for ten Corporal E launches in late 1945 and early 1946. The end of World War II in 1945 led to a less stringent schedule, however. JPL designed the airframe for the missile, which Douglas Aircraft of Santa Monica then produced. JPL also designed the rocket engine and some other parts, while Sperry Gyroscope of Great Neck, New York, supplied and installed the initial guidance-and-control system, usually referred to as an autopilot.[51]

In 1944 von Kármán estimated that a rocket with a range of 30 to 40 miles would be necessary to serve as a prototype for a later missile. He thought such a vehicle would need an engine with 20,000 pounds of thrust and 60 seconds of burning time. Experience at JPL to that point indicated that the only rocket type yet developed that could meet von Kármán's specifications was a liquid-propellant vehicle burning red fuming nitric acid and aniline. JPL had already tested engines using these hypergolic propellants and proved them to be reliable. The fuel—initially 80 percent aniline mixed with 20 percent furfuryl alcohol—could be used as a coolant for the engine, since JPL had considerable experience with that type of cooling, and a mild steel could be the material for the engine. Thus, such a propulsion system could be adapted to the Corporal E's requirements. Early plans called for centrifugal, turbine-driven pumps to feed the propellants. Since Aerojet had a turborocket under development, JPL thought it could take advantage of the nearby rocket firm's experience and adapt it to a pump for the Corporal. This general design became Corporal F, which was never completed, while Corporal E used air pressurization, as had the WAC Corporal, air being favored over nitrogen because of its more general availability at test sites, which were a long way from JPL.[52] The use of White Sands Proving Ground and other distant launch and test sites constituted a major difference from the convenient arrangement at Peenemünde, where test launches could take place next to the development facilities, although some were later carried out further east.

The first major design for a Corporal E engine was a 650-pound mild-steel version featuring helical cooling passages. Such a heavy propulsion device followed four unsuccessful attempts to scale up the WAC Corporal B engine to 200 pounds. None of them passed their proof testing. In the new 650-pound engine, the cooling passages were machined to a heavy outer shell that formed a sort of hourglass shape around the throat of the nozzle. The injector consisted of eighty pairs of impinging jets that dispersed the oxidizer (fuming nitric acid) onto the fuel. In direction, velocity, and diameter, these streams were similar to those in the WAC Corporal A. The injector face was a showerhead type with orifices more or less uniformly distributed over it. It mixed the propellants in a ratio of 2.65 parts of oxidizer to 1 of fuel. Silver solder attached the outer shell of the combustion chamber to an inner shell. When several of these heavyweight engines underwent proof testing, they cracked and the nozzle throats eroded as the burning propellant exhausted out the rear. But three engines with the inner and outer shells welded together proved to be suitable for flight testing.[53]

The first Corporal E with this heavyweight engine was launched from the Army's White Sands Proving Ground in New Mexico on May 22, 1947. The rocket was 39 feet 2⅜ inches long, with a diameter of 30 inches in the center section. It carried 1,768 pounds of fuel and 4,668 pounds of oxidizer. Its intended range was 60 miles, and it actually achieved a range of 62½ (in one account, 64¼) miles. The second launch occurred on July 17, and the rocket failed to achieve enough thrust to rise significantly until 90 seconds of burning had reduced the weight to the point that it flew a very short distance. On November 4, round 3 was more successful, but its propellants burned for only 43 seconds instead of 60 before the engine quit. This reduced its range to just over 14 miles. Both it and round 2 (which one observer called the "rabbit killer" because it flew along the ground) experienced burn-throughs in the throat area, the helical cooling coils proving inadequate to their purpose.[54]

Deciding that, in addition to these flaws, the engine was too heavy, the Corporal team set out to design a much lighter-weight engine. Several versions combining features of the WAC Corporal B and 650-pound Corporal E combustion chambers all suffered burnout of the throat area during static tests. The upshot was a redesigned engine weighing about 125 pounds. This engine stemmed in part from an examination of the V-2, revealing that its cooling passages were axial (with no helix angle; that is, they took the shortest distance around the combustion chamber's circumference). Analysis showed the advantage of that arrangement, so JPL adopted it. The lab also made the inner shell of the new engine corrugated and the outer shell smooth. It changed the shape of the combustion chamber from semispherical to essentially cylindrical, with the inside diameter reduced from 23 inches to 11.09 and the length reduced slightly, thereby cutting the weight and also making for a smaller area to be cooled and a smaller, thus lighter, injector.

It took two designs to achieve a satisfactory injector, the first having burned through on its initial static test. The second injector had fifty-two pairs of impinging jets angled about 2.5 degrees in the direction of (but located well away from) the chamber wall. Initially the Corporal team retained the mixture ratio of 2.65. But static tests of the axially cooled engine in November 1948 at the ORDCIT test station in Muroc, California (in the Mojave Desert above the San Gabriel Mountains and considerably north of JPL), showed that lower mixture ratios yielded higher characteristic velocities and specific impulses as well as smoother operation. Accordingly, the mixture ratio was reduced first to 2.45 and then to 2.2. Later still, the propellant was changed to stabilized fuming nitric acid (including a very small amount of hydrogen fluoride) as the oxidizer and aniline–furfuryl alcohol–hydrazine (in the percentages of 46.5, 46.5, and 7.0 respectively) as the fuel. With this propellant, the mixture ratio shifted further downwards to 2.13 because of changes in the densities of the propellants. The resultant engine, made of mild steel, provided high reliability. Its success rested primarily upon its "unique configuration, wherein the cool, uncorrugated outer shell carrie[d] the chamber pressure loads, and the thin inner shell, corrugated to form forty-four axial cooling passages, [was] copper-brazed to the outer shell." The inside of the inner shell (the combustion chamber inside face) was plated with chrome to resist corrosion from the propellants.[55]

Until August 1944 the focus of the Corporal effort had been upon "propulsion, launching, and design problems," as Clark Millikan had written to the Sperry Gyroscope Company. But he added, "in the over-all picture the problem of remote control occupies a very essential role." Caltech, he went on, had "agreed to accept the responsibility for this phase of the work as prime contractor," but "the actual development work" would be carried out by subcontract. Negotiations led to a purchase order in February 1945 for ten control systems, which on Sperry's recommendation involved pneumatic components.[56]

A report written in 1958 stated, "The purpose of this early guidance and control system was merely to provide sufficient control to allow a demonstration of the propulsion system and to maintain the missile trajectory within the confines of the [White Sands] Proving Ground," so there is no need to describe it in detail. The report characterized it as "an elementary all-pneumatic autopilot designed by Sperry Gyroscope Company, and a crude form of manual override guidance from the ground." The system worked adequately for the first and third launches, already described, but on the fourth launch on June 7, 1949, the rocket veered to the left of vertical almost right away and began a roll about 15 seconds after launch. Consequently, 23 seconds into its flight, the launch control officer initiated radio cutoff. Apparently vibration adversely affected the mechanical autopilot. This was the first flight with the lightweight, axially cooled engine, and indications were that it, the airframe, and the rest of the propulsion system worked satisfactorily. This led to a decision to develop a tactical guidance-and-control system.[57]

The decision created difficult problems for JPL. On the one hand, the need for a tactical missile of reasonably long range was becoming critical to the Army. On the other hand, JPL initially was not given firm specifications for the missile. The background to this untenable situation was that, in the late 1940s, the Army was the only service that did not have atomic weapons. Early on, atomic warheads had been too large for any but strategic uses, but in 1947 and 1948, atomic testing had shown that the weapons could be smaller, lighter, and cheaper. Thus they could counter the large numbers of troops and tanks available to the Soviet bloc in Europe, a significant problem for the North Atlantic Treaty Organization.

With the possibility of war with the Soviets looming in planners' thinking, Col. Holger N. Toftoy, who was responsible for the Army guided missile program, asked JPL's Louis Dunn and William Pickering in September 1949 if the Corporal could be reliably guided and controlled. Toftoy obviously saw it as a possible delivery platform for an atomic weapon. The two managers answered affirmatively. The Army did not yet have authority to develop a tactical missile with an atomic warhead, but in late 1949 that service decided to proceed with development of a suitable guidance system. Only later, in December 1950, did approval for the Corporal to carry an atomic warhead come from the Department of Defense. Meanwhile, on January 18, 1950, the chief of Army Ordnance directed JPL to expedite the development of the Corporal as an interim tactical missile. It later became a full-fledged tactical missile, but for some time after 1949 the military specifications for it remained "in a constant state of flux."[58]

There was thus a sense of urgency about the development but, at first, no formal approval to develop it or specific military characteristics upon which to base the design. By mid-1949 the airframe and propulsion systems were already in an advanced and satisfactory state of development, so they could serve as an established technological base around which the guidance and control system could be developed. Indications were that the control would have to be accurate to no more than a half-mile circular probable error (CPE, usually referred to as circular error probable, CEP).[59] This meant that 50 percent of the warheads had to land within a circle having a radius of half a mile. Later the requirement changed to a ±500-foot error in range and ±100 feet in cross-range deviation (azimuth).[60] Still later it changed again. Thus the accuracy JPL sought to achieve was a moving target, with the military specifications not finally available until 1953, very late in the research and development cycle. When the specifications were finally laid down, they exceeded the capabilities of the system that was being developed.

The Army's desire to get the system developed as fast as possible probably led to JPL's developing the guidance and control system in-house. It also dictated the use of existing components. This resulted in a radio command system in conjunction with a World War II radar, the SCR-584, both ground-based, instead of an inertial system. The radar would track the missile in azimuth, or lateral deviations from the intended path. It was not accurate enough to do this, but the problem was overcome by locking the radar in azimuth and by making other modifications including a different dish and refinements of some circuits. The radar would also track the missile in elevation in conjunction with a transponder aboard the missile. Computers converted data from tracking into commands to an autopilot using an electronic amplifier to control a pneumatic servo system that actuated the flight-control surfaces—both jet vanes in the rocket exhaust and aerodynamic control surfaces on its four fins. The Corporal E team sent the commands to the missile by using a pulse-coding system that General Electric had developed for the Hermes project (see chapter 4). A radio command unit, also a GE development, interpreted the pulses at the missile end of the system.[61]

The autopilot did not rely simply upon commands from the ground but also got signals from gyros and accelerometers on board the missile. Since the onboard system was not sufficiently accurate, however, it could be overridden by the ground control system governed by the radar. Additionally, JPL adopted a Doppler system like that on the V-2 for cutoff of propulsion, but in contrast to the Germans, the American lab used the tracking radar in addition to the Doppler station to obtain data on the Corporal's position and velocity. Then the computers, developed from a Reeves general-purpose analog computer used as a beginning point, processed the data from both the Doppler system and the radar and sent signals to the missile. Unlike the German system, that for the Corporal did not handle the cutoff in two stages. Instead, after having been prepared automatically by a timer, the radio signal from the cutoff computer on the ground actuated propulsion-system shutoff circuits, a blasting cap, and a solenoid. The blasting cap detonated and released gases under high pressure that caused a spring-loaded piston to operate a latch that caused the propellant valve to close in 0.008 seconds. This cut off airflow to the propellant tanks, and hence propulsion, allowing the air remaining in the pressurized tank to operate the control servomechanisms for the rest of the flight.[62]

A further refinement to the guidance and control system was an arrangement to provide control even after engine cutoff. This was totally different from the arrangement for the V-2, which had no such range correction once propellant cutoff occurred. The goal of the supplemental system was to eliminate errors in range that might result: (1) from the way the propellant shutoff actually occurred, (2) from winds and other atmospheric conditions, or (3) from other causes occurring after propulsion cutoff. The correction in range resulted from measurements of the missile's trajectory near its peak. A range-correction computer calculated the error at projected impact based on the trajectory measurement, and the radar system transmitted a correction to the missile to be stored and applied during the final 20 seconds of flight, once the Corporal had reentered the atmosphere and there was sufficient external air pressure for the aerodynamic control surfaces to realign the trajectory. Range-correction electronics in the missile sent commands to the servo for the pitch-control surfaces to initiate a maneuver to correct for the predicted impact error.[63]

The Corporal team could not test this range-correction system until late in the flight program after earlier tests had proved the capabilities of the other major controls. Meanwhile, to develop such an elaborate system, the JPL guidance and control staff under Pickering had grown from a small part of the project in mid-1949 to half of the entire organization by 1953. Starting in 1949, this staff replaced the all-pneumatic control system in the Sperry arrangement with a JPL-designed electro-pneumatic autopilot. Static tests of the system before the first flight test had shown that vibration was causing mechanical failures, but the first flight tests with parts of the developing system occurred on July 11, 1950. The system, flown on this fifth round of Corporal E flight tests, consisted of an electronic automatic pilot commanding pneumatic servomotors. There were radar control overriding the autopilot for the rising portion of the trajectory, Doppler cutoff, and integrating rate gyros controlling the trajectory after it reached its peak. The flight reached a range of only 51.2 miles because a failure of an air coupling reduced propellant flow rates, but it was considered a success. It marked the end of the phase of Corporal as a test vehicle for development of components. After this, the team shifted to development of a tactical weapon capable of hitting within a few hundred feet of an impact point. For five more rounds of testing, the vehicle retained its designation as Corporal E. Then it evolved into Corporal I with such intermediate designations as RTV-G-2 and XSSM-A-17 for its preliminary development and production phases, respectively.[64]

Round 6 of the Corporal E firings took place on November 2, 1950. The missile experienced multiple failures, landing 35.9 miles downrange, about 35 miles short of projections. Later static tests revealed problems with a propellant regulator that caused overrich mixture ratios on both rounds 5 and 6. Failure of a coupling had again resulted in loss of air pressure for both rounds. The radar beacon to provide overriding guidance in azimuth operated satisfactorily until failure of a flight-beacon transmitter some 36 seconds into the flight. The Doppler beacon never went into operation to cut off propellant flow at the proper moment because the missile failed to achieve the velocity prescribed for shutting off the propellant flow, and also because the Doppler beacon itself failed at 24 seconds after liftoff. As a final blow, all electronic equipment failed, apparently because of the extreme vibration the rocket developed.[65]

While JPL wrestled with these problems, Douglas Aircraft, having built the first ten Corporal airframes, received an order for twenty more, not including the equipment for the control system. These twenty units, to be tested in conjunction with JPL's tactical guidance-and-control system, were designated Corporal I's. By the end of 1950, JPL had learned that the expected accuracy would be ±500 feet in range and ±100 feet in azimuth for an overall range of 26 to 75 nautical miles (about 30 to 86 statute miles). Meanwhile, round 7 of the Corporal E flew in January 1951. It landed downrange at 63.85 miles, 5 miles short of the targeted impact point. This flight was the first to demonstrate propellant shutoff and also the first to use a new multicell air tank and a new air-disconnect coupling. These two design changes increased the reliability of the propulsion system significantly. The guidance and control system generally performed well, but there continued to be a number of malfunctions of individual components. For example, the missile began to roll at 40 seconds into the flight when an electronic connection linking the autopilot to the central power supply failed. Also, the ground radar supplied the control system with erroneous information that accounted for two of the five miles of shortfall.

Round 8 on March 22, 1951, did somewhat better, hitting about 4 miles short of the target, but round 9 on July 12 landed 20 miles beyond the target because of the failure of the Doppler transponder and the propellant cutoff system. Perhaps in consequence, the final round of Corporal E was never employed. But the Corporal team had learned from the first nine rounds how little they understood about the flight environment of the vehicle, especially vibrations that occurred when it was operating. They began to use vibration test tables to improve the design. They also decided to test individual components before installation. This testing resulted in changes of suppliers and individual parts; it also led to repairs before launch and redesigns in the case of multiple failures of a given component.[66]

The next twenty airframes built by Douglas incorporated delta-shaped rear fins instead of the trapezoidal fins used earlier on both the Corporal E and the WAC Corporal. These improved the maneuverability of the vehicle. Homer Stewart related in an interview in 1982 that he first did the testing for the delta shape in an informal way. He said that about 1947 the designers had apparently already decided to change from the trapezoidal design (illustrated in a study Stewart had done in 1944 on stability estimates for the Corporal) to the delta. One day when he was home looking after his young son, who had the flu, he made a balsa-wood model of the new delta configuration he and others had been planning. Somewhat like Kurzweg at Peenemünde, he used weights to adjust the model's center of gravity and tested it for about two hours, amusing his sick child in the process. He moved the center of gravity until he found the proper balance between stability and instability. He said that with this crude method, he came up with an answer that was very close to the one later provided by a wind tunnel about a year later.[67]

Stewart did not say which wind tunnel confirmed his crude tests, but after April 1949 JPL had a 12-inch-diameter supersonic tunnel designed by Allen Puckett that provided uniform flow from Mach 1 to Mach 3 by means of a flexible nozzle. It was followed in 1951 by a tunnel of 20 inches capable of speeds up to Mach 4.8. These tunnels provided new data on skin friction that missile designers found invaluable. The largest uncertainty in the design of the Corporal had been the drag, and now data on drag and heat transfer could be obtained. Before the 12-inch supersonic tunnel was available, engineers had to estimate aerodynamic moments and forces that would operate on the Corporal as then conceived by using linearized theory. Later the engineers determined the form drag with the help of two-dimensional Prandtl-Glauert corrections to published experiments. They arrived at viscous drag by means of the Kármán-Schönherr equation for calculating incompressible turbulent skin friction. But they had to use approximation in their trajectory calculations. In short, they used aerodynamic theory to calculate roughly the way the Corporal would be affected in flight until they could get experimental and flight-test data. The first flight test's data indeed "appeared to be in fair agreement [with the calculations] in view of the accuracy of measurement," though some base-pressure measurements were higher than estimated.

Finally in September 1950 the Corporal team conducted the first series of supersonic wind tunnel tests to develop the configuration of the Corporal with the delta-shaped fins. These afforded a much smaller margin of static stability than before, to enable greater maneuverability for the missile. In July of 1951, models of the Corporal were tested in the 20-inch tunnel to verify the data from the 12-inch tunnel and to extend the airspeed to Mach 3.67.[68]

On October 10, 1951, round 11 (flight 10, since round 10 had not been launched) with the new delta fins was the first to have the basic configuration of Corporal I. Unfortunately, the frequency regulator for the central power supply failed on takeoff, causing the missile to follow nearly a vertical trajectory. Range safety cut its flight short so that it would impact between White Sands and the city of Las Cruces, New Mexico. It was followed by round 12 on December 6. This Corporal I featured "all missile components except the range computer and azimuth programmer," meaning that it did have a prototype radar, Doppler, and computers.

Before these launches, the Army had invited several companies in the summer of 1951 to bid on production contracts as prime contractors. Firestone Tire and Rubber Company of Los Angeles won the bid for the missile, ultimately building 320 of them. Gilfillan Brothers, also of Los Angeles, subsequently got the contract to produce the ground guidance equipment. Ryan Aeronautical Company in San Diego manufactured the engines for both the Douglas- and the Firestone-built versions, but apparently not as a prime contractor. The Firestone missiles did not contain all of the guidance equipment, because JPL still did not have the command unit and the range correction unit developed fully enough. According to an Ordnance Corps report of 1958, JPL received the Firestone missiles and disassembled them for inspection. It rebuilt them and administered preflight testing before sending them to White Sands for the actual flight tests. Then it sent comments to the manufacturer to help improve factory production. As Clayton Koppes reported, the three prime contractors (including JPL) failed to work together effectively. Problems resulted.

For early flight tests, the JPL staff involved with the Corporal traveled to White Sands for each launch. In 1951, as flight tests became more frequent, JPL created a flight operations and test section at White Sands to carry out the research launches and conduct JPL's business at the proving ground. This section became resident in New Mexico early in 1952. During that year JPL launched twenty-six Corporal rounds, including the first ten of the Firestone lot as well as sixteen produced by Douglas. The first two rounds had a difference in range of 1,400 feet, but one of these missiles had an error in azimuth of only 23 feet at a range of 69 miles. The first Firestone missile flew on August 7. Overall, the launches during 1952 satisfactorily demonstrated the experimental ultra-high-frequency Doppler system and the azimuth programmer. Already by January 1952 two sets of prototype ground equipment were built, apparently by JPL, since Gilfillan still was not under contract. The first full-scale field test with tactical ground equipment took place in

September 1952. The basic elements for Corporal I were nearly all in place at that time.[69]

The Corporal I flight evaluations continued through 1953 and into 1954. Inflight failures of components continued, especially those of the propellant shutoff circuit. Other continuing problems with the Corporal involved the command unit, which GE had used on the Hermes (see chapter 4). This proved adequate as an experimental unit, but it overheated when used for more than 10 minutes (including time on the ground) and was affected by vibrations in the flight environment. The JPL team used various work-arounds through round 28. In April 1952 JPL contracted with Motorola Research Laboratory in Phoenix, Arizona, to design a new command unit. Called the CU-54, the new device differed completely from the older unit and was comparatively free of overheating problems. JPL modified it to prevent jamming and make it more rugged. Motorola and Gilfillan then competed to build the more rugged version, designated CU-54A. Motorola won the competition because of the superior performance of its design under vibration.

There were also problems with gyroscopes. In 1949, the Corporal had begun to use gyros manufactured by Kearfott Company of Cliffton, New Jersey. Instead of Sperry's air-driven rotors for the gyroscopes and pneumatic pickoff devices on an earlier system, Kearfott's gyros were driven by electric motors and employed electrical pickoffs, although the fin actuators on the servo systems remained pneumatic. With both the Sperry and Kearfott gyros, there were problems, so in 1950 JPL switched to gyros from Schwien Engineering Company of Van Nuys, California. It was the Schwien gyro that became standard for the Corporal, although it came to be manufactured by Clary Multiplier Corporation of San Gabriel, California, using the Schwien design with the latter's permission. Further flight testing showed that even the Schwien gyro was subject to effects from vibration, and the ones built by Clary suffered from poor workmanship, including dirt and even screws and nuts in the bearings.[70]

Because of such problems and resultant engineering changes to correct them, a second production order to Firestone for Corporal missiles in late 1954 carried a redesignation of the missile as Corporal II (official designation, XM2E-1). Separate contracts with Gilfillan in 1953 and 1955 called for improving the reliability of electronic components. The basic guidance concept remained unchanged, but Gilfillan redesigned components, aided by technical advice and field evaluation from JPL. From December 1953 through 1954, JPL rebuilt and launched Corporal I's to serve as prototypes for the Corporal II electronic equipment. One major change from Corporal

I to Corporal II was in the Doppler system. Its operating frequency was increased from a fixed figure of 38 megacycles to the ultra-high-frequency range of 450 to 480 megacycles. More minor improvements to the radio link improved tactical operations.

The first Corporal I with components modified by JPL, making it a Corporal II prototype, landed 234 meters (about 768 feet) short and 116 meters (about 381 feet) to the right, well within the final tactical requirements of 300 meters (about 984 feet) circular probable error. This occurred in a launch on October 8, 1953. The first production Corporal II launched on October 28, 1954. It landed 43 meters (roughly 141 feet) short and 169 meters (about 555 feet) to the right, again well within specifications. The Corporal IIs used the Gilfillan electronic components, rather than ones manufactured by various companies under subcontract to Firestone for the Corporal I's. Overall, the circular probable error for all Corporal II rounds fired by JPL was 350 meters (about 1,148 feet), 50 meters in excess of the desired CPE. JPL retained technical control of the Corporal program throughout 1955, relinquishing it in 1956 while continuing to provide technical assistance to Firestone and Gilfillan.[71]

Besides the less-than-ideal accuracy of Corporal II overall, there were problems with propellant shutoff during firings of the missile by Army field forces. Fact-finding investigations and informal discussions conducted by Firestone, Gilfillan, the field forces, Army Ordnance, and JPL led to greater care by field forces personnel in following operational procedures. This eliminated shutoff problems. Since 1951–52, JPL had gained greater understanding of the flight environment and the effects of vibration upon components, especially electronic ones, and redesigns had followed. Nevertheless, because the missile failed to meet specifications laid down by the Army (however late in the design cycle and however unrealistic under the circumstances), the Army Field Forces board recommended abandoning the Corporal. Many using organizations, on the other hand, decided that the advantages to be gained from field experience with a guided missile outweighed the shortcomings of the Corporal as a weapon. It was, after all, the first surface-to-surface, long-range guided missile to be fired by American troops. So the Corporal was declared operational in 1954, and in January 1955 the 259th Field Artillery Missile Battalion deployed to Europe with the Corporal I. Eight Corporal II battalions replaced it during 1956 and the first half of 1957.[72]

As deployed, the Corporal was 45.4 feet long and 30 inches in diameter, with a fin span of 6.1 feet. Its range was listed as 86 nautical miles (99 statute

Table 3.1. Comparison of V-2 and Corporal Missiles

	V-2	Corporal[a]
Length (feet)	46	45.4
Diameter (feet)	5.4	2.5
Launch weight (pounds)	28,440	11,200
Thrust (pounds)	55,100	20,000
Range (miles)	200	99
Specific impulse (lbf-sec/lbm)	210	221

Sources: Dornberger, *V-2*, xvii–xviii; Schilling, "V-2 Rocket Engine," 285; Nicholas and Rossi, *Missile Data Book*, 3–5; Ordway and Wakeford, *International Missile and Spacecraft Guide*, 3.

a. Neither Nicholas and Rossi nor Ordway and Wakeford specifies whether these figures are for Corporal I or II, but presumably the latter.

miles). Its velocity was Mach 3.3. Its specific impulse as given by one source was 221 lbf-sec/lbm. Table 3.1 shows how these figures compared with those for the V-2.

Obviously, the Corporal was less powerful and had a shorter range than the V-2, although the effective range of the German missile was actually more like 165 miles. On the other hand, the Corporal's propulsion system had a higher specific impulse than the V-2's. It is hard to compare the accuracy of the two, given the difference in range, but probably the Corporal was more accurate. In some respects, such as the axial nature of the cooling system and the use of Doppler radar for propellant cutoff, the Corporal had borrowed from the V-2. In most respects, however, the American missile represented an independent development, in some cases one that independently adopted features developed at Peenemünde too late to incorporate them into the V-2. These included a showerhead injector and the use of hypergolic propellants. Both had been developed for the Wasserfall antiaircraft rocket, and a single injector plate later became a standard element in the construction of the rockets designed in Huntsville.[73]

Like the V-2, the Corporal was not a notably successful weapons system. Its flaws in this regard, however, can be explained in significant part by its being introduced as a research vehicle and only later converted to a tactical weapon—one, moreover, for which military specifications were not available until far into the research and development program. Another factor in the Corporal's shortcomings was that many of the contractors were unprepared to meet the stringent standards that missiles required. Also, JPL was inexperienced in working with contractors. Under the circumstances, the conclusion of an Ordnance Corps report written in 1958 and published at JPL is not far from the mark: "The primary objectives of the program were realized."

Among the program's achievements was the research done in circuit design, which advanced technology in that field. Before Corporal, little was known in the United States about the effects of vibration on electronic equipment. The vibration tables used to test these effects may have been the first effective simulators of the flight environment in that area. Subsequently, both testing for the effects of vibrations and analysis of components and systems for reliability became standard practice in missile development. According to the Ordnance Corps report, by 1958 JPL telemetry equipment had also become standard on most U.S. test ranges. The experience of Corporal and its problems also offered several lessons for future rocket and missile development. One was that "technical system responsibility and unchallengeable authority for action must be clearly assigned at the earliest possible date"; another was that "required military characteristics must be defined at the outset, with a realistic regard for the limitations of the system" and for the probability of those characteristics being achieved within the available time frame. These were important lessons for both JPL and the Army,[74] although they would not be followed in development of the Sergeant missile.

Another specific achievement of the Corporal program that certainly deserves mention is the axially cooled, lightweight engine. Although the idea for the axial direction of cooling flow came from the V-2, the overall engine was a significant and original achievement. It was both light and efficient, and while there seems to have been no direct use of its design in subsequent engines, it is likely that propulsion engineers learned something of their art from it. Similarly, although the guidance and control system had lots of problems, it was a sophisticated development. Since it was a command-guidance type of system, it had little direct application to future systems, which moved as quickly as was practical to inertial and celestial technologies. But it did mark a step forward in some respects, and presumably guidance-and-control experts learned from it. For example, although computers used to process data from both Doppler and radar sources in order to send corrective commands to the missile were based on the Reeves general-purpose analog computer, that off-the-shelf device required further development to be used for the Corporal. According to Bragg's history of the Corporal, at least, JPL developed a precision operational amplifier to serve as the "heart" of the special-purpose computer. It marked "an improvement over commercially available items and a definite contribution to computer development."

As early as 1953, transistors began to replace vacuum tubes in the electronic components of the Corporal system. This was not entirely path-breaking. Transistors had been developed elsewhere,[75] but their use this early shows that JPL was conversant with new technological developments and quick to adopt them. Another notable achievement of the Corporal was occasioned by the decision to stick with a pneumatic servo system when most people at the time thought only hydraulic systems were suitable for actuating control mechanisms. A pneumatic system required a precision valve to control the air. It had to rotate rapidly in proportion to the small direct-current output of an amplifier. Such a valve was ready in 1950 and constituted, according to Howard Seifert, a major contribution to the guidance art.[76]

Despite these achievements, Clayton Koppes's summation of the program seems valid: "Corporal was at best an interim weapon. The army's demand for speed, research uncertainties, and contract problems had produced a system Pickering characterized as a lash-up of existing equipment and designs."[77]

Management and Influence

In the course of the Corporal's development, the management structure at JPL had changed dramatically. When Malina succeeded von Kármán as director in December 1944, he did so only in an acting capacity. Von Kármán recommended the "acting" in his title, and also recommended that Louis G. Dunn be assistant director, with JPL further directed by an executive board headed in an acting capacity by Clark Millikan, who would also be the new acting director of GALCIT. Caltech approved all of these proposals. According to Clayton Koppes, there had been "coolness" between von Kármán and the younger Millikan that Malina and Dunn inherited, making their relations with the board difficult, although, following von Kármán's recommendation, both were also members of the board.[78]

It does appear that there was some friction between Clark Millikan and Malina, but it is difficult to know how much to make of it. According to Malina's first wife, Liljan, the source of the difficulties was their politics. Malina was somewhat militantly liberal, and Millikan was highly conservative. She said it bothered Millikan that Malina was so egalitarian and desirous of making the world a better place. Perhaps this would not have mattered to Millikan had Malina not been prone to say things that offended his conservative friends. Indeed, Malina may already have been suspected of Communist leanings. There were investigations of his friends in 1950, but already in 1945, when Malina's marriage to Liljan was breaking up, the FBI had searched through literature in his house. He said later that he and others were concerned about the development of missiles and atomic bombs.

He went on two missions to Europe for the War Department in 1944 and 1946, and while there in 1946 he had looked into the formation of the United Nations Educational, Scientific, and Cultural Organization (UNESCO). He was anxious to work for world peace, so in 1946 he took a leave of absence from JPL. He allowed his security clearance to lapse and went to work for UNESCO in 1947, ultimately becoming an artist but also a promoter of international cooperation in astronautics.[79]

Dunn succeeded Malina as acting director of JPL on May 20, 1946, although Malina remained on the board for some time. Dunn became director (no longer acting) on January 1, 1947, in effect succeeding von Kármán. A conservative from South Africa, Dunn had ideas about racial equality and other issues that were quite different from Malina's, but the two men had been friends who simply avoided discussing the issues they disagreed on. Dunn was much more formal than his predecessors. Whereas things were pretty relaxed under Malina, Dunn brought more structure and discipline to JPL. He was also cautious, hence concerned at the growth of the lab during his tenure. From 385 employees in June of 1946, the number grew to 785 in 1950 and 1,061 in 1953, causing Dunn to create division heads above the section heads who had previously reported to him directly. There were four divisions by September 1950, with William Pickering heading the one on guided-missile electronics.

Pickering came to function as the project manager of Corporal in everything but name. In August 1954, Dunn resigned from JPL to take a leading role in developing the Atlas missile for the recently established Ramo-Wooldridge Corporation (see chapter 7). At his suggestion, Caltech appointed Pickering as his successor. A New Zealander by birth, Pickering continued the tradition of having foreign-born directors at JPL. Easier to know than the formal Dunn, Pickering was also less stringent as a manager. Whereas Dunn had favored a project-based structure, Pickering reinstated organization by discipline. He remained as director until 1976. Howard Seifert, who had come to GALCIT in 1942 and worked with Summerfield on liquid-propellant developments, characterized the three JPL directors by reference to an incident when some mechanics cut off the casual Malina's necktie because *he* was too formal. Seifert said they would never have cut Dunn's tie off without losing their jobs, and they would not have cut off Pickering's either, but he would not have fired them for that offense alone. He added that Dunn had a rigid quality but undoubtedly was extremely capable.[80]

Despite all the changes in personnel and management from Malina and von Kármán through Dunn to Pickering, and despite the differences in the

Figure 12. Three early JPL directors *(left to right)*: William H. Pickering; Theodore von Kármán, cofounder and first director; and Frank J. Malina, cofounder and successor to von Kármán as (acting) director. All were active in the development of early U.S. rocketry. Courtesy of NASA/JPL-Caltech.

personalities and values of these individuals, one constant seems to have been an organization that was not highly structured or well suited to dealing with outside industry and the design/fielding of a weapon system, as distinguished from a research vehicle. Even Dunn's project organization seems not to have been well suited to the kind of systems engineering that soon became common in missile development.[81]

It may well be, however, that the rather loose organization that characterized JPL in this period was conducive to the kinds of innovations achieved by Parsons in his virtual invention of castable composite propellants and by Summerfield in his discovery (separate from that of Thiel) of how to design a liquid-propellant engine. Incidentally, both Parsons and Summerfield, along with Tsien, left JPL in the course of the development of the Corporal missile. Parsons left in 1944 and worked at Aerojet from 1941 until 1946, apparently overlapping at the two places for about four years. Thereafter he held a series of jobs and devoted increasing time to exotic pursuits involving the use of drugs, the occult, and even black magic. Finally, on June 17, 1952, he was packing for a trip to Mexico when he dropped a container with fulminate of mercury in it. It exploded, killing the inventive but unusual chemist.[82] It is interesting to speculate, in view of both Goddard's and Oberth's interests in the occult, whether there might not be some connection between such proclivities and inventiveness, at least in the field of rocketry.

Martin Summerfield had served as the assistant chief engineer of the GALCIT rocket research project from 1941 to 1943. Then he had gone over to Aerojet for two years on a special engine project, after which he spent 1945–49 as head of the rocket research division of JPL (redesignated the rockets and materials division in the July 3, 1946, organization chart). In 1949 he joined the Princeton University faculty, teaching aerospace engineering there until 1978. He served as Astor Professor of Applied Science at New York University in 1979–80, and then went to work full-time at the Princeton Combustion Research Laboratories, where he had become part-time chief scientist in 1975 and president in 1978. From 1951 until 1962 he was editor in chief of the *Journal of the American Rocket Society*, and he continued to do research in combustion processes and gas dynamics, fuels and propellants, solid propellant ignition and combustion, and related fields. Because of the investigations of people with whom he and Malina held political discussions, Summerfield lost his clearance in 1952, but it was restored after a few years. Thus, with a hiatus, he continued to make contributions to rocket propellant and combustion research.[83]

Tsien left Caltech in August 1946, turning over the research analysis section to Homer J. Stewart and accepting a faculty position at MIT. He did not find his former school a comfortable place, although he was promoted to full professor in 1947 with almost unprecedented speed. Tsien was harsh in criticism of senior colleagues who he felt lacked theoretical rigor. His faith in mathematical analysis in the von Kármán tradition set him apart even at this elitist institution. He returned to Caltech in 1949 but fell under the same FBI suspicion that cost Summerfield his clearance. Tsien ultimately returned to his native China, where he played an important role in Chinese rocket development.[84]

These were all critical players in early rocket development at JPL, but as will be seen in succeeding chapters, many other talented engineers remained in the growing JPL organization, including Stewart, Dunn, and Pickering, who carried on the work the pioneers had started. The very capable Clark Millikan formally succeeded von Kármán as director of the Guggenheim Aeronautical Laboratory and as chairman of the JPL board on March 11,

1949.[85] As will also be seen, many people from JPL besides Louis Dunn later showed up in positions of importance on other missile and rocket projects, carrying with them much that they had learned in their work at JPL as well as their talents. Thus in a variety of ways, some of them incalculable, the early work at JPL contributed to U.S. rocketry, even though JPL itself essentially got out of the rocket propulsion business in the late 1950s in favor of spacecraft development.[86]

One very tangible way that JPL contributed to U.S. missiles and rockets was its early work in hypergolics. This technology transferred to Aerojet, which later became the contractor for the Titan II, a missile that used storable liquid propellants that ignited on contact. The same technology was also used in the Titan III and Titan IV liquid stages. These launch vehicles thus were direct beneficiaries of the early hypergolic propellants Malina learned about in part from the Navy in Annapolis. This was a significant contribution from both indigenous American research efforts during World War II.[87]

Aside from what will be discussed later, there was one other way in which JPL influenced American rocket and missile development. In 1943 von Kármán organized what Malina described as "the first graduate course in jet propulsion engineering in the U.S.A." It was taught by the staff at GALCIT and JPL, plus some engineers from local industries, for the rest of the war, with its lectures published by the Air Technical Service command in 1946. The course was repeated in 1948, supplemented that year by the Army Ground Force's Officer's Guided Missile Course. According to Karl Klager, an expert in propellant chemistry, and his coauthor, Albert O. Dekker, it "helped to generate expertise in the scientific-engineering community as well as the military."[88]

4

From Pompton Lakes to White Sands

Other Liquid-Propellant Rocket Developments, 1930–1954

Goddard, with his small band of technicians, and Malina's Suicide Squad were far from the only early developers of liquid-propellant rockets in the United States.

The American Rocket Society

Among the many others, undoubtedly the most important in the period before World War II were the individuals associated with what became, in 1934, the American Rocket Society. This organization, first called the American Interplanetary Society, had its birth on April 4, 1930, in a third-floor apartment on West 22nd Street in New York City. There, eleven men and a lone woman created the society in the home of G. Edward Pendray and his wife, Leatrice, who were both journalists. Interestingly, in view of the science-fiction connections of other rocket enthusiasts, nine of the twelve founders either wrote science fiction or were associated with Hugo Gernsbach's *Science Wonder Stories*, a magazine of the day specializing in adventure stories. The group chose as its president David Lasser, managing editor of Gernsbach's magazine and an MIT graduate, with Pendray as vice president.[1]

Although Goddard became a member of the Society, Pendray wrote that, characteristically, "Members of the Society could learn almost nothing about the technical details of his work." As a sort of substitute, in 1931 the Pendrays went to Europe and witnessed the static test of a small liquid-propellant engine at the Raketenflugplatz, reporting the experience in May to the Society back in New York. Aside from the people he influenced who came to the United States as immigrants from Germany, this was another example of the ways in which Oberth, who inspired the people at the Raketenflugplatz, also contributed indirectly to American rocketry. Soon Society members were testing their own rockets. They had the usual share of failures and partial successes, deciding that they needed to develop an engine that would operate reliably and not burn through before they launched any more whole rockets. As Pendray reported, "The Society then began a long, often discouraging, but finally successful series of motor development tests." This work "finally culminated in . . . a practical liquid-cooled regenerative motor designed by James H. Wyld."[2]

Wyld had earned a degree with high honors in mechanical engineering from Princeton in 1935 and then studied electrical engineering under a fellowship in 1935–36. While working on the rocket engine, he was employed from January 1937 to February 1938 at the Linde Air Products Company where, among other things, he learned about gas flow, thermal stress, heat transfer, and the theory of combustion at high temperatures. He had read extensively about aerodynamics, including the effects of compressibility and supersonic flow. This reading encompassed the works of von Kármán, Prandtl, Busemann, and the American aerodynamicists Lyman Briggs and Hugh L. Dryden. In his study of rocket engines, he seems to have included the work of Harry W. Bull of Syracuse, New York, also a Society member who had written in its journal about his combustion-chamber tests in 1932 involving a small engine featuring regenerative cooling around the nozzle throat only. Wyld was also aware of the work on regenerative cooling done by the Austrian Eugen Sänger and of Oberth's discussion of the subject.[3]

In 1935 Wyld witnessed a series of static tests of rocket engines in which an explosion hurled a fragment into a female bystander's elbow, causing a compound fracture. Far from deterred by the danger, he was stimulated to undertake a study of rocket engines and came up with a design for a regeneratively cooled propulsion system that would permit firings of longer duration than before without burn-throughs. This became the first American engine to apply regenerative cooling to the entire combustion chamber. Built in 1938, it was among three engines tested at a new stand of the Society at New Rochelle, New York, on December 10, 1938. It burned steadily for 13.5 seconds and achieved an exhaust velocity of 6,870 ft/sec.[4]

The first test of Wyld's engine was limited in duration by an oxygen shortage at the testing site, but at the American Rocket Society's proving stand in Midvale, New Jersey, there were longer tests on June 8, June 22, and August 1, 1941. The significance of this engine was twofold. First, it led directly to the founding of America's first rocket company, Reaction Motors, by Wyld and three other men who had been active in the Society's experiments. Second, it was from Wyld that Frank Malina learned about regenerative cooling for the engines developed at what became JPL.[5]

Reaction Motors and the Naval Engineering Experiment Station

Encouraged by repeated static tests of Wyld's engine lasting up to 40 seconds, Lovell Lawrence Jr., another of the founders of Reaction Motors, contacted Dr. George Lewis, director of research for the National Advisory Committee for Aeronautics. Lewis put Lawrence in contact with the U.S. Navy's Bureau of Aeronautics, which eventually sent a representative in November 1941 to see a static test of the regeneratively cooled engine. The test was successful and the representative reacted enthusiastically, leading to the incorporation of the company on December 16, 1941, only nine days after the Japanese attack of Pearl Harbor precipitated American entry into World War II.[6]

As it happened, another engine besides Wyld's that had been tested in New Rochelle on that same December day in 1938 was a tubular model, also regeneratively cooled, built by a midshipman from the Naval Academy named Robert C. Truax. Truax's engine was erratic in its operation, but his life became intertwined with the activities of Reaction Motors and was significant in its own right. As a high school student, the future midshipman had read all the literature he could find on rocketry. Entering the Naval Academy in June of 1935, he spent what little spare time he could find in the rigorous Annapolis schedule designing a sounding rocket with regenerative cooling for the throat area of the nozzle. He scrounged parts and obtained use of an old lathe to build it. He then tested it at the Naval Engineering Experiment Station across the Severn River, using compressed air and gasoline as the propellants. Finally he was able to get 25 pounds of thrust and several minutes of running time, with the results reported in the American Rocket Society's journal in April 1938. Subsequently, despite Navy regulations to the contrary, he was able to substitute gaseous oxygen for compressed air as an oxidizer. He continued his development into 1939, when he graduated on June 1 and was commissioned an ensign. After sea duty, he reported to the Navy's Bureau of Aeronautics from April to August 1941 at "the first jet propulsion desk in the Ship Installation Division." There he was responsible for looking into jet-assisted takeoff for seaplanes. In response to his own recommendation that a JATO system be developed, he was soon reassigned to Annapolis, where he headed a jet propulsion project at the Naval Engineering Experiment Station.[7]

Assigned to his project were a Mr. Robertson Youngquist and three ensigns, Ray C. Stiff, J. P. Patton, and William Schubert. In developing a JATO for the PBY seaplane—their specific assignment—they recognized the need for a combustion chamber that was regeneratively cooled. They selected red fuming nitric acid and gasoline as propellants and began tests to determine atomizer (injector) criteria, the optimal volume of the combustion chamber, and rates of heat transfer. After some successful tests, an expanded staff began designing an engine with 1,500 pounds of thrust in late spring 1942. Tested in June, the engine exploded.

Meanwhile Ensign Stiff, who held a bachelor's degree in chemical engineering from Texas Technical College (later Texas Tech University) and a master's in chemical engineering from the University of Texas, had discovered that aniline and other chemicals ignited spontaneously with nitric acid. This information became critical to JPL's efforts to develop a liquid-propellant JATO unit. The explosion of the engine Truax's Annapolis group was developing also forced it to resort to aniline as a fuel despite the greater availability of gasoline. The result was a JATO device designated DU-1 (for droppable unit number 1). In spring 1943, two 1,500-pound-thrust DU-1 units suspended from the wing struts on either side of a PBY shortened its takeoff distance by an average of 60 percent.[8]

Before this, in a related development, Reaction Motors had entered into a contract with the Navy in early 1942 to deliver the existing Wyld engine, develop and demonstrate a similar 100-pound-thrust engine burning aviation gasoline and liquid oxygen, develop a 1,000-pound-thrust engine with the same propellants, and demonstrate sequential starts of the last engine with throttling from full to half thrust. The founding partners sent off the Wyld engine and received $5,000 for it, allowing them to rent buildings at Pompton Lakes, New Jersey, buy a truck, and hire a few more employees, bringing the total to nearly twenty by the end of 1942.

When they tested the Wyld engine with gasoline instead of its designed fuel, alcohol, it promptly burned through. A redesign with a copper nozzle replacing the previous aluminum one showed that gasoline and liquid oxygen could indeed be made to work in a regeneratively cooled combustion chamber. Then began the process of scaling up to the required 1,000-pound-thrust engine. The nozzle was again replaced, with a stainless steel one. Lawrence designed throttling valves that were remotely controlled by electric motors and gears. But the process of scaling up yielded the same sorts of delayed ignition, improper propellant mixing, and the like that had plagued German developers and those at JPL. Nevertheless, the new firm delivered the throttleable engine within the 180 days stipulated in the Navy contract.[9]

A follow-on contract with the Navy required Reaction Motors to deliver a 3,000-pound-thrust engine to provide assisted takeoff of a Martin PBM patrol bomber. For this engine, another of the partners, John Shesta, developed an injector system in which the oxygen was fed from about forty holes in an upper injector plate. These aligned with holes in a lower plate that fed the gasoline in the opposite direction. The two sets of injectors propelled the gasoline and oxygen in cones that intersected to produce good mixing. However, scaling up produced problems with heat transfer that required adjustments in the circulation of the fuel for cooling. The solution was helical winding of "ribbons" that carried the gasoline around the combustion chamber, slowing it down for better heat transfer. A trial-and-error process to place the cooling passages at the strategic angles and lengths for proper heat transfer finally solved the problem.

The resultant engine weighed 75 pounds, with the propellants fed by nitrogen pressure bottles. The combustion chamber was about 6 feet long and 6 inches in diameter. A test on May 6, 1943, yielded 24 seconds of high thrust, the maximum registering at 3,180 pounds. There were subsequent igniter problems, but the company achieved a long run estimated at over a minute on May 14, 1943, this becoming the standard burn time. In November 1943, Reaction Motors got a small building at the Naval Engineering Experiment Station in Annapolis, and on January 12, 1944, the larger JATO unit got its test on a Martin PBM flown by Marine Capt. William L. Gore. The test was successful, but the Navy never put any of the Reaction Motors JATOs into operational use, probably because of the difficulty of using liquid oxygen, which boiled off if stored for any length of time. The storable hypergolic propellants used by Truax and Aerojet were much better suited to operational use.[10]

Consequently, the Navy's Bureau of Aeronautics issued a contract to Aerojet for an improved version of the Navy experimental station's DU-1 and a separate pump-fed JATO for permanent installation on a PB2Y3 four-engine seaplane. Now-Lieutenant Truax and Lieutenant (j.g.) Stiff received orders for duty at the Aerojet plant in California to oversee the two projects. Stiff was responsible for the droppable unit, of which about a hundred were built, while Truax worked on the pump-fed unit, which never went into production. Ultimately the Coast Guard used some of the JATO units Stiff oversaw (designated 38ALDW1500) for offshore rescue work. Once he completed his five years of Navy service, Stiff joined Aerojet as a civilian engineer. He rose to be vice president and general manager of Aerojet's liquid rocket division in 1963 and then, in 1972, president of the Aerojet Energy Conversion Company. In 1969 he was made a Fellow of the American Institute of Aeronautics and Astronautics (into which the American Rocket Society had merged) for "his notable contributions in the design, development and production of liquid rocket propulsion systems, including the engines for Titan I, II, and III."

Meanwhile, Truax returned to the Engineering Experiment Station, where under the leadership of Lt. William Schubert the team developed an engine for the Gorgon missile, using Robertson Youngquist's pintle valve to solve a problem of hard starts. Reaction Motors constructed the Gorgon using the station's specifications for this engine that burned mixed nitric/sulfuric acid and monoethylanaline. Gorgon was not a single missile but a family of missiles, some using this engine and some using other kinds of jet propulsion. Its various technical designs led to a number of other tactical missiles, including the Lark, for which the Engineering Experiment Station provided the engine specifications and Reaction Motors produced the engines.[11]

The experimental station in Annapolis grew to include twenty officers with engineering backgrounds and about a hundred sailors and civilian technicians. After the end of the war, it moved and became part of the Pilotless Aircraft Unit at the Naval Air Station, Mojave, California. In 1946 that unit moved from its desert location to the Naval Air Facility, Point Mugu, on the coast, where on October 1, 1946, the Naval Air Missile Test Center came into existence. This developed into the Navy's primary facility for testing and evaluating guided missiles and target drones. On June 16, 1958, the Naval Missile Center, as it was then called, became the headquarters for the Pacific Missile Range, where many ballistic missiles would be launched.[12]

In the interim, Reaction Motors increasingly found that rocket testing at its location in Pompton Lakes in northeastern New Jersey roused the ire of neighbors. With the help of the Navy, the young company moved in mid-1946 to the U.S. Naval Ammunition Supply Depot at Lake Denmark, further west in the middle of northern New Jersey. There it had a testing area of 25 acres instead of 3.5 acres at Pompton Lakes. Moreover, the noise was shielded from neighbors to some extent by hills and to a greater extent by "thousands of acres of government property acquired specifically for ammunition storage and testing." However, while the company had expanded its payroll to 120 in the middle of 1946, it had accepted a number of fixed-price contracts that caused its development costs to exceed its income. In 1947 and 1948, Laurence S. Rockefeller invested $300,000 in the firm. Then in 1953 the Mathieson Chemical Corporation bought a controlling interest in Reaction Motors' outstanding stock, later merging with Olin Industries to form the Olin Mathieson Chemical Corporation. The number of employees climbed to 669 in 1955 and 1,639 in early 1958. Then on April 30, 1958, the Thiokol Chemical Corporation, which had become a major producer of solid-propellant rocket motors (see chapter 5), merged with Reaction Motors, which then became the Reaction Motors Division of Thiokol.[13]

Although this merger looked beneficial for both parties, in retrospect it proved ill-fated for Reaction Motors. While Aerojet was able to work effectively from the beginning with both solid and liquid propellants, the two types had significantly different requirements. Solid-propellant engineers focused largely upon the propellant. Casings had their own problems and development issues, as did control methodologies, segmenting, and the like. But propellants and their chemistry were primary. In liquid-propellant rocket development, on the other hand, once the specific propellants for a given vehicle were chosen, the issues were really with injectors, ignition, plumbing fixtures like tubing and valves, and propellant delivery systems, whether pressure feeding or pumps. Thus the concerns of the two partners in the merger were quite dissimilar. Moreover, as a small division of a much larger firm, Reaction Motors had to compete with other components of Thiokol for funding and management attention.[14]

Between the mid-1940s and the late 1950s, Reaction Motors had developed rocket engines for the X-1 and D-558-2 experimental research aircraft, followed by the throttleable engine for the X-15 rocket research airplane that eventually flew to the edge of space in the 1960s and reached a speed of Mach 6.7. In addition, the firm had developed an engine for the Viking sounding rocket, which became the starting point for the Vanguard launch vehicle, and engines for the Air Force's MX-774 program, which provided technology for the Atlas missile and space-launch vehicle. After the merger with Thiokol, it developed some engines for tactical missiles like the Sparrow III and the Bullpup. It developed some auxiliary (small) rocket engines for the space program, notably the Surveyor vernier. But limited by its location, where spreading postwar suburban housing had resurrected complaints about noise, Reaction Motors had never been able to produce many rockets with large production runs or large engines beyond the size of the X-15 power plant.[15]

As profits declined, personnel had to be cut, and prospects looked dim at the beginning of the 1970s when the Department of Defense was reducing its research and development budget, with funding for liquid-propellant rocket technology falling more than that for solids. In 1970 Thiokol decided to discontinue working in the liquid-propellant field, and in June 1972 Reaction Motors ceased to exist.

Despite its ultimate failure as a business, the organization had shown considerable innovation and had made some lasting contributions to American rocketry besides Wyld's regenerative cooling. A second important legacy was the "spaghetti" construction for combustion chambers, invented and named by Edward A. Neu Jr. Neu applied for a patent in 1950, and received it in 1965, but had developed the concept earlier. It involved preforming cooling tubes so that they became the shell for the combustion chamber when joined together. They created a strong yet light chamber. The materials used for the tubes and the methods of connecting them varied, but the firm used the basic technique on many of its engines up through the LR99 for the X-15. By the mid-1950s, other firms picked up on the technique or developed it independently. Rocketdyne used it on the Jupiter and Atlas engines, Aerojet on the Titan engines. Later, Rocketdyne used it on all of the combustion chambers for the Saturn series, and today's space shuttle main engines still use the concept.[16]

Another contribution of Reaction Motors, according to former project engineer Maurice E. "Bud" Parker, who later worked for Thiokol in Huntsville, Alabama, was that "the propulsion specialists at the Marshall Space Flight Center studied the LR-99 design features and control system which made the system fail-safe and man-rated, and later applied this knowledge to the design of the Saturn engine where man-rating was to be all important for the upcoming Apollo program." Even though Reaction Motors never won any competitions to build large engines for intercontinental ballistic missiles, the firm did build some components for them. For example, the company manufactured a number of "fast-acting shutoff valves for engine staging" for the Atlas and some valves for the Titan II and III.[17] Thus, although Reaction Motors itself ultimately was not successful in the world of business, it made important contributions to rocket technology.

Projects Paperclip and Hermes

A larger contribution to American rocketry came, as already seen in part, from bringing the V-2 and most of the leading Germans who had developed it to this country after World War II. As the Soviet offensive against Germany pushed west, by January 1945 the Germans at Peenemünde could hear the sounds of Russian artillery. Orders came down to von Braun to evacuate Peenemünde and move to the area of the Mittelwerk underground production facilities near Nordhausen in the Harz Mountains. Eventually he and many of his associates ended up in the Bavarian Alps, where they surrendered and underwent interrogation by British and American intelligence and technical specialists. Others had remained in the vicinity of Nordhausen and came to the attention of the Allied forces there.[18]

U.S. Army Ordnance was especially interested in learning more about missile technology from the German engineers under von Braun, from technical documents the Germans had created in the process of their research and development, and from V-2 missiles themselves. Besides contracting with JPL to conduct research that ultimately resulted in the Corporal missile, Ordnance had also initiated a contract on November 15, 1944, with the General Electric Company (GE) to perform research and development on guided missiles. In conjunction with this effort, called project Hermes, the Ordnance Department sought to provide GE's engineers with captured V-2s to be studied and test fired. Both the department and its contractor had sent people to Germany in hopes of bringing back many of the Germans, their documents, and some V-2s.[19]

Dr. Richard W. Porter was General Electric's project engineer for Hermes from 1944 to 1953. He had earned a B.S. in electrical engineering at the University of Kansas and a Ph.D. in the same discipline from Yale. He did not know much about rockets at the start of the project, which raises the question of why the Army hired GE to do the job. According to Art Robinson, Porter's deputy, what attracted the Army was the firm's "broad technological base." Porter himself said that it would have been logical to put people working on jet engines in charge of Project Hermes, but those individuals were all heavily occupied with their jet-engine work. So he assembled an initial team of five people, and went to Germany. Before they went, they engaged in intensive study of rocketry, even managing to get Robert Goddard to come to GE and share his knowledge before his death—a rare departure from his usual practice.

For the Army, Col. Holger N. Toftoy headed the Ordnance technical intelligence teams charged with finding and evaluating captured enemy equipment and weapons. Working under him were Majors James P. Hamill and Robert Staver. According to Porter, "we ended up with quite an extensive set of technical documents and data including significant research reports, field service and operational manuals and handbooks. In addition to this, we acquired more than a dozen freight carloads full of developmental components and research equipment in Bavaria, as well as shiploads of production V-2 parts taken from the underground factory at Nordhausen."[20]

Under the auspices of Project Overcast, which essentially became Project Paperclip in 1946, von Braun and a handful of his associates flew to the United States on September 18, 1945. They were followed by three groups of about 118 others who had worked at Peenemünde. These arrived in the United States by ship between November 1945 and February 1946, ending

Figure 13. After World War II ended in 1945, Wernher von Braun led about 120 of his colleagues, who at Peenemünde had developed the V-2 rocket for the German military, to the United States as part of Operation Paperclip. During the following five years the team, under contract to the U.S. Army, worked on high-altitude firings of the captured V-2 rockets at White Sands Proving Ground in New Mexico and on guided missile development at Fort Bliss, Texas, where this photo was taken. Courtesy of NASA.

up at Fort Bliss in Texas, across the state border from White Sands Proving Ground in New Mexico, where the V-2s would be launched after assembly.[21]

The purposes of bringing the Germans to the United States, according to Toftoy, were "rapid dissemination" of their "knowledge and experience" through further interrogation by American scientific and technical personnel; assembly and firing of the V-2s to learn about their "design, functioning, flight characteristics, etc."; and a continuation of the Germans' development work. The anticipated interrogations by "scientists and engineers from all interested services" as well as from educational institutions and firms beginning to engage in missile development did occur. The Germans helped Porter's team in preparing to launch V-2s. Finally, they also worked on a prototype for the Hermes II missile, which was a two-stage vehicle with rocket and ramjet components, as well as on other Hermes missiles.[22]

The area to which the Germans working at White Sands had come was much dryer and more barren than Peenemünde. The one common feature was sand, but sand with a difference. The sand at the new site often whipped up into a windblown fury. "The New Mexico sandstorm is an implacable enemy of rocket work," wrote Milton Rosen later about his own work at White

Sands in launching Viking rockets. “You can see it coming in the distance, a long whitish cloud crawling close to the ground. It strikes with unbelievable force, sometimes as a steady gale, sometimes in gusts,” and can destroy a rocket if it penetrates to the critical parts. All that could be done was to protect the vehicle with rubberized covers.[23]

Despite such hazards, on November 22, 1946, the Germans signed the first of a series of one-year contracts. Before that, a number of them had set to work with their General Electric counterparts under shorter-term contracts, sorting through the V-2 parts from three hundred freight cars that carried them to New Mexico, arriving in August 1945. The components had traveled via rail from Nordhausen to Antwerp, where seventeen Liberty ships took them to New Orleans. After another rail journey to Las Cruces, they proceeded on flatbed trucks across the mountains to White Sands, which had just been established as a proving ground on July 9, 1945. As might be imagined, the journey was not kind to many of the components, which had already been in bad shape from weather, looting, and German attempts at demolition. The first Germans started working at White Sands in the fall of 1945. They numbered some thirty-nine by March of 1946 and were indispensable to the GE personnel in identifying parts and components, repairing them, and assembling them into complete missiles.[24]

General Electric and Project Hermes

The backgrounds of the individual Germans ranged from science and engineering to technical. They outnumbered the GE people at first; the total number of GE employees at White Sands did not reach thirty-four until late 1947. But gradually the proportion of Germans declined until, by spring 1947, there were none left at White Sands. GE people took over the work, assisted increasingly by Army personnel who learned much about handling large missiles from the experience. Meanwhile the Germans had passed on their extensive knowledge to the Americans. At first there was a language barrier, but the Peenemünders worked to improve their English.[25]

At the beginning of the V-2 program at White Sands, the tentative goal had been to launch twenty-five missiles. Those running the project had naïvely assumed that this could be done by about ten GE engineers assisted by the Germans and Army personnel. As the program went forward, the number of missiles to be launched increased twice, and the need became clear for more time than originally anticipated between launches to analyze the data. The number of GE personnel remained at roughly thirty-four from the end of 1947 to the end of the project for GE in 1951, with the breakdown approximately as shown in table 4.1.

Table 4.1. Numbers and Types of GE Personnel for Project Hermes at White Sands

Type	Number
Engineers and engineering assistants	17
Mechanical shop personnel	11
Electrical shop personnel	3
Clerks and administrators	3
Total	34

Source: L. D. White, "Final Report," 43, NHRC.

The Germans told the GE and military personnel that when missiles had been stored for an extended period, there was a large increase in failures during flight. So when repairs were found necessary, testing was repeated. Then the group performed two overall tests before a missile left the White Sands assembly building (completed in February 1946; before that, a Quonset hut had been used). Once a V-2 was at the launching site (completed in time for Tiny Tim tests for the WAC Corporal on September 26, 1945), the team performed one full test before the day of launch and another just before the propellants were loaded.[26]

Ultimately the V-2 program envisioned firing as many as a hundred rockets, although not all of them were launched. Since mostly untrained people had gathered the parts in Germany, many were not available in sufficient quantity to assemble all the V-2s needed, especially as the components and subassemblies were often not in good repair. Components that the program had to procure from sources in the United States included the mixing device (computer), gyroscopes, the tail structure, piping, the oxygen vent valve, wiring, and the main electrical distributor. Most of the U.S. components were basically the same as the items they were replacing, although use of American materials and standards made them less than carbon copies. Obviously, however, the need to assemble and test the V-2s before launch, as well as to find commercial sources in the United States to build replacements, contributed to the learning experiences of a variety of Americans involved.

The replacement mixing devices were built by GE. These had electrical circuits that were essentially identical to their German counterparts, but their mechanical construction was very different. The German units used solid copper wire, which did not work well in tests, especially vibration tests. GE personnel substituted wire with many strands, which held up better. There was also a different power supply with reduced voltage, which lowered the chances of voltage breakdown. And there was a new mechanical design to facilitate use of standard American components. GE-built computers flew in more than half of the launches and produced far fewer failures than the German mixing devices, whose problems mostly stemmed from deterioration of components rather than poor design.

Most of the German gyros were in such bad shape that they had to be rebuilt. A firm named Waldorf and Kearns built the American replacements, which were not identical but "seemed to be at least as good as the German models in almost every respect." There were other changes, but one final one will suffice to give a sense of the differences between the European and the American V-2 equipment. The German servomotors used to move the jet vanes were barely adequate to their purpose when new. The ones transported to New Mexico were no longer new and showed signs of wear and leakage of hydraulic fluid. GE modified an aircraft motor to replace the German unit. The new version had more than adequate power. These changes in the various components of the control system resulted in much better performance of the system. Initially, 46 percent of the V-2s had exhibited steering problems; for those with American control components, the percentage dropped to 5.[27]

V-2 Launches

By March 15, 1946, the V-2 team at White Sands was ready to conduct a static test of an entire missile at a test stand built into the side of a cliff. The Americans had awarded a contract for the construction of the stand on October 26, 1945. Rated for 90,000 pounds of thrust, it was completed by March and continued in use until 1949. The first V-2's engine burned for 57 seconds, preparing the way for the first flight.[28]

The team launched this V-2 on April 16, 1946, a month and a day after the successful static test. Shortly after liftoff, the missile began to roll. It soon entered a violent spin, turning to the east because of the prevailing wind. Using an emergency system developed by the Naval Research Laboratory, the launch team cut off propulsion 19 seconds after liftoff. The missile reached an altitude of 18,000 feet and a distance from the launch site of 5.3 miles. When it hit, the propellant tanks were roughly two-thirds full, so they exploded. Presumably this and the impact itself precluded determining the cause of the missile's erratic behavior by examinating the wreckage. Since the missile carried no telemetry equipment, the only information for analysis of the failure came from photography and visual observations.

It seemed likely from the way the missile flew that the cause was a broken jet vane. This could have been caused either by defects in the vane or by an igniter striking the vane as it was expelled by expanding gases from the combustion chamber. After this failure, the V-2 team instituted a variety of measures to ensure that vanes did not break on future launches. These included X-rays of the vanes, load tests, torque tests of the vanes' mounting threads, and even cardboard covers to protect the vanes against being struck by an igniter. (The cardboard burned off once the engine reached full thrust.) These measures apparently worked, since none of the other missiles behaved in the same way.[29]

This flight was unusual in not having telemetry, which normally was the best source of information about causes of failures, the other two being optics and recovery of parts and onboard recording devices. The information from telemetry typically included turbine speed, acceleration and velocity, chamber pressure, and the movements of the jet vanes. But the six telemetry channels could not begin to cover all possible causes of failure. Since recovery of parts and recorders not monitored by telemetry was always problematical and optics supplied limited information, investigations into subsequent failures usually could yield the general nature of a problem but not its precise cause.[30]

Undeterred by the failure of the first launch, the U.S. Army–GE–German team did not cancel an invitation for the press to cover flight 2 on May 10, 1946. The optimism this suggested bore fruit in a generally successful launch, with the V-2 reaching an altitude of 70 miles. Impact was 31 miles north of the launch site after an onboard integrating accelerometer cut off the engine. The angle of tilt following launch was 4 degrees steeper than planned, but performance was otherwise essentially flawless. There were seventy-one further launches from White Sands under either GE or Army auspices, through September 19, 1952. (The last GE flight was on June 28, 1951, with the ensuing five conducted by the Army alone.) Depending upon the criteria of failure, some 52 to 68 percent of the flights conducted under GE auspices were successful, but many of the failures still yielded useful information.[31] In addition, they taught the participants a lot about the missiles through the investigations that ensued.

Scientific Experiments

Scientific experiments quickly became a very important part of the V-2 flights, for which the Army had a number of specific goals directly related to future development of large missiles. Besides training for military personnel and contractors in the preparation and launching of such missiles, Army Ordnance planned to use the V-2s to gather aerodynamic data including boundary-layer transition, drag, and heat transfer as well as data on ballistic trajectories. The Army also envisioned operational testing of missile components including ramjets, and flight testing of staged rockets.[32]

But beyond gathering engineering data, Toftoy had a wider vision for the V-2s. In September 1945 he had assumed direction of the Army's guided missile program in Washington, D.C., and he assigned Lt. Col. James B. Bain to look into interservice cooperation, including use of the V-2s for scientific research. Bain located a variety of groups in both the Army and the Navy, including their contract laboratories, that were interested in using rockets for upper-atmosphere research. On January 16, 1946, at the Naval Research Laboratory (NRL), forty-one people from a dozen different institutions met and established a V-2 Rocket Panel, which ultimately became the Upper Atmosphere Rocket Research Panel. The V-2 panel was chaired by Ernst H. Krause of NRL and, when he left in 1947, James A. Van Allen of the Applied Physics Laboratory of Johns Hopkins University became chair. The details of its operations are beyond the scope of this history, but basically the Army used the panel to allocate research space on flights not already dedicated to the Army's own research.[33]

The NRL developed the telemetering system used at White Sands for the V-2 flights. It also designed "warheads" to carry experiments, with the Naval Gun Factory manufacturing them, since the original V-2 warhead casing was unsuitable for such use. As the experiment program continued, there were further variations in the casings. Initially the experiments weighed significantly less than the high explosive carried in the operational V-2 during the war, so lead weights were added to maintain the missile's designed center of gravity for aerodynamic stability. Later both the experimental payloads and the cases (hence the forward airframes of the V-2s) became increasingly large. This pushed the overall weight of the V-2s significantly over the design weight, as shown in table 4.2. Complete figures are not available, but the percentage of missiles with major contour modifications rose to 80 in 1950 and to 100 in 1951.[34]

Another modification necessitated by the experiment program was a method of detaching the warhead section from the rest of the missile. This was done after the missile reached its greatest height, so that the warhead with the experiments would not fall as rapidly as the main body's aerodynamic shape caused it to do. On early flights, the impact produced a large crater, virtually precluding recovery of either experimental payloads or—when flights failed and investigators were seeking causes—missile parts. After trying various methods of separating the warhead, the V-2 team found it most effective to place a small amount of explosive next to the four longitudinal members of the control chamber behind the payload casing. When the V-2 was on its downward trajectory, the explosion was set off remotely

Table 4.2. Changes to V-2s for U.S. Experiments

	Load Added to Design Weight (pounds)	Increase in Payload (percent)	Percentage with Major Contour Modifications
1946	150	6.8	0
1947	400	18.2	7
1948	527	23.8	41
1949	1,036	47.0	75

Source: U.S. Army Ordnance Corps/GE, "Hermes Guided Missile Research," 2, NHRC.

at an altitude depending on experimental requirements, usually about 40 miles above the desert floor. The separated pieces of the missile would fall more gently, generally permitting recovery of experimental equipment in reasonably good shape, although the nose section was not always recovered, causing some experimenters to place their experiments in the rear section of the missile.[35]

The areas explored by these experiments ranged from atmospheric physics to cosmic radiation measurements and included studies of the solar spectrum and ionospheric physics, among other disciplines. As the NRL's Homer Newell wrote, "By the time the last V-2 was fired in the fall of 1952, a rich harvest of information on atmospheric temperatures, pressures, densities, composition, ionization, and winds, atmospheric and solar radiations, the earth's magnetic field at high altitudes, and cosmic rays had been reaped." Also, the experience helped to give birth to a new technical culture and to the field of space science that blossomed further after the creation of the National Aeronautics and Space Administration in 1958.[36]

Much more significant for the development of launch vehicle technology than the results of the scientific experiments was simply the practice Americans got in launching the V-2s, operating them in flight, and analyzing the failures. One notable flight was the first night launch on December 17, 1946. Round 17, as it was counted, reached an altitude of 116 miles, highest of the series. Round 19 on January 23, 1947, followed its flight plan less well but, for the first time, the onboard telemetry system worked perfectly. Thus, despite an irregular liftoff with a roll beginning soon afterwards, the program counted this flight a success.[37]

Bumper WAC

One of the major contributions to missile and launch vehicle technology by Project Hermes came from the Bumper WAC project. This combined the V-2s with WAC Corporal B rockets in a two-stage configuration. According to Frank Malina, the idea for the combination originated with Martin Summerfield. Another account, not necessarily incompatible, holds that there was a discussion of using the two vehicles as a "step-rocket" among Colonel Toftoy, Richard Porter of GE, and Clark Millikan from JPL at the June 13, 1946, launch of a V-2 at White Sands. Millikan then had JPL undertake studies of both the feasibility of the project and the performance it might produce, resulting in a report by Homer J. Stewart. At the suggestion of Toftoy, Wernher von Braun and Ludwig Roth at Fort Bliss produced a similar study. Army Ordnance then authorized the project in October 1946.

Its purpose was to test vehicle separation at high speeds, to attain speeds and altitudes greater than were otherwise possible, and to investigate such phenomena as aerodynamic heating at high speeds within the atmosphere. Since a number of groups were already looking into launching satellites, and since more than one rocket stage commonly came to be used for that purpose, these matters became important.[38]

GE had overall responsibility for the Bumper project as part of the Hermes effort. It coordinated among the agencies involved, assembled the V-2s and WACs as modified for two-stage launch, performed preflight checkout tests, and handled the launch itself. JPL did the required theoretical investigations and designed the second stage from the WAC Corporal B, plus its separation system. Douglas Aircraft built the second stage, then designed and manufactured the special parts on the V-2 required for the mission.[39]

The engineers decided to place the WAC in a submerged position in the V-2's nose. Separation would be achieved through ignition of the second-stage engine. This required the installation of a conical, heavy-steel blast deflector within the shortened instrument compartment, with ducting to allow the rocket blast to escape. The compartment would also contain guide rails for the Bumper WAC, as the modified WAC Corporal was called. Doors would seal the nose cavity while the V-2 (or Bumper) climbed, then open as the WAC engine fired. At this time, the V-2 engine had to stop firing.

The sequence would begin as a normal launch of the V-2. Once near the peak of the missiles' ascent, the first stage's integrating accelerometer would signal the engine to adjust from the 25-ton- to the 8-ton-thrust propulsion stage. Then the second integrating accelerometer signal, which normally would cut off the 8-ton stage of engine thrust, would instead actuate the start of the WAC engine and the opening of the duct doors so its thrust could escape. As the Bumper WAC engine fired, its exhaust would burn through an electrical wire, actuating the V-2's cutoff. The WAC would have no control system but would be stabilized through spin imparted by two solid-propellant motors firing in opposite directions around the vehicle's circumference at the center of gravity. The solid motors' igniters would be actuated by a switch as the V-2 and Bumper WAC separated.[40]

To further ensure stability for the Bumper WAC, JPL increased its fin area by about 50 percent from the WAC Corporal B's configuration and used four rather than just three fins to further enhance stability. Since the Bumper WAC occupied most of the Bumper's instrument compartment, a small amount of instrumentation went into the WAC's nose cone. This included a Doppler receiver-transmitter and a small telemetry system that sent back information on the skin temperature of the WAC's nose. Also telemetered were the performance of the Bumper, the effects of changes in the V-2's center of gravity resulting from the presence of the WAC second stage, and data about the second-stage ignition and separation.[41]

On May 13, 1948, the first Bumper WAC vehicle launched with only a solid-propellant motor in the second stage, which otherwise had the same center of gravity, weight, and structure as the liquid-propellant Bumper WAC. This test successfully demonstrated separation. On flight 2 with the same configuration on August 19, 1948, the Bumper vehicle's turbines speeded up excessively at 33 seconds after launch, and the modified V-2 never reached the speed necessary for the integrating accelerometer to start the engine cutoff and inaugurate the firing of the WAC motor. Nevertheless, telemetry provided valuable information about the configuration of the two stages flying together.[42]

The next two rounds of the Bumper WAC program featured the WAC in its regular configuration, except that it only had 32 rather than 45 seconds of propellant, with ballast providing the normal weight and center of gravity. On the first flight (round 3 of the Bumper launches) on September 30, 1948, the Bumper performed as designed but the WAC exploded during ignition. Apparently the problem resulted from the low pressure at the separation altitude. There had been tests on the ground with a small vacuum container placed around the exhaust end of the WAC to simulate ignition at high altitude. This failed to simulate actual conditions. Recognizing this after the failure, the project engineers placed a diaphragm over the exhaust nozzle to retain higher atmospheric pressure in the combustion chamber. The firing of the engine broke the diaphragm. This change solved the problem. The following flight (round 4) on November 1, 1948, saw erratic steering on the Bumper and another failure attributed to a broken alcohol feed line resulting from the dynamic loads the modified airframe imparted to the V-2.

Despite these disappointments, GE launched Bumper WAC round 5 on

February 24, 1949, with a full 45 seconds of propellant loaded into the second stage. All the planned sequences worked, and at 98,813 feet (18.7 miles) separation occurred. The spin-motors at the WAC's center of gravity imparted 420 rpm of spin, and the second stage reached a reported altitude of 244 miles and a maximum speed of 7,553 ft/sec. These were the highest speed and the greatest altitude reached by a rocket or missile to that date. The highly successful launch demonstrated the validity of theory that a rocket's velocity could be increased with a second stage. Also shown was a method of igniting a rocket engine at high altitude, offering a foundation for later two- and multiple-stage vehicles. Aerodynamic heating data were obtained up to Mach 6, and telemetering worked to the maximum altitude, proving that communication could pass through the ionosphere.[43]

The final Bumper launch at White Sands took place on April 21, 1949. There was a premature cutoff of propulsion for the V-2 stage at 48 seconds after launch. Investigators concluded that the configuration produced too much vibration, evidently the cause of the failures for Bumpers 2 and 4 as well.

The final two launches of the Bumper program occurred at the newly activated Joint Long Range Proving Grounds in Florida, which later became Cape Canaveral Air Force Station and had previously been the site of the Banana River Naval Air Station until taken over by the Air Force on September 1, 1948. The Coast Guard did not open a small patch of land for missile use until February 1950, and Bumper 7 would be the first launch attempted from the mosquito-infested, sparsely settled area. The launchpad was a 100-square-foot swatch of concrete poured onto the sand and equipped with tunnels and hatches for equipment. The gantry was composed of painters' scaffolding with plywood platforms. A former dressing shack for swimmers became a launch control facility after it was suitably surrounded by sandbags. The reason for moving the launches to the Atlantic coast of Florida was the plan to send the Bumper WAC on an almost horizontal path to collect aerodynamic data at very high speeds.[44] The White Sands range was not large enough for this to take place in New Mexico.

The V-2 stage with the WAC Corporal on its shoulders would launch vertically but was supposed to turn to 22 degrees from the horizontal for separation. Burnout was set to occur at 125,000 feet, with the peak altitude intended to be 160,000 feet over the ocean. Used to launching in the dry desert, the GE-Army-JPL launch crew failed to anticipate the effects of Florida's humidity. They watched with a hundred media representatives as the V-2 failed to launch on July 19, 1950. The moist, salty air had condensed

Figure 14. A V-2 rocket carrying a WAC Corporal, noticeably more slender than the first stage. This Bumper WAC, as it was called, launched on July 24, 1950, from Cape Canaveral, Florida, as part of Project Bumper. NASA Historical Reference Collection, Washington, D.C. Courtesy of NASA.

inside the vehicle and prevented it from firing by impairing the valve-control electromagnet. Delays in the launch had left the very cold liquid oxygen in the tank for seven hours, promoting the condensation. Bumper 8 launched successfully on July 24, 1950, but the WAC failed to accelerate after separation. After drying and rechecking in a hangar, Bumper 7 successfully launched on July 29, 1950. Excessive precession of the pitch gyro caused a separation angle closer to the horizontal than intended—a mere 10 degrees—and a separation altitude of only 48,000. There the drag forces were much greater than in the less dense atmosphere of 125,000 feet, so the WAC reached a speed of only 3,286 mph. Although this was reportedly the highest speed yet achieved in Earth's atmosphere, and although the optical and telemetry collection was successful, the vehicle did not yield the high-speed data the experimenters had hoped for.[45]

Operations Sandy and Pushover

While preparations were being made for the Bumper WAC project, six GE engineers assisted in launching a V-2 from the fantail of the aircraft carrier USS *Midway*. The launch took place on September 6, 1947, and the missile headed off at an angle from the vertical of 45 degrees even though the ship had almost no pitch or roll. After it nearly hit the bridge and then straightened out momentarily, it began tumbling. This test, called Operation Sandy, could hardly have been encouraging for those in the Bureau of Aeronautics who wanted the Navy to develop ballistic missiles for shipboard use. Even less so were other tests in which two V-2s loaded with propellants were knocked over on purpose aboard the *Midway* in part of what was called Operation Pushover. They detonated, as the Navy expected. Finally, at White Sands on a mocked-up ship deck, a V-2 with propellants burning was also knocked over on December 3, 1948. This produced a huge blast that cracked structural supports. The deck itself ruptured, with alcohol and liquid oxygen running through the hole and igniting. This later provided reinforcement for the Navy's opposition to liquid propellants in what became the Polaris program.[46]

Tactical Hermes Missiles

A detailed account of the various tactical missiles worked on by GE and the former Peenemünders would be out of place here. Except for one solid-propellant version, they appear not to have had a large influence on missile and launch vehicle technology. They began as weapons projects but became simply test vehicles in October 1953. Hermes A-1, patterned after Wasserfall, was a 25-foot-long antiaircraft test vehicle converted in 1946 to a surface-to-surface mission. A few of them flew from White Sands, and V-2s tested their basic guidance-and-control subsystems in flights during 1948. The A-2 featured both liquid- and solid-propellant designs. The solid-propellant version had four flight tests and will be discussed in chapter 5.

The Hermes II V-2/ramjet combination had four flight tests, but none in which the ramjet stage was fully functional. The first, designated Hermes II round 0, featured a "dummy ram wing" on the nose of a V-2 that GE had modified. Launching on May 29, 1947, it experienced control problems 4 seconds after liftoff because of a faulty gyroscope and landed south of the Mexican city of Juarez. According to Fred Ordway and Mitchell Sharpe, range safety officer Ernst Steinhoff did not direct fuel cutoff because he did not want the residual propellant to cause a big fire. Steinhoff expected the missile would clear El Paso and Juarez, as it did. The incident nevertheless led to considerable tightening of range safety. There were three more Hermes II flights, but eventually the missile merged with another ramjet project, Hermes B, and both were cancelled by 1954.[47]

The A-3 was the final Hermes missile. Its development began late in 1947, accelerated in 1951, and ended in 1954. It was an ambitious tactical missile, but its characteristics were continually revised. An A-3A (RV-A-8) had seven test flights in 1953 and early 1954, while the A-3B (XSSM-A-16) flew six times in 1954. It reached a speed above Mach 3 and a range of 61 nautical miles (70 statute miles). The B-model had sea-level nominal thrust of 22,600 pounds and an alleged specific impulse at sea level of 242 lbf-sec/lbm, much higher than the V-2 using the same liquid oxygen and alcohol propellants. It also featured an inertial guidance system corrected by radio and radar command. But because the missile was terminated with so few flight tests, it is not clear how much influence it had upon launch vehicle or missile technology, particularly as the Germans had left the Texas–New Mexico area in 1950 and GE, after the cancellation of Hermes, met limited success in obtaining government contracts and soon dissolved its Guided Missile Department. At its highest level of activity in 1952, Project Hermes had employed about 1,250 people by one account, a number that declined to 450 in 1954.[48]

Legacies

According to a GE publication written in 1965, the engineers who worked on Project Hermes "formed a unique nucleus of talent that was fully realized when they took over many top management slots in General Electric missile and space efforts in later years." Since GE did have a role in the guidance and control system for Atlas (secondary, however, to the role of the Arma Division of American Bosch Arma Corporation) and many of the fleet ballistic missiles starting with Polaris (in conjunction with the Draper Laboratory at MIT), this claim is plausible. In fact, the firm claimed that work on the Hermes A-3 "led directly to the Atlas guidance system." GE also later supplied the first-stage engine for the Vanguard launch vehicle and the second stage for the short-lived Atlas-Vega vehicle, among others. William R. Eaton, who was part of the Hermes team, later became the GE manager of the Mississippi Test Facility where Saturn launch vehicles were tested. Finally, once the Hermes project ended in 1954–55, the people who worked on it shifted their focus to a ballistic reentry vehicle for the Air Force.

Additionally, the A-1 had featured a new type of injector with doublet streams for fuel and oxidizer that impinged on one another from alternating holes drilled in multiple concentric circles in the face of the injector. George Sutton paid a visit to GE in the later 1940s and took information about the injector design back to North American Aviation, where he then worked. North American modified the design for use in larger-diameter injectors for other projects. Thus it does seem that the firm's experience with Hermes made a contribution (via missiles in the case of Polaris) to future launch-vehicle technology.[49]

Even apart from GE's somewhat limited role in this technology, clearly there was a legacy of considerable importance from Project Hermes, including the firing of the V-2s and the related transfer of the Germans under von Braun to this country. But opinions have been mixed about how significant the legacy really was. Alan J. Levine in his book *The Missile and Space Race* took a skeptical view: "While some things were learned during Hermes, the program was not too useful." John R. London, who focused on the V-2 launches but also covered ramjet research in an article on White Sands, was more positive when he stated, "A better understanding of high-speed aerodynamics, ballistics, and other aerospace technologies was gained, allowing the successful development of a variety of aerospace vehicles and systems. The technical legacy of the White Sands V-2 program advanced American space technology."[50]

Even more emphatic was Julius H. Braun, who had worked on the project while serving in the Army. He spoke of "a massive technology transfer to the U.S. rocket and missile community" from the V-2. "There was a steady flow of visitors from industry, government labs, universities and other services," he observed, and continued:

> Propulsion experts from the [German] team traveled to North American Aviation in Southern California to assist in formulating a program to design, build and test large liquid propellant rocket engines. This program led to the formation of the NAA Rocketdyne division. In a similar manner, guidance specialists provided the basic concepts around which the NAA Autonetics division ultimately evolved. AC Spark Plug Division, IBM, RCA, and many other organizations also developed guidance and control skills by expanding on the V-2 experience.

Braun opined that "The rapid exploitation and wide dissemination of captured information" from the Germans "saved the U.S. at least ten years during the severe R&D cutbacks of the postwar period." He listed a great many U.S. missiles and rockets that "incorporated components . . . derived from the V-2 and its HERMES follow-on programs."[51]

As Braun suggests, the V-2 technology was a starting point for many efforts by both the Germans and U.S. engineers to develop more advanced technology for the rockets and missiles that followed. But it was just that, a starting point: U.S. engineers did not simply copy the V-2 technology but went beyond it. Even in assembling the V-2s for firing in this country, GE engineers and others had to develop modifications to German technology in making replacement parts. Firms that contracted to make the parts undoubtedly learned from the effort. GE engineers also learned a great deal from working on the various tactical rockets that were part of the Hermes effort.

While many visitors surely picked up a great deal of useful information from the Germans, as did the GE engineers themselves, Martin Summerfield, who questioned von Braun and a Mr. J. Paul and Professor Carl Wagner about the rocket engine for the V-2 on April 19, 1946, evidently picked up little of use from the interchange. Summerfield had already learned from his own research at JPL the kinds of technical lessons that the Germans imparted. Others from JPL had a good chance to look over the V-2s while testing the Corporal and firing the Bumper WAC. According to Clayton Koppes, "They concluded there was relatively little they wanted to apply to their projects."[52] Of course, as discussed in chapter 3, even the Corporal did borrow some technology from the V-2, but much else had been developed independently.

Other visitors to the Germans came with less experience in rocketry and probably, as Braun says, benefited greatly from their visits. Later chapters will show that bringing the Germans, their documents, and their V-2s (disassembled as they were) did provide a great impulse to U.S. rocketry, as Toftoy and others in Army Ordnance had foreseen and intended. Projects like the Bumper WAC added to the fund of engineering data. But as Braun had mentioned, funding for rocket and missile development was limited in the immediate post–World War II United States. Firing the V- 2s had cost about $1 million per year through 1951, but until the American government and people became alarmed about a threat from the Soviet Union with its missiles and potential warheads, there would not be a truly sustained, major effort in the United States to go beyond the technologies already developed by JPL, Reaction Motors, Aerojet, GE, and the Germans.[53]

One other legacy lay in changing popular perceptions. At the start of the Hermes project, most Americans probably thought no more about rocketry than the father of GE team member Joe Hoffman, who joined the effort in late 1946 and commented later, "The public had no interest or knowledge about rocketry. . . . My own father was stunned when he heard I was doing work in the field. He just couldn't believe that a big, reputable, successful company like General Electric would be fooling around with rockets."[54]

Soon, however, articles about the rocket launches at White Sands began to appear in newspapers and popular magazines. As early as May 27, 1946, an article with photos appeared in the popular *Life* magazine with the headline "U.S. Tests Rockets in New Mexico." Over the next couple of years, similar articles appeared in *Parade*, the Sunday *Washington Star*, the Sunday *New York Herald Tribune*, the *New York Times*, and *Popular Mechanics*. In October 1950, *National Geographic* published "Seeing the Earth from 80 Miles Up" by Clyde T. Holliday with accompanying photos.[55]

According to John R. London, "The American imagination was captured" by such articles, although he goes too far in saying "The V-2 program in America provided a belief in the exploration of space, and a vision of how it might be accomplished, to the American people."[56] At least the program had not done this to very many people by late 1949 when a Gallup poll indicated that only 15 percent of Americans believed "men in rockets" would "be able to reach the moon within the next 50 years."[57] Joe Hoffman's father may not have become part of that 15 percent, but it is likely that he no longer was incredulous that a "successful company like General Electric would be fooling around with rockets."

5

From Eaton Canyon to the Sergeant Missile

Solid-Propellant Rocket Developments, 1940–1962

Until the mid-1950s, the major efforts in the evolution of large surface-to-surface missiles in the United States focused on liquid propellants. This changed in the mid- to late 1950s with the Polaris and Minuteman programs. Those two missiles inaugurated a major shift in ballistic-missile (although not space-launch) technology away from liquids. After their successful development, most intermediate range and intercontinental ballistic missiles came to use solid propellants, freeing up many liquid-propellant missiles for modification and use as launch vehicles. In the meantime, as an array of achievements in solid-propellant rocket technology prepared the way for the breakthrough that enabled the success of the Polaris and Minuteman, they also led toward the use of solid-propellant boosters for a variety of launch vehicles—notably Titan III, Titan IV, and the space shuttles, but also Scout and upper stages for Delta.

World War II and Rockets

During World War II, at the beginning of recent solid-propellant development,[1] the vast majority of rockets produced for use in combat employed extruded "double-base" propellants usually consisting mostly of nitrocellulose and nitroglycerine.[2] These were limited in size by the nature of the extrusion process used at that time. In "wet" extrusion, nitrocellulose was suspended in a solvent, swelling into a doughlike composition, and was then forced (extruded) through dies to form it into masses of propellant, called grains. The small masses of propellant produced by this process had to be thin, so the solvent could evaporate, and the elasticity of the resultant grain was too low for bonding large charges to the motor's case. With a "dry" or solventless process, there were also limitations on the size of the grain as well as a greater risk of explosion. These factors pointed to the need for castable propellants.

But before a truly viable process for producing large castable propellants could be developed, the United States, being at war, needed an assortment of rockets to attack such targets as ships (including submarines), enemy fortifications, gun emplacements, aircraft, tanks, and logistical systems. The development of these weapons did not lead directly to any launch vehicle technology, but the organizations that developed them later played a role in furthering that technology. Two individuals provided the leadership in producing the comparatively small wartime rockets with extruded grains.

One was Clarence Hickman, who had worked with Goddard on rockets intended for military applications during World War I. He then earned a Ph.D. at Clark University and went to work at Bell Telephone Laboratories. After consulting with Goddard, in June 1940 Hickman submitted a series of rocket proposals to Frank B. Jewett, president of Bell Labs as well as a division chairman in the recently created National Defense Research Committee. The upshot was the creation of Section H (for Hickman) of the NDRC's Division of Armor and Ordnance. Hickman's section had responsibility for researching and developing rocket ordnance. Although Section H was initially located at the Naval Proving Ground at Dahlgren, Virginia, it ended up working largely for the Army.

Hickman favored wet-extruded double-base propellants because they had shorter burning times than dry-extruded ones. He and his associates worked with this type of propellant first at Dahlgren and then, in the course of the war, at the Navy Powder Factory at Indian Head, Maryland, and finally at the Allegany Ordnance Plant, Pinto Branch, on the West Virginia bank of the Potomac River west of Cumberland, Maryland. There at the end of 1943 they set up Allegany Ballistics Laboratory (ABL) as a rocket development facility operated for Section H by George Washington University. By using traps, cages, and other devices to hold the pieces or columns of solvent-extruded double-base propellant, they helped develop the bazooka antitank weapon, a 4.5-inch aircraft rocket, JATO devices with less smoke than those produced by Aerojet using John Parsons's asphalt-based propellant, and a recoilless gun, among other weapons.[3] ABL later became an important producer of upper-stage motors for missiles and rockets.

Hickman's counterpart on the West Coast, and the other leader in the development of U.S. rockets for World War II use, was physics professor Charles Lauritsen of Caltech. Lauritsen was vice-chairman of the Division of Armor and Ordnance (Division A) of the NDRC, and in that capacity he had made an extended trip to England to observe rocket developments there. The English had found a way to make solventless double-base propellant by dry extrusion. This required extremely heavy presses but yielded a thicker grain than the wet-extruded propellant, offering higher propellant loading and the longer burning time that Lauritsen preferred.

Convinced of the superiority of this kind of extrusion and believing that a larger rocket program was needed than Section H could provide with its limited facilities, Lauritsen argued successfully for a West Coast program. Caltech then set up operations in Eaton Canyon in the foothills of the San Gabriel Mountains northeast of the campus near Pasadena. It operated from 1942 to 1945 and expanded to a 3,000-person effort involving research and development of rocket motors, pilot production thereof, development of fuses and warheads, and both static and flight testing. The group produced an antisubmarine rocket 7.2 inches in diameter, a 4.5-inch barrage rocket, several retro-rockets (fired from the rear of airplanes at submarines), 3.5- and 5-inch forward-firing aircraft rockets, and the 11.75-inch-wide Tiny Tim rocket, which produced 30,000 pounds of thrust and weighed 1,385 pounds. (This last item was the rocket later used as a booster for the WAC Corporal.)

By contrast with Section H, Section L (for Lauritsen) served mainly the requirements of the Navy. In need of a place to test and evaluate the rockets being developed at Eaton Canyon, in November 1943 the Navy established the Naval Ordnance Test Station (NOTS) in the sparsely populated desert region around Inyokern well north of the San Gabriel Mountains. Like the Allegany Ballistics Laboratory, NOTS was destined to play a significant role in the history of U.S. rocketry, mostly with tactical rockets but also with ballistic missiles and launch vehicles.

One early contribution was the development of the 5-inch White Whizzer rocket by members of the Caltech team who had already moved to NOTS but were still under direction of the university rather than the Navy. By about January 1944, combustion instability had become a problem with the 2.25-inch motors for some of the tactical rockets. These motors used tubular, partially internal-burning charges of double-base propellant. Radial holes drilled in the grain helped solve severe pressure excursions, presumably by allowing the gas produced by the burning propellant to escape from the internal cavity. A young researcher named Edward W. Price, who had not yet received his bachelor's degree but later became one of the nation's leading experts in combustion instability, suggested creating a star-shaped perforation in the grain for internal burning. He thought this might do a better job than the radial holes in preventing the oscillatory gas flow that was actually causing the charges of propellant to split. He tested the star perforation, and it did produce stable burning.

In 1946 Price applied this technique to the White Whizzer, which featured a star-perforated, internal-burning grain with the outside of the charge wrapped in plastic to inhibit burning there. This geometry allowed higher loading of propellant than the previous design, which had channels for gas flow both inside and outside the grain. And since the grain itself protected the case from the heat in the internal cavity, the case could be made of lightweight aluminum, affording better acceleration. Ground-launched about May 1946, the White Whizzer yielded a speed of 3,200 ft/sec, then a record for solid rockets. The internal-burning, aluminum-cased design later appeared in the 5.0-inch Zuni and Sidewinder tactical missiles. Internal burning also came to be a feature of a great many other solid rockets, including ballistic missiles and stages for launch vehicles. This apparently was the first use of such a grain design in the United States.[4]

Castable Composite Propellants

Meanwhile, the first known castable solid propellant was the composite developed by John Parsons at the Guggenheim Aeronautical Laboratory at Caltech back in 1942 for use in JATO motors.[5] Containing asphalt as a binder and potassium perchlorate as an oxidizer, this propellant—known as GALCIT 53 after the lab's acronym—did not have a particularly impressive performance compared, for example, with the double-base composition called ballistite. But it retained its performance at temperatures down to 40°F far better than the compressed black powder Parsons had been using in JPL's JATO motors. At lower temperatures, however, GALCIT 53 cracked. It also melted in the tropical sun and was very smoky when burning. This last characteristic limited the takeoff of follow-on aircraft using JATO units on a single runway because the smoke restricted visibility. These shortcomings led researchers at GALCIT and its successor, JPL, to search for an elastic binder with storage limits beyond GALCIT 53's extremes of -9°F and 120°F. In particular, a young engineer named Charles Bartley, who was employed at JPL from June 1944 to August 1951, began examining synthetic rubbers and polymers, eventually hitting upon a liquid polysulfide compound designated LP-2, which worked as a solid-propellant binder. It was made by the Thiokol Chemical Corporation for sealing aircraft tanks and other applications.[6]

Like many innovations, LP-2 was discovered by inadvertence. In 1926 a physician named Joseph C. Patrick, who found chemistry more interesting than medicine, had been seeking an inexpensive way to produce antifreeze from ethylene using sodium polysulfide as a hydrolyzing agent. Instead of antifreeze, his procedure yielded a synthetic rubber, which led him to cofound a firm named Thiokol that marketed the material for gaskets, sealants, adhesives, and coatings (the polysulfide polymer being resistant to weather, solvents, and electrical arcing). Then in 1942, Patrick and an employee named H. L. Ferguson found a way to make the first liquid polymer that included no volatile solvent and yet could be cured to form a rubberlike solid. It found use during World War II in sealing not only fuel tanks in aircraft but also gun turrets, fuselages, air ducts, and the like.[7]

Before learning of LP-2, Bartley and his associates at JPL had tried a variety of moldable synthetic rubbers to use as binders and fuels, including Buna-S, Buna-N, and Neoprene. Neoprene worked best, both as a binder and as a fuel, but molding it required high pressures, like the extrusion process used with double-base propellants, which made the production of large propellant grains impractical. Meanwhile, Thiokol chemists seeking new applications for their polymer had begun to release technical data about it. At a meeting of the American Chemical Society, Bartley asked about a liquid that would polymerize to a solid elastomer, a rubberlike substance. A Dr. Frank M. McMillan who represented Shell Oil in the San Francisco area had found out about Thiokol's product and shared the information with Bartley, who acquired small quantities of LP-2 from Walt Boswell, the Thiokol representative for the western United States.[8]

With encouragement from Army Ordnance and the Navy, Bartley—joined by John I. Shafer, a JPL design engineer, and H. Lawrence Thackwell Jr., whose expertise lay in aircraft structures—began in 1947 to develop a small rocket designated Thunderbird, which had a 6-inch diameter. The three researchers used it for testing whether polysulfide propellants could withstand the forces of high acceleration that a large launch vehicle might also encounter. Bartley had already found that an end-burning grain of polysulfide propellant did not produce steady thrust but burned faster at first and then leveled off. He suspected that accelerated burning along the case was reshaping the grain into a cone, a hypothesis he confirmed by quenching the flame partway through the burn.

To solve the problem of unsteady thrust, the three JPL engineers adopted a grain design developed in Great Britain in the late 1930s. Like Ed Price's White Whizzer, it featured an internal-burning, star-shaped cavity that protected the case from excess heat because all of the burning was in the middle of the propellant grain. It also provided a relatively constant level of thrust because, as the star points burned away, the internal cavity became a cylinder with roughly the same surface area as the initial star. When Bartley read about the star design in a British report and asked Shafer to investigate

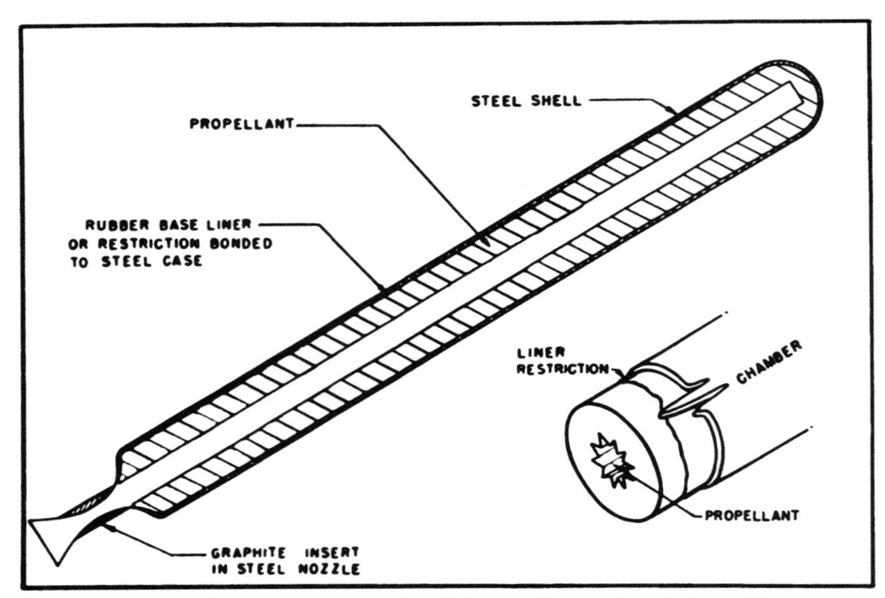

Figure 15. Schematic drawing of an early solid-propellant rocket design with an internal-burning, star-shaped cavity in a propellant grain bonded to the case. Note the graphite insert in the nozzle to protect it from the hot exhaust gases. From the redacted version of Thackwell and Shafer, "Applicability," approved for use by the NASA Management Office at JPL. Courtesy of NASA/JPL-Caltech.

it, Shafer found that the government-owned, contractor-operated Allegany Ballistics Laboratory had used the British design in the uncompleted Vicar rocket and a scaled-down version named the Curate. Using equations from the ABL report on the projects, Shafer developed a number of star designs beginning in 1947. Combining a polysulfide propellant with the star design and casting it in the case so that a bond formed, the team under Bartley produced the successful Thunderbird rocket that passed its flight tests in 1947–48.[9]

While this rocket was being tested, another significant development for composite propellants was occurring—the replacement of potassium perchlorate as an oxidizer by ammonium perchlorate, which offered higher performance and less smoke. Apparently the Thunderbird used a propellant with the designation JPL100, which contained a mixture of ammonium perchlorate and potassium perchlorate in a polysulfide binder.[10] Already in 1947, however, JPL had developed a propellant designated JPL118 that used only ammonium perchlorate as an oxidizer together with polysulfide as the binder and a couple of curing agents. Although this propellant had yet to be fully investigated in 1947,[11] by mid-1948 it provided a specific impulse of 198 lbf-sec/lbm at sea level using an expansion ratio of 10 for the rocket nozzle.[12] This was still relatively low compared with the typical performance of double-base propellants (about 230 lbf-sec/lbm), but it was higher than the 185 lbf-sec/lbm for the asphalt–potassium perchlorate propellant and the 190 lbf-sec/lbm for JPL100.[13]

Aerojet likewise began using ammonium perchlorate in its Aeroplex (polyester polymer) propellants in 1948 in order to increase specific impulse and reduce smoke. Funded by the Navy Bureau of Aeronautics to develop a basic understanding of the production and employment of solid propellants, Aerojet increased the specific impulse of its ammonium perchlorate propellants to 235 lbf-sec/lbm, but its Aeroplex binder was not case bondable, leading the firm to switch in 1954 to a polyurethane propellant that was.[14]

In the interim, Thiokol sought to sell its polymer to Aerojet and another manufacturer of rockets, the Hercules Powder Company, but both rejected Thiokol's product because its 32 percent sulfur content made it a poor fuel. Thiokol sales representative Joseph Crosby then asked the Army if it wanted to allow the promising developments with the Thunderbird rocket to end. The answer, in the spring of 1947, was Army Ordnance's encouragement for Thiokol to go into the rocket business itself. In early 1948 Thiokol set up its budding rocket operations in a former ordnance plant in Elkton, Maryland, and in April of 1949, without abandoning the Elkton location, it moved to the Army's new Rocket Research and Development Center at Redstone Arsenal in Huntsville, Alabama.[15]

The Falcon Missile

About this time, under contract with the Army, Thiokol produced a T-40 motor intended for use as a JATO unit. As a propellant it used JPL100 (rechristened T10 by Thiokol) in a case-bonded motor design. Also in 1949, Thiokol designed the T-41 motor for the Hughes Aircraft Company's Falcon missile under development for the Air Force. This was a shorter version of JPL's Thunderbird motor. It began production at Elkton, then moved to Huntsville, where a larger version called the T-42 evolved from it.[16]

According to Air Force Col. Edward N. Hall, who was highly instrumental in promoting the development of solid propellants, the Falcon tactical, air-to-air missile contributed "quality control techniques for rubber-base propellants, design data for case-bonded grains, [and] aging characteristics of rubber-based propellants" to the evolving store of knowledge about solid-propellant technology.[17] It appears that Thiokol did not make these contributions on its own. JPL provided considerable assistance in an early example of technology transfer. In October of 1947, Charles Bartley of JPL was present at a meeting of Thiokol personnel, representatives of Army Ordnance, and the Navy's Bureau of Aeronautics to discuss the kind of work Thiokol was expected to do in the further development of polysulfide propellants. The next day, Bartley met again with Thiokol personnel to relate JPL's experience with polysulfide-perchlorate propellants.[18]

In early 1949 a trip report by a Thiokol employee discussed a visit to JPL's Solid Rocket Section, of which Bartley was the chief. The report covered such matters as the grease used for extracting the mandrel placed in the curing propellant to create the internal cavity and igniters that employed black powder or a special igniter powder. Other subjects discussed were grinding ammonium and potassium perchlorate, combining them with the liquid polymer in a vertical mixer, pouring the propellant, preparing the liner for the combustion chamber, and testing. Also reported was a visit to Western Electrochemical Company, which supplied the perchlorate. The document concluded with some recommended changes to Thiokol's operating procedures at Elkton.[19]

As helpful as JPL's assistance was to Thiokol, however, the kinds of contributions Hall mentions seem to have come also from work done independently at Thiokol's Elkton and Huntsville plants. For example, Thiokol discovered that the size of perchlorate particles was important in motor operation and propellant castability, so it introduced a micromerograph to measure particle size. To reduce the deleterious absorption of moisture by the perchlorates, it installed air conditioning in the grinding rooms. It determined the optimal mixing time for the propellant. And it began using a Teflon coating in place of the barium grease that JPL had used to extract the mandrel from the middle of the cast propellant after it cured. This change was necessary because, even when the grease-affected part of the propellant was sanded after extraction, some grease enfolded into the propellant, causing weak areas. Thiokol also introduced a "temperature-programmed cure cycle," pressurized curing, and a method of casting that eliminated propellant voids resulting from shrinkage and air bubbles.[20]

The Sergeant Test Vehicle

While these developments were occurring at Thiokol, JPL was wrestling with a test vehicle named the Sergeant, not to be confused with the later missile of that name. Army Ordnance had authorized its development. It was a sounding rocket with a diameter of 15 inches, which was quite large for the day, although only an inch larger than an aircraft rocket developed at NOTS in 1945 and dubbed the Big Richard. Designed with an extremely thin steel case of 0.065 inches and a star-shaped perforation, the Sergeant test vehicle was expected to attain an altitude of up to 700,000 feet (130-odd miles) while carrying a 50-pound payload. Static tests with a thicker case in February 1949 showed that a polysulfide grain of that diameter could function without deformation. But when the JPL researchers, including Bartley, Shafer, and Thackwell, shifted to the thinner case, the result was twelve successive explosions, the last on April 27, 1950.

At this point, JPL director Louis Dunn cancelled the project for the sounding rocket and cut back all solid-propellant work at the laboratory to basic research. The researchers soon determined that the explosions had two causes: a chamber pressure that was too high for the thin case, and points in the star-shaped cavity that were too sharp, promoting cracks. An easy solution would have been a more conservative (thicker) case and rounded points in the star configuration. With Dunn's cancellation of the project, Thackwell transferred to Thiokol's Redstone Division in Huntsville, Alabama, taking his knowledge of solid-propellant rocketry with him.[21]

The RV-A-10 Missile

About this time, Thiokol had teamed up with General Electric in the Hermes Project to produce a solid-propellant missile, much larger than the Sergeant sounding rocket, that was initially known as the A-2. The project operated on a shoestring and was eventually cancelled, but not before it achieved significant progress in solid-propellant technology.[22]

Army Ordnance's original requirements for the A-2 were to carry a 500-pound warhead as far as 75 nautical miles, but the requirements changed to a payload of 1,500 pounds, necessitating a motor with a diameter of 31 inches. Thiokol became the contractor for the motor, with development starting in May 1950. The timing allowed the project to take advantage of work on the Sergeant sounding rocket and of Larry Thackwell's experience with it. By December 1951 a static test of the 31-inch motor had been successful. Then, between January 1952 and March 1953, there were twenty more static tests at Redstone Arsenal and four flight tests of the missile at Patrick Air Force Base in Florida. In the course of the program, the missile came to be designated the RV-A-10. Unanticipated problems with nozzle erosion and combustion instability arose but were solved.[23]

The four flight tests achieved a maximum range of 52 miles (on flight 1) and a maximum altitude of 195,000 feet or 37 miles (flight 2) using a motor case 0.20 inches thick and a propellant grain employing a star-shaped perforation with broadened tips. The propellant, which was designated TRX-110A and included 63 percent ammonium perchlorate by weight as the oxidizer, took advantage of an Air Force–sponsored project (MX-105) entitled Improvement of Polysulfide-Perchlorate Propellants that had begun in 1950 and issued its final report, written by Thiokol employees, in May 1951. Meanwhile, on test motor 2 a propellant designated T13, which combined ammonium perchlorate (66.97 percent by weight) with a polysulfide designated LP-33 (29.72 percent), achieved a sea-level specific impulse at 80°F of 196.2 lbf-sec/lbm. It also, unfortunately, demonstrated combustion instability. The TRX-110, which had a slightly lower specific impulse but no combustion instability, was substituted.[24]

To arrive at the blunter-tipped star perforation, Thiokol drew not only on JPL's experience with the Sergeant test vehicle, which Thackwell shared, but on photoelastic studies of grains performed at the company's request by the Armour Institute (later renamed the Illinois Institute of Technology). Together with a thicker case wall than JPL had used, this modification eliminated JPL's problems with cracks and explosions.[25] TRX-110 proved to have insufficient initial thrust, but a change in the size of the ammonium perchlorate particles, from a coarse-fine mixture to consistently fine, yielded higher initial thrust and a more consistent thrust over time—another desirable trait. Meanwhile, the Thiokol-GE team gradually learned about the thermal environment to which the RV-A-10 nozzles were exposed. Design of the nozzles evolved through subscale and full-scale motor tests employing a variety of materials and techniques for fabrication. Among the choices, an SAE 1020 steel with carbon inserts proved to be the best, while a roll weld proved superior to casting or forging.[26]

Another problem was the appearance of cracks and voids in the large grain for the RV-A-10 when it was cured at atmospheric pressure—probably the cause of a burnout of the liner on motor 2. The solution was twofold: in the pouring and in the curing. Thiokol personnel poured the first two mixes of propellant into the motor chamber at a temperature 10 degrees Fahrenheit hotter than normal, with the last mix 10 degrees cooler than normal. Then they cured the propellant under 20 psi of pressure with a layer of liner material laid over it to prevent air from contacting the grain. Together, these two procedures eliminated the voids and cracks.[27]

With these advances in the state of the art of producing solid-propellant motors, the RV-A-10 by the time of its flight testing in February–March

1953 became the first known solid-propellant rocket motor of such a large size—31 inches in diameter and 14 feet 4 inches long—to fly. Among its other apparent firsts were scaling up the mixing and casting of polysulfide propellants to the extent that more than 5,000 pounds of it could be processed in a single day; the routine use of many mixes in a single motor; the use of an igniter rolled into coiled plastic tubing, called a jelly roll, to avoid the requirement for a heavy closure at the nozzle end to aid in ignition; and the first use of jet vanes inserted into the exhaust stream of a large solid-propellant rocket to provide thrust vector control.

As recently as December 1945, the head of the Office of Scientific Research and Development during World War II, Vannevar Bush, had stated, "I don't think anybody in the world knows how [to build an accurate intercontinental ballistic missile] and I feel confident it will not be done for a long time to come." Many people even in the rocket field did not believe that solid-propellant rockets could be efficient enough or burn long enough to serve as long-range missiles. The RV-A-10 was the first rocket to remove such doubts from at least some people's minds. Arguably, it provided a significant part of the technological basis for the entire next generation of missiles from Polaris and Minuteman to the large solid boosters for the Titan IIIs and IVs and the space shuttle,[28] although many further technological developments, including significant improvements in propellant performance, would be necessary before they became possible.

The Sergeant Missile

Meanwhile, the first major application of the technologies developed for the RV-A-10 was the Sergeant missile, for which JPL began planning in 1953 under its ORDCIT contract with Army Ordnance. JPL submitted a proposal for a Sergeant missile in April 1954, and on June 11 the Army's chief of ordnance programmed $100,000 in fiscal year 1955 research and development funds for it. At the same time, he transferred control of the effort to the commanding general of Redstone Arsenal. Using lessons learned from the liquid-propellant Corporal missile, JPL proposed a co-contractor for the development and ultimate manufacture of the missile. In February 1956 a Sergeant Contractor Selection Committee unanimously chose the Sperry Gyroscope division of Sperry Rand Corporation for this role, based on JPL's recommendation and Sperry's capabilities and experience with other missiles, including being prime contractor for the Sparrow I air-to-air missile system for the Navy. Meanwhile, already on April 1, 1954, the Redstone Arsenal had entered into a supplemental agreement with the Redstone Division of Thiokol to work on the solid-propellant motor for the Sergeant, and the program to develop the missile began officially in January 1955.[29]

There is no need for a detailed history of the Sergeant missile here. It took longer to develop than planned and was not operational until 1962, by which time the Navy had completed the much more capable and important Polaris A1 and the Air Force was close to fielding the far more significant and successful Minuteman I. JPL director Louis Dunn had warned in 1954 that if the Army did not provide for an orderly research and development program for the Sergeant, "ill-chosen designs" would "plague the system for many years."

This warning became a prophecy. The Army did not provide consistent funding, and then insisted on a compressed schedule. The problem was complicated by differences between JPL and Sperry and by JPL's becoming a NASA instead of an Army contractor in December 1958. The result was a missile that failed to achieve its mandated in-flight reliability of 95 percent. While it met a slipped ordnance support readiness date of June 1962, it was classified as a limited-production weapons system until June 1968. On the plus side, it was equal to its predecessor, Corporal, in range and firepower while being only half as large and requiring less than a third as much ground support equipment. Its solid-propellant motor could be gotten ready for firing in a matter of minutes instead of the hours required for the liquid-propellant Corporal. An all-inertial guidance system on board Sergeant made it virtually immune to known enemy countermeasures, whereas Corporal depended on a vulnerable electronic link to guidance equipment on the ground.[30]

The Sergeant motor, designated JPL500, was a modification or direct descendant of the RV-A-10's motor. The latter, using the TRX-110A propulsion formulation, employed 63 percent ammonium perchlorate as an oxidizer, whereas the Sergeant motor's propellant had 63.3 percent of that oxidizer and 33.2 percent LP-33 liquid polymer in addition to small percentages of a curing agent, two reinforcing agents, and a curing accelerator. At a nozzle expansion ratio of 5.39, its specific impulse was about 185 lbf-sec/lbm, considerably lower than the performance of Polaris A1. It employed a five-point star grain configuration, a case of 4130 steel at a nominal thickness of 0.109 inches (roughly over half the thickness of the RV-A-10 case), and a nozzle that, like the RV-A-10's, used SAE 1020 steel with a graphite nozzle-throat insert.[31]

Although it was not at the forefront of solid-propellant technology by the time of its completion, the Sergeant did make contributions to the devel-

Figure 16. Aerial view of the Jet Propulsion Laboratory on January 1, 1961, set against the backdrop of the San Gabriel Mountains in southern California. Courtesy of NASA/JPL-Caltech.

opment of launch vehicles—but, intriguingly, only through a scaled-down version of itself. The next chapter will discuss Project Orbiter, which provides the background for these contributions. Here let it suffice to say that JPL had scaled down Sergeant motors from 31 to 6 inches in diameter for performing tests on various solid propellants and grain designs. By 1958 it had performed static tests on more than three hundred of the scaled-down motors and had flight-tested fifty of them, all without failures. These reliable motors became the basis for upper stages in reentry test vehicles for the Jupiter missile (called Jupiter C) and in the launch vehicles for Explorer and Pioneer satellites, which used modified Redstone and Jupiter C missiles in the first stage.[32]

These contributions aside, the Sergeant missile was the culmination of a series of developments in solid-propellant rocketry that prepared the way, in part, for future advances. It apparently did not, itself, make significant contributions to the technology. Moreover, it marked the end of JPL's role in rocketry per se. Delays in its development can be blamed in part on the Army's failing to heed Dunn's warning and provide consistent funding, in part on conflicts between JPL and Sperry, and perhaps primarily on Sperry's own problems. JPL's other commitments both to the Army and to NASA after the end of 1958 also impeded its work on Sergeant, however, and it bore some of the blame for ill feelings with Sperry. While JPL arguably improved its systems engineering from Corporal to Sergeant, this improvement benefited spacecraft rather than launch vehicles. For after Sergeant, JPL shifted its focus to building such spacecraft and to engaging in other space-related efforts to the exclusion of work on terrestrial rockets and missiles.[33]

Conclusions

The developments discussed in this chapter marked significant progress in solid-propellant technology. From fairly modest technical beginnings at the start of America's participation in World War II, solid-propellant rocketry witnessed several important advances. John Parsons's path-breaking discovery of an asphalt-based composite propellant at JPL led to Charles Bartley's adaptation of Thiokol's liquid polysulfide for use as a binder for composite propellants that was far superior to asphalt. Ed Price's use of an internal-burning, star-shaped cavity on the White Whizzer and JPL's apparently independent employment of similar technology from England via the Allegany Ballistics Laboratory laid the groundwork for later uses of internal-burning, case-bonded charges in missiles and launch vehicles.

JPL's involvement in Thiokol's becoming a major contractor for missiles and rockets ironically provided competition for the Caltech lab's own industrial creation, Aerojet. Both would come to play major roles in developing and producing propulsion units for a great variety of launch vehicles. Finally, Thiokol's work on the RV-A-10 began the process of scaling up the size of solid-propellant motors that would culminate in the truly gargantuan boosters for the Titan IV and the space shuttle.

Before the jump from the RV-A-10 to those huge solid-rocket motors could take place, however, further technological developments would be necessary. These will be discussed in chapter 9 and the sequel to this book. Meanwhile, important progress was also occurring in liquid-propellant technology for rockets.

6

Redstone, Jupiter C, and Juno I, 1946–1961

Following the development of the Redstone missile, rocket technology in the United States moved from the prehistory of launch vehicles to their actual history. A modified Redstone became the first stage of America's initial launch vehicle, the Juno I, and Redstone marked a significant improvement in several technologies for such vehicles, including propulsion, materials, structures, and guidance and control. It also represented an early example of technology transfer between armed services, as the Redstone engine developed from one of the engines North American Aviation had designed for the Air Force's Navaho missile, which never went into production.

Redstone

The Redstone began its development as the Hermes C1, which General Electric envisioned as a three-stage vehicle. However, GE shelved its original conception in 1946 because at that time it had insufficient data to design such a missile. Not until October 1950 did Army Ordnance direct the firm to proceed with a feasibility study of Hermes C1 and to work with the Germans under von Braun to that end. Meanwhile in April 1948 Colonel Toftoy, still heading the Rocket Branch in the Army Ordnance Department, recommended the establishment of a rocket laboratory, and on November 18 the chief of ordnance announced the reactivation of Redstone Arsenal in Huntsville, Alabama.[1] This became the site for most of the development of the Redstone missile as well as the source of its name.

The Chemical Warfare Service had set up an installation called the Huntsville Arsenal in 1941 to make mortar and howitzer shells. Later that year the Army Ordnance Corps began constructing the Redstone Ordnance Plant to produce munitions. Redesignated Redstone Arsenal in 1943, it and Huntsville Arsenal together employed about 20,000 people at the peak of production during World War II, but Redstone Arsenal went into standby status in 1947. Following its reactivation in 1948 as the research and development center for rockets that Toftoy had proposed, it incorporated the deactivated Huntsville Arsenal in 1949.[2]

On October 28, 1949, the secretary of the Army approved a proposal to move the guided missile group—formerly the Ordnance Research and Development Division Sub-Office (Rocket)—from Fort Bliss, Texas, to Redstone Arsenal. There it became the Ordnance Guided Missile Center on the site of the former Huntsville Arsenal. The Army officially established the center on April 15, 1950. It then took about six months for the people and equipment to be transferred, the Germans now having their wives and families with them. The green countryside, reminiscent of Germany, appealed to most of them, and the development of activities associated with rocketry began the conversion of the Watercress Capital of the World, as Huntsville had been called, into what residents called the Space Capital of the World. In the process, the population grew from 16,000 in 1950 to 48,000 in 1956 and 136,102 by 1970. Of course, the Germans were not the only ones who moved from the western desert to Redstone Arsenal. Also included were some 800 civil servants, a group of GE employees, and about 500 military personnel.[3]

In September 1950, just months after the beginning of the Korean War, Army Ordnance directed the transfer of responsibility for Hermes C1 to the Ordnance Guided Missile Center, giving it responsibility for design, manufacture, and testing. Back on July 10, soon after North Korean troops invaded South Korea, Army Ordnance had told the center to study the possibility of developing a tactical missile with a 500-mile range and a circular error probable of 1,000 yards to support the Army Field Forces. No additional funding immediately accompanied this requirement, despite the war and the fact that the Soviet Union had successfully tested an atomic bomb in 1949. The center designated the 500-mile missile the Hermes C1, just one of many designations it would carry until it officially became the Redstone on April 8, 1952.[4]

Despite their recent move from Fort Bliss, low funding, and the difficulty of creating satisfactory offices, laboratory space, and test stands from the facilities of the former Huntsville Arsenal, the personnel under von Braun quickly completed a preliminary study. Von Braun served as project engineer and put together a report, which he presented to the Department of Defense's Research and Development Board in the fall of 1950. Although the preliminary study did not designate any one type of propulsion system for the 500-mile missile, it did point out that the XLR43-NA-1 liquid-propellant rocket engine, developed by North American Aviation for the Air Force's Navaho missile project, was available and could be mass produced as early as late summer 1951. By the end of January 1951, the preliminary study group had concluded that the North American engine should be used to develop the missile in conjunction with an inertial guidance system supplemented by radio guidance. "We favored this method," von Braun later wrote, "because it was simple, reliable, accurate—and available."[5]

In February 1951, Colonel Toftoy changed the payload for the projected missile from somewhere between 1,500 and 3,000 pounds to 6,900 pounds. He recognized that this would reduce the range severely. During a visit that month by Kaufman T. Keller, a former chairman of Chrysler Corporation whom the administration had named director of guided missiles in the office of the secretary of defense, the Ordnance Guided Missile Center agreed that the missile would be developed in twenty months following receipt of funding estimated to be $18 million for the first 4 of 100 Hermes C1s, as they were still called. Keller subsequently agreed to reduce the requirement to 75 research-and-development missiles, with 12 of them completed by May 1953.[6]

Army Ordnance sent the Center its initial funding of $2.5 million only on May 1, 1951. Until then, all the engineers at Redstone Arsenal could do was continue preliminary design. Once development began in May, it continued for seven and a half years until the flight test of the last research-and-development vehicle. For most of this time, the developers operated without formally established military characteristics, the goal being simply the delivery of a warhead weighing 6,900 pounds. Proposed military characteristics did appear in 1954, but as the Redstone developed further, these changed. As of September 30, 1951, preliminary design characteristics showed a missile with 75,000 pounds of thrust, a specific impulse of 218.8 lbf-sec/lbm, and a range of 178 miles, remarkably close to the final achievements of the missile.[7]

Meanwhile, the Ordnance Guided Missile Center had asked North American Aviation for a program to modify the XLR43-NA-1 engine for use on the Redstone. A letter contract with NAA on March 27, 1951, provided 120 days of research and development to make that engine comply with the Ordnance Corps' specifications and to deliver a mock-up and two prototypes of the engine, which would then be designated NAA 75-110, referring to 75,000 pounds of thrust operating for 110 seconds. Supplemental contracts in 1952 and 1953 increased the quantity of engines to be delivered and called for their improvement. These contracts included only nineteen engines, with subsequent power plants purchased by the prime contractor, Chrysler, through subcontracts. About this same time, in a series of reorganizations from August 1951 to September 1952, the Ordnance Guided Missile Center became the Guided Missile Development Division.[8]

North American Aviation and the Redstone Engines

The story of how North American Aviation had developed the XLR43-NA-1 illustrates the ways in which launch vehicle technology developed in the United States. NAA came into existence in 1928 as a holding company for a variety of aviation-related firms, including an airline. It was hard hit by the depression after 1929, and General Motors acquired it in 1934, hiring James H. "Dutch" Kindelberger as its president. Dutch was a pilot in World War I and an engineer who worked for Donald Douglas before moving to NAA. Described as "a hard-driving bear of a man with a gruff, earthy sense of humor—mostly scatological—[who] ran the kind of flexible operation that smart people loved to work for," he reorganized NAA into a manufacturing firm that built thousands of P-51 Mustangs for the Army Air Forces in World War II, as well as the B-25 Mitchell bomber and the T-6 Texas trainer. With the more cautious but also visionary John Leland Atwood as his chief engineer, Dutch made NAA one of the principal manufacturers of military aircraft, with its workforce rising to 90,000 at the height of production before it fell to 5,000 shortly after the end of the war. Atwood succeeded Kindelberger as president in 1948, when Dutch became chairman of the board and General Motors sold its share of the company. The two managers decided to continue their pursuit of the military aircraft market in the much less lucrative climate of the postwar, when many of their competitors were shifting to commercial airliners.[9]

Despite the drop in business, NAA had lots of money from wartime production, and Kindelberger determined to hire a top-notch individual to head a research laboratory he would fill with quality engineers in the fields of electronics, automatic control with gyroscopes, jet propulsion, and missiles. He selected William Bollay, the former von Kármán student whose lecture had gotten the Suicide Squad started. Following receipt of his Ph.D. in aeronautical engineering at Caltech, Bollay had joined the Navy in 1941 and been assigned to Annapolis, where the Bureau of Aeronautics (BuAer) was working on experimental engines including the JATOs Robert Truax was developing. At war's end, Bollay was chief of the Power Plant Development Branch for BuAer. As such, he was responsible for turbojet engines, and NAA was then developing the FJ-1 Fury, destined to become one of the Navy's first jet fighters. Bollay came to work for NAA during the fall of 1945 in a building near the Los Angeles airport, where he would create what became the Aerophysics Laboratory.[10]

On October 31, 1945, the Army Air Forces' Air Technical Service Command invited leading aircraft firms to bid on studies of guided missiles. NAA proposed a 175–500-mile surface-to-surface rocket it called Navaho (for *N*orth *A*merican *V*ehicle *A*lcohol plus *H*ydrogen peroxide and *O*xygen). The proposal resulted in a contract on March 29, 1946, for an experimental missile designated MX-770. Other contracts for the missile followed. The Navaho ultimately evolved into a complicated project before its cancellation in 1958. It included a rocket booster and ramjet engines with lots of legacies passed on to aerospace technology, but for the Redstone, only the rocket engine that evolved to become NAA 75-10 is relevant.[11]

Apparently North American did not originally intend to manufacture the engine. An employee from the early days recalled that NAA was "forced into the engine business—we had the prime contract for Navaho and couldn't find a subcontractor who would tackle the engine for it, so we decided to build it for ourselves." NAA's plans for developing the engine began with the German V-2 as a model, but soon the firm's engineers envisioned "an entirely new design rated at 75,000 pounds thrust" as compared with about 56,000 for the V-2. Already in the spring of 1946, Bollay and his associates visited Fort Bliss to conduct numerous interviews with Germans who had worked on the V-2, and held subsequent conversations with such people as von Braun, Walter Riedel, and Konrad Dannenberg. By the middle of June 1946, Bollay's team began redesigning the V-2 engine with the aid of drawings and other documents obtained from the Peenemünde files captured in Germany. That summer Colonel Toftoy arranged for NAA to obtain two V-2 engines, and in September the firm secured the loan of a complete V-2.[12]

The North American engineers did not limit themselves to consulting the Germans and their documents but conferred as well with JPL, General Electric, Bell Aircraft Corporation, the National Advisory Committee for Aeronautics laboratory in Cleveland (later Lewis Research Center), the Naval Ordnance Test Station at Inyokern, and Aerojet about various aspects of rocket technology. Also, the development engineer for the 75,000-pound-thrust engine was George Sutton, who joined NAA in 1946 as a rocket propulsion engineer, bringing with him more than three years of experience working on rockets at Aerojet.[13]

In these ways the NAA engineers acquired much of the available fund of knowledge about rocketry that had been accumulating slowly. By October 1947 the Aerophysics Laboratory staff had grown to 500-odd people. This necessitated a move to a plant in nearby Downey in July 1948. By the following fall, the engineers had taken apart and reconditioned a V-2 engine, examining all of its parts carefully. The team had also built the XLR41-NA-1, a rocket engine like the V-2 but using American manufacturing techniques, American design standards, some improved materials, and various replacements of small components. Then, by early 1950, NAA had redesigned the engine to feature a cylindrical shape, replacing the ellipsoid contour of the V-2, which had produced efficient propulsion but was hard to form and weld. Bollay's people kept the V-2's propellants, 75-percent alcohol and liquid oxygen. But in place of the eighteen-pot design of the V-2 injectors, which had avoided the combustion instability the Germans encountered when they tried impinging injector plates, the NAA engineers developed two types of flat injectors—a doublet version in which the alcohol and liquid oxygen impinged on one another to achieve mixing, and a triplet, wherein two streams of alcohol met one of liquid oxygen. They tested subscale versions of these injectors in small engines fired in the parking lot. Their methodology was purely empirical, showing the undeveloped state of analytical capabilities in this period, and they encountered combustion instability. But they found that the triplet type of injector provided slightly higher performance due to improved mixing of the propellants.[14]

Meanwhile, NAA searched for a location where it could test engines larger than those fired in the parking lot. It found a location in the Santa Susana Mountains northwest of Downey (and all of Los Angeles) in Ventura County. There the firm got a permit in November 1947 to test its engines. NAA leased the land and built rocket-testing facilities in the rugged area where Tom Mix had starred in western movies, using company funds for about a third of the initial costs and Air Force funding for the rest. By early 1950 the first full-scale static test could be made on XLR-43-NA-1, as the new engine was designated.[15]

Full-scale engine tests with the triplet injector revealed severe combustion instability, so the engineers reverted to the doublet injector, which partly relieved the problem. Although the reduced combustion instability came at a cost of lower performance, the XLR-43-NA-1 still outperformed the V-2, enabling use of a simpler and less bulky cylindrical combustion chamber that looked a bit like a farmer's milk can with a bottom that flared out at the nozzle. The engine delivered 75,000 pounds of thrust at a specific impulse 8 percent better than the V-2 and with a 40 percent reduction in weight. The new engine retained the double-wall construction of the V-2 with regenerative and film cooling. Tinkering with the placement of the igniter and injecting liquid oxygen before the fuel solved the problem with combustion instability. The engine used hydrogen peroxide–powered turbopumps like those on the V-2 except that these—smaller, lighter, and faster—also provided higher pressures in the combustion chamber.

Like the V-2, the XLR43-NA-1 began ignition with a preliminary stage in which the propellants flowed at only some 10 to 15 percent of full combus-

Figure 17. Redstone missile engine, a modified and improved version of the Air Force's Navaho cruise missile engine of the late 1940s. The U.S. engine used a cylindrical combustion chamber in place of the bulky, nearly spherical V-2 combustion chamber. Courtesy of NASA.

tion rates. If observation suggested that the engine was burning satisfactorily, it was allowed to proceed to "main stage" combustion. To enable the engineers to observe ignition and prestage transition, von Braun suggested that NAA engineers roll a small Army-surplus tank to the rear of the nozzle. By looking at the combustion process from inside the tank, engineers could see what was happening, enabling them to reduce problems with rough starts by changing sequencing and improving purges of the system in a trial-and-error process. Through such methods, the Air Force's XLR43-NA-1 Navaho engine became the basis for the Army Redstone missile's NAA 75-110.[16]

Having supervised the development of this engine and the expansion of the Aerophysics Laboratory to a complex with about 2,400 people on staff, Bollay left North American in 1951 to set up his own company, which built Army battlefield missiles. In 1949 he had hired Samuel K. Hoffman, whose background included design engineering for Fairchild Aircraft, Lycoming

Manufacturing, and the Allison Division of General Motors. Hoffman then worked his way up from project engineer with the Lycoming Division of the Aviation Corporation to become its chief engineer, responsible for the design, development, and production of aircraft engines. In 1945 he became a professor of aeronautical engineering at his alma mater, Penn State, the position he left in 1949 to head the Propulsion Section of what became NAA's Aerophysics Laboratory.

As Hoffman later recalled, Bollay had hired him for his practical experience building engines, something many of the brilliant but young engineers working in the laboratory did not possess. Hoffman now succeeded Bollay. Meanwhile, they had overseen the development of a significantly new rocket engine. Although it used the V-2 as a starting point and bore considerable resemblance to the cylindrical engine developed at Peenemünde before the end of the war, it had advanced substantially beyond the German technology and provided greater thrust with a smaller weight penalty. Moreover, it marked the beginnings of another rocket-engine manufacturing organization, the future Rocketdyne division of NAA,[17] which was destined to be the foremost producer of rocket engines in the country.

Meanwhile, development of the NAA 75-110 engine for the Redstone missile did not stop in 1951. Improvements continued through seven engine types, designated A-1 through A-7. Each of these engines had fundamentally the same operational features, designed for the same performance parameters. They were interchangeable, requiring only minor modifications in their tubing before installation in the Redstone missile. All except A-5 flew on Redstone tests between August 20, 1953, and November 5, 1958, with A-1 being the prototype and A-2, for example, having a liquid-oxygen-pump inducer added to prevent cavitation—bubbles forming in the oxidizer, causing lower performance of the turbopump and even damage to hardware as the bubbles imploded, releasing high energy.[18]

During the course of these improvements, Chrysler had become the prime contractor for the Redstone missile, being issued a letter contract in October 1952 and a more formal one on June 19, 1953. Thereafter, it and NAA undertook a product improvement program to increase engine reliability and reproducibility. An example of the results is provided by table 6.1, a comparison of the numbers of components in the pneumatic control system for the A-1 and A-7 engines, used respectively in 1953 and 1958.[19] Obviously, the fewer the components needed to operate a complex system like a large missile, the fewer the problems with its launching and trajectory. Thus this threefold reduction in components on a single system for the engine must have contributed significantly to reliability.

Table 6.1. Pneumatic Control System Components of Redstone Engines

	A-1 Engine	A-7 Engine
Regulators	4	1
Relief Valves	2	1
Solenoid Valves	12	4
Pressure Switches	6	2
Check Valves	3	0
Test Connections	4	1
Pneumatic Filters	0	1
Total	31	10

Source: Stadhalter, "Redstone: Built-In Producibility," NASM.

This was especially true since Rocketdyne engineers (as they became in 1955) tested each new component design both in the laboratory and in static firings before qualifying it for production. They also simulated operating conditions at extreme temperatures, levels of humidity, dust, and the like, since the Redstone was scheduled for deployment and use by the Army in the field. Static engine tests showed a reliability exceeding 96 percent, a remarkably high figure in view of two facts: About 50 percent of Redstone engine components were produced, at least in part, by outside suppliers. And these parts had to be built to a higher standard than those used in conventional aircraft, because the stresses of an operating rocket engine were greater. To ensure reliability, Rocketdyne subjected all welds in stressed components to radiographic inspection. Then the Army required a minimum of four static engine tests to prove that each new model worked satisfactorily before acceptance of the propulsion system. Two of these tests had to last 15 seconds each, while a third was for the full rated duration—presumably 110 seconds, although according to Chrysler's publication on the Redstone, the engine ultimately produced 78,000 pounds of thrust for a duration of 117 seconds.[20]

Redstone Fuselage

In contrast with the arrangement for the engine, the Guided Missile Development Division did the preliminary design of the fuselage for the Redstone in-house. William A. Mrazek, who had worked on the V-2 and Wasserfall at Peenemünde and arrived at Fort Bliss in 1946, was responsible for designing the structure, in which aluminum propellant tanks essentially constituted the center unit. The tanks were pressurized to prevent collapse as the propellants flowed out into the combustion chamber. This arrangement marked a major change from the V-2 with its sheet steel skin enveloping longerons, stringers, and frames. It was made possible in the Redstone by the fact that, in another departure from the V-2 design, the nose cone and warhead with the guidance and control system would separate from the rest of the missile after engine cutoff. This made it unnecessary for the center unit to withstand the aerodynamic heating and dynamic pressures of reentry into the atmosphere. But the nose section did have to undergo forces of reentry without premature detonation of its warhead or damage to the terminal guidance and control system—another new feature—so it had a skin of alloy steel attached to bulkheads, rings, and stringers. The aft unit just behind the warhead section was of similar construction. It carried explosive screws for separation from the center unit following engine cutoff and two air vanes for terminal guidance following reentry.

The tail unit had a riveted aluminum structure with four stabilizing fins and air rudders. It had to withstand the heat of the rocket exhaust, and Mrazek coordinated closely with NAA to integrate the engine into the structure. Carbon jet vanes extended into the exhaust stream, as in the V-2. For the whole design, Mrazek was responsible for the structural calculations as well as the structural and materiel research. For the preliminary design, he did this before performing any wind tunnel tests. Instead he relied upon data from wind tunnel work for similarly configured missiles. Once the preliminary design was ready, the Ordnance Corps contracted with the Reynolds Metals Company on July 18, 1952, to design, build, and assemble the fuselage components. Reynolds remained the subcontractor once Chrysler became the prime contractor. Given the hasty preliminary design, there were naturally changes as the missile reached maturity, including adding nine inches to the center section, eliminating four inches from the tail section, and making other alterations to the tail.[21]

Redstone Guidance and Control

As with the fuselage, the Guided Missile Development Division began developing the guidance and control system for the Redstone in-house. Many of the contributors were Germans who had worked on various aspects of the V-2 guidance and control devices either at Peenemünde or at German firms or technical institutes. They had continued their work at Fort Bliss as part of a guidance and control group under Professor Theodor Buchhold, who had helped with the guidance system of the V-2 from the Technical Institute of Darmstadt, where he and Professor Carl Wagner had developed a complicated cutoff device for the German missile. A number of young engineers from the United States became members of the team, and eventually, under contract, the Ford Instrument Company, a Long Island division of Sperry Rand, also contributed, especially in manufacturing but also in final design and modification.[22]

Although the initial proposal for what became the Redstone missile had included supplemental radio guidance in addition to inertial guidance, the group around von Braun decided to pursue a pure inertial system. When he was still in Germany at Kreiselgeräte in 1944, Fritz Mueller had done some preliminary work on air-bearing gyros to replace those spinning on ball bearings, the kind used on the V-2. The advantage of the air bearings was less friction and torque, hence greater accuracy and longevity without unwanted precession from friction. Mueller even built a laboratory model of the SG-66 stabilized platform using air bearings.[23]

At Fort Bliss, Mueller built up a small gyro laboratory with five or six people, some German and some American, supplemented by military personnel when needed for experiments or construction. These researchers did further work on air-bearing gyros, also making them synchronous (so they ran at a constant speed) and symmetrical (so that they did not shift their centers of gravity when materials changed in temperature while operating). They developed more accurate servo and pickoff systems for an improved stabilized platform, and Mueller was responsible for developing a so-called internal-gimbal system to replace the gimbals on the exterior of the platform. The new arrangement can be compared with a universal joint in an automobile, and it made the structure much smaller, with the components more accessible to engineers and mechanics. The platform featured three single-degree-of-freedom gyros (operating in only one axis each) for stabilization in space.

Much of this work was already completed when the group moved to Huntsville, but it continued there, especially after efforts began to focus on the specific needs of the Redstone missile. Buchhold devised the overall scheme for the Redstone guidance and control system, using an analog guidance computer that American engineer J. S. Farrior had developed. The resultant system included two improved pendulous integrating gyro accelerometers (PIGAs), suggested by Mueller, to be mounted to the stabilized platform, called the ST-80. These PIGAs also had air bearings and provided signals indicating acceleration laterally and longitudinally to analog computers for determining engine cutoff and to provide information about displacement. The computers then combined the various data inputs from the gyros to send appropriate signals to the jet vanes and air rudders as well as the engine.[24]

By December 1951, Buchhold's group had finished the design and built prototypes for roughly 85 percent of the guidance and control equipment for the missile. After looking into possible contractors, it initiated the process of contracting with the Ford Instrument Company to design, modify, improve, and simplify the group's designs and to manufacture a prototype of the complete guidance and control system for the Redstone. Before this inertial guidance system was completed, the missile itself was ready for flight testing. So the Germans found drawings of the LEV-3 control system used in the V-2 and had Ford Instrument make nine of them for use in the early Redstone tests.

Since the early flights were intended to test the engine and the rocket's structure, this expedient did not pose a problem and actually had the advantage of permitting testing of the ST-80's components while the LEV-3 controlled the rocket. On three launches between August 20, 1953, and May 5, 1954, the Redstone flew with only the LEV-3 for control. The ST-80 system rode along as a "passenger" on flights 4 and 5 (August 18 and November 17, 1954). Then on flights 7 and 8 (April 20 and May 24, 1955) the ST-80 provided lateral guidance while the LEV-3 furnished control. Thereafter through flight 21 (December 18, 1956) the LEV-3 controlled some flights while the ST-80 system furnished full guidance and control on others. After flight 21 the ST-80 provided full guidance and control on all flights.[25]

One other new feature on the Redstone as compared with the wartime V-2s was the use of transistors in addition to magnetic amplifiers in its guidance and control system. Buchhold had decided not to use any vacuum tubes because of their unreliability, so the Redstone began to employ transistors at about the same time as the Corporal. The transistors greatly improved the guidance and control computers over those used on the V-2. The Redstone system included one computer for range, which received data from the range accelerometer and sent signals to a separate computer that calculated engine cutoff at between 96 and 107 seconds after launch, depending on the desired distance. Another computer received information from the lateral accelerometer and sent signals to the control computer, which combined that information with the range computer's and provided signals to the control devices. From cutoff to reentry, the only control came from air jets, but during reentry and thereafter, stored data about deviations from the predetermined trajectory allowed the control computer to send corrections to the control surfaces on the nose section.[26]

The Redstone guidance and control system clearly marked a major step forward in the evolution of guidance and control technology for rockets and missiles. Although it evolved from the work done on the V-2, like the engine and fuselage it also reflected many improvements. According to J. S. Farrior, the Redstone guidance and control system was "the first accurate inertial ballistic missile guidance system to complete a flight evaluation program," which it did in 1954. It then provided the basis for the development, with further improvements and miniaturization, of the Jupiter and Pershing guidance and control systems.[27]

Flight Testing

The first flight test of the Redstone was on August 20, 1953, at the Atlantic Missile Range at Cape Canaveral, using a missile built at Redstone Arsenal rather than at the Chrysler plant, with a LEV-3 control system, also called an autopilot. Present at the launch were several NAA engineers including Dieter Huzel, who provided a report of the event. Huzel had worked at Peenemünde and Fort Bliss, but instead of accompanying his German friends to Huntsville, he went to work in 1950 for NAA, where he was soon assigned as development engineer for the Redstone engine. He thus added his knowledge to that being acquired by the engineers at Rocketdyne.

Huzel reported that the missile had launched and flown normally for 40 seconds until it entered cloud cover. But problems soon developed. Flight data indicated unexpectedly high temperatures at the rear of the missile about the time the Redstone entered the clouds. Still, the power plant continued to function normally. At the 53rd second, telemetering was temporarily disturbed, but the speed of the turbopump remained normal up to the 80th second, indicating continued normal functioning of the engine. Then the turbopump speed fell off along with propellant flow rates for about four seconds, whereupon the flow rates of the alcohol and oxygen moved in opposite directions, only to return to normal at the 98th second of flight. Cutoff of the engine by radio signal (necessitated by use of the LEV-3 instead of the ST-80 system) occurred at the 105th second. Huzel speculated that disturbances in the guidance and control system, possibly starting with mechanical devices in the tail section, caused the missile to go off course and begin slow somersaults at the 80th second, resulting in power-plant problems. When the fins stabilized the vehicle on a downward trajectory, he said, the power plant revived, indicating that it "operated entirely satisfactor[ily] during its first launching." The guidance and control group, on the other hand, blamed the power plant for the failure.

With the second launch approaching, von Braun insisted that all personnel involved in the launch recount their experiences until an explanation emerged. Finally a technician from the guidance area recalled that he had tightened every screw he could reach before closing the guidance compartment. But one of these screws was for a balancing trim potentiometer involved in roll control. By tightening the screw after it had already been adjusted, the technician added a roll bias to the system so that the missile rolled beyond the ability of the vanes to stabilize it. Once that became known, the launch team corrected the problem (apparently by covering the potentiometer to prevent tampering after final adjustment). Since the goals of the first flight were proper engine operation, integrity of the airframe, responses to control inputs, and registering of the environment inside and around the missile, the launch actually was a success from the standpoint of research and development, although it was recorded as a failure. Not only did telemetered data from the flight prove valuable, but the flight showed that human error by someone trying to do a good job could cause problems and that absolute honesty about such errors was often critical to determining causes of failure.[28]

Initially, the Ordnance Guided Missile Center had planned to design, develop, and manufacture the Redstone entirely in-house. As speed of development became more important, it was apparent that capabilities and facilities for doing this rapidly enough were lacking. Even after a decision to rely upon industry to provide major assemblies and components, the Guided Missile Development Division (as it became) still wanted to assemble the missiles. This not only followed the Army arsenal tradition but was considered necessary, given the early stage of rocket development and the perception that most American firms were not yet prepared to do the job correctly. Also, as Redstone Arsenal saw the situation, if its military and civilian engineers were to perform technical supervision over work done by contractors, they had to do not only research and development but, sometimes at least, pilot production. On April 1, 1952, however, Army Ordnance disapproved "manufacture and assembly of [Redstone] missiles beyond that required to get a prime contractor successfully operating." Oddly, then, the contract with Chrysler came long after contracts for key components were awarded to NAA, Reynolds, and Ford. Complications with getting plant facilities for Chrysler contributed to a compromise arrangement. Except for the components produced by subcontractors, the Guided Missile Development Division manufactured and assembled missiles 1–12 and 18–29. Chrysler produced and assembled missiles 13–17 and 30 onwards.[29]

Of the 37 missiles fired through November 1958, just 13 experienced malfunctions, for an almost 65 percent success rate. One, missile number 3, exploded on the launchpad. Only 12 of the 37 flew specifically for the Redstone program. The other 25, carrying the designation Jupiter A, provided information to assist with the Jupiter program. Thus these tests served at the same time to increase the reliability of the Redstone and to help with

Table 6.2. Comparison of Corporal, Sergeant, and Redstone Missiles

	Corporal	Sergeant	Redstone
Length (feet)	45.4	34.5	69.3
Thrust (pounds)	20,000	50,000	78,000
Range (miles)	99	75–100	175
Payload (pounds)	1,500	1,500	6,305
Number produced	1,101	475	120
Unit cost	$293,000	$1,008,000	$4,266,000

Sources: Bullard, *Redstone,* 70, 92, 100, 162–65; von Braun, "Redstone, Jupiter, and Juno," 110–11; Nicholas and Rossi, *Missile Data Book,* 3-5 to 3-9.

the development of the Jupiter. The test program led to improvements in the propulsion and guidance-and-control systems as well as the fuselage. At the end of research and development, the Redstone had a length of 69 feet 4 inches, a diameter of 70 inches, a trajectory ranging from 50 to 175 miles, an accuracy of 300 meters (984 feet) circular error probable, and a payload of 6,305 pounds. It was deployed by the Army on June 18, 1958, and deactivated in June of 1964, replaced by the faster and more mobile Pershing, also developed at Redstone Arsenal. As missiles go, it was fairly inexpensive, costing $92.5 million for research, development, testing, and evaluation, as compared to $172.1 million for the Sergeant but only $50.1 million for the Corporal. Per missile, Redstone cost $4,266,000, however, compared with $1,008,000 for Sergeant, which had a larger production run, and $293,000 for Corporal, which was produced in much greater numbers, as shown in part in table 6.2.[30] Obviously, the Redstone had much higher performance in terms of range, thrust, and payload than its older cousins among Army missiles, hence the higher per-missile cost.

Project Orbiter and Competition with Vanguard

Cost aside, Redstone certainly contributed more innovations to American rocketry than did either Corporal or Sergeant, although JPL's overall contributions were certainly important. Redstone also held the distinction of becoming the first stage of the United States' first true launch vehicle, with scaled-down Sergeants forming the next three stages. How Redstone and these by-products of the Sergeant program came to be modified into a launch vehicle is a complicated story, elements of which remain something of a mystery to this day.

By 1954, with the Redstone's development well along, von Braun had begun to think about using the missile to help launch a small satellite. The same year, Lt. Cdr. George Hoover at the Air Branch of the Office of Naval

Figure 18. Redstone missile on the launchpad at Cape Canaveral on May 15, 1958. Courtesy of NASA.

Research (ONR), who had long been interested in launching satellites, concluded that recent improvements in rocketry had reached the stage where this was possible. The two men and others met in Washington, D.C., on June 25, 1954, and von Braun proposed to launch a satellite using a Redstone with larger propellant tanks for extended burning time, plus three stages consisting of solid-propellant Loki antiaircraft missiles. The ensuing developments were complex, but they resulted in Project Orbiter, supported by the Army and ONR, for the launching of a satellite as part of the United States' participation in the International Geophysical Year (IGY), a period from July 1, 1957 to December 31, 1958, proposed by the International Council of Scientific Unions for scientific research, which came to include the launch of satellites.[31]

While elements of the Department of Defense were considering von Braun's proposal, written in September 1954, the Naval Research Laboratory submitted a separate satellite-launching proposal, Project Vanguard, involving use of NRL's sounding rocket, Viking, as a projected first stage. Suddenly the Navy found itself supporting both Orbiter and Vanguard. The decision lay essentially in the hands of the assistant secretary of defense for research and development, Donald A. Quarles, who referred the two proposals plus a half-hearted and informal one from the Air Force to an ad hoc committee for evaluation. Homer J. Stewart of Caltech and JPL headed this eight-person Committee on Special Capabilities, which began looking at the proposals in May 1955. The details of the deliberations, committee makeup, and proposals need not detain us here. But for reasons that have been debated, a majority voted for Vanguard, with two people, one of them Stewart, casting minority votes for Orbiter. The process was messy and convoluted, but the decision went to Vanguard. Even a revised proposal by the Army and a second vote by the Stewart committee on August 23, 1955, failed to sway the majority. Vanguard remained the United States' launch vehicle for the IGY.[32]

Jupiter C

Even though the decision for satellite launch had gone against Project Orbiter, von Braun later said, "We at Huntsville could not scrap the satellite idea, and we did not scrap our satellite-oriented hardware." Instead it became the "basis for the Jupiter C (Composite Reentry Test Vehicle)." The term Jupiter C is something of a misnomer. The Jupiter missile enjoyed a higher priority for firing at Cape Canaveral than the Redstone, so the Army Ballistic Missile Agency (ABMA)—activated at Redstone Arsenal on February 1, 1956, and given responsibility for the Redstone and Jupiter missiles—had designated Redstone missiles used to test Jupiter components as Jupiter As. This practice carried over to the Jupiter C.[33]

The purpose of Jupiter C was to test a scaled-down nose cone for the Jupiter missile, which would travel three times as fast as the Redstone (Mach 15.1 as compared with Mach 5) and thus would be subjected to much higher

Figure 19. Key members of the von Braun team at the Army Ballistic Missile Agency *(left to right)*: Ernst Stuhlinger, director, Research Projects Office; Helmut Hoelzer, director, Computation Laboratory; Karl L. Heimburg, director, Test Laboratory; Ernst Geissler, director, Aeroballistics Laboratory; Erich W. Neubert, director, Systems Analysis Reliability Laboratory; Walter Haeussermann, director, Guidance and Control Laboratory; Wernher von Braun, director, Development Operations Division; William A. Mrazek, director, Structures and Mechanics Laboratory; Hans Hueter, director, System Support Equipment Laboratory; Eberhard Rees, deputy director, Development Operations Division; Kurt Debus, director, Missile Firing Laboratory; and Hans H. Maus, director, Fabrication and Assembly Engineering Laboratory. Courtesy of NASA.

aerodynamic heating upon reentry into the atmosphere. This problem of reentry heating for missiles traveling at such high speeds was a major concern. The Air Force favored using enough metal on the nose cone to absorb the heat, at the cost of a heavier launch weight. The Army had been experimenting with ablative nose cones, employing a material that gradually vaporized, dissipating the heat as the material converted from a solid into a gas. One of the bases for these experiments, which had been ongoing since 1953, was research at Cornell University using shock tubes. The Navy had funded this research beginning in 1946. Another source of information for the ablative nose cones was an attempt to find a better substance for jet vanes on the Redstone than graphite. A fiberglass-reinforced plastic proved quite unsuitable for jet vanes because it eroded severely from ablation, but it showed promise as a nose-cone material. Engineers at Redstone Arsenal began testing substances for nose cones in the jet exhaust of rockets. They ended up, serendipitously, using the fiberglass-reinforced plastic from the Redstone jet-vane experiment on the Jupiter C, which tested the concept in actual flight.[34]

For the Jupiter C, the Redstone was modified, essentially as von Braun had proposed for Project Orbiter, by elongating the missile to hold larger fuel and oxidizer tanks. Also, at the suggestion of Rocketdyne, the fuel changed from alcohol to Hydyne (unsymmetrical dimethylhydrazine and diethylene triamine).[35] This increased the specific impulse by about 15 percent, raising the thrust of the first stage to 83,000 pounds from the nominal 75,000 pounds of the Redstone at this time. The oxidizer remained liquid oxygen. The substitution of Hydyne for alcohol did not require any major change in the engine hardware, because the two fuels had the same density. To supplement the added performance from the Hydyne, the larger tanks increased the burning time of the engine from 121 to 155 seconds. Such tanks were made possible by the substantial reduction in weight between the heavy warhead for the Redstone missile and the upper stages for Jupiter C. The longer engine burn required the addition of a second hydrogen peroxide tank to run the turbopump the additional 34 seconds.[36]

In a further modification, the Redstone's nose section required bolstering to accommodate the upper stages. The rockets used in these upper stages were not the Lokis that von Braun had proposed for Project Orbiter. To aid in drawing up that proposal, Army Ordnance had asked JPL for its input. JPL's report recommended the use of scaled-down Sergeant rockets, developed for tests of the Sergeant missile. These 6-inch-diameter scale models of the Sergeant were the product of many tests in which they had proved reliable, and in September 1955, JPL began designing clusters of them for use in the Jupiter C flight program.

JPL clustered eleven of the motors to constitute stage 2 and, inside the eleven, three more for stage 3. Both sets were placed in a tub that was attached to a shaft connected with electric spin motors in the Redstone's instrument compartment. The spin system, developed by the Aerophysics Development Corporation under contract to ABMA, rotated the tub before the launch, with the spin rate increased in flight. This spinning provided ballistic stability and also eliminated problems from any difference in thrust between individual scaled-down Sergeants.

The particular arrangements for the spinning of the stages had to be intricate to avoid resonance between the spin frequency and the bending frequencies of the first stage. If this resonance had occurred, it would have resulted in violent vibrations that could have shaken the vehicle to pieces. The problem was complicated by the fact that the bending frequencies of the missile increased as the propellants dissipated in burning. To keep the spin frequency frjom coinciding with the bending frequency, the system spun the tub containing the upper stages at 550 rpm before the launch. About 70 seconds after liftoff, a governor gradually changed the speed to 650 rpm. After 155 seconds of flight, the speed again increased to 750 rpm, where it stayed until the stages were fired in sequence.

Perhaps remembering the problem with ignition at altitude of the WAC portion of the Bumper WAC, the JPL engineers designed an igniter that would fire in a vacuum. Then they sealed each motor so that it would retain atmospheric pressure. Finally, they sealed the igniter in a container that would maintain the pressure at sea level. This triple insurance added a small weight penalty but proved to be effective.[37]

The first Jupiter C missile launched without a nose cone on September 20, 1956. It landed 3,350 miles from Cape Canaveral with all stages functioning properly. This set a distance record lasting until the launching of intercontinental ballistic missiles. More important, it showed that the multistage vehicle worked as designed. The second launch occurred on May 15, 1957. This vehicle carried a scaled-down Jupiter nose cone. A problem with the guidance system caused the final stage to fly beyond the target area, so the nose cone could not be recovered. However, telemetry showed that it had survived the rigors of reentry without problems. On August 8, 1957, the third Jupiter C carried its nose cone 1,160 miles downrange. The Navy recovered it, giving evidence that the ablation principle worked, and an ablative coating soon became the standard method of protecting reentry vehicles from aerodynamic heating.[38]

Juno I

Less than two months later, on October 4, 1957, the Soviet Union shocked the world—especially the United States—by successfully launching Sputnik 1, the world's first artificial satellite. Not really a large object, it had a diameter of 22 inches and weighed 184 pounds. But it made the 3.5-pound, 6-inch satellite Vanguard was preparing to launch look tiny by comparison. A quirk of fate placed secretary of defense–designate Neil H. McElroy at Redstone

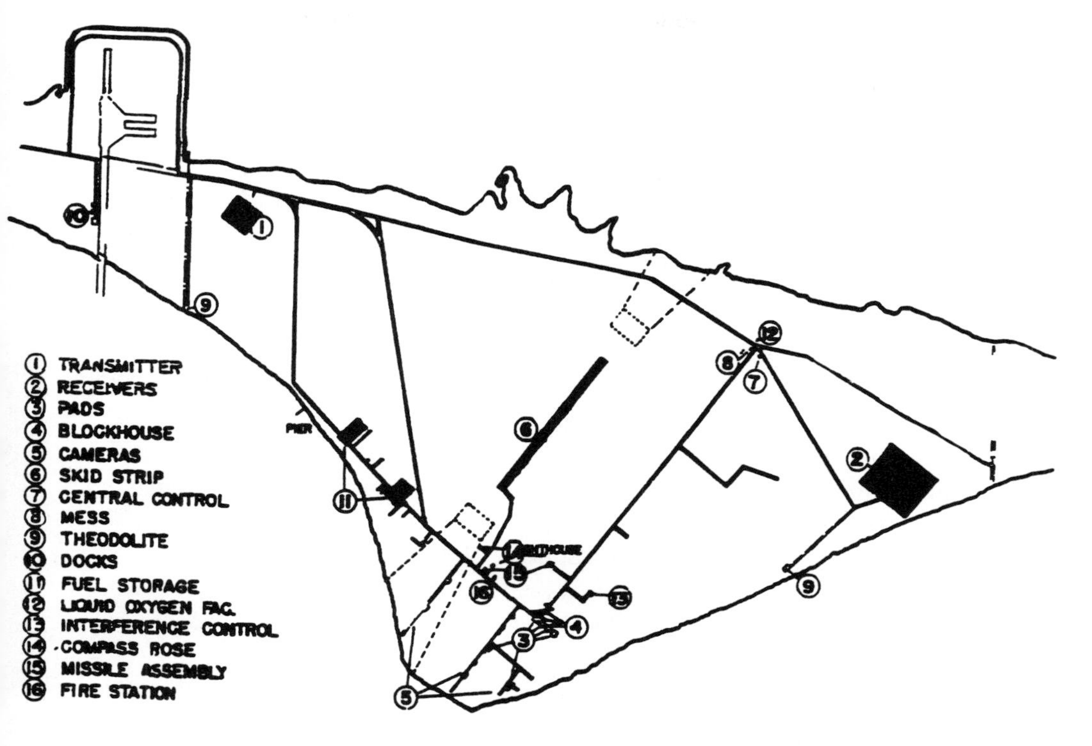

Figure 20. Undated map of Cape Canaveral showing early facilities at the launch site. Official U.S. Air Force photo, courtesy of the 45 Space Wing History Office, Patrick AFB, Fla.

Arsenal the night of Sputnik's launch. Accompanied by Army Secretary Wilbur M. Brucker and a couple of Army generals, Lyman Lemnitzer and James Gavin, he was at a cocktail party talking with Wernher von Braun and Maj. Gen. John B. Medaris, commander of the Army Ballistic Missile Agency, when Medaris's public affairs officer reported the Soviets' success.

Von Braun had already expressed his feelings about such a satellite in his September 1955 proposal for Project Orbiter. "The establishment of a man-made satellite, no matter how humble, would be a scientific achievement of tremendous impact," he had written. "Since it is a project that could be realized within a few years with rocket and guided missile experience available *now*, it is only logical to assume that other countries could do the same. It would be a blow to U.S. prestige if we did not do it first." He had also contributed to a series of articles in the widely circulated *Collier's* magazine and been advocating space flight on such national television programs in the early to mid-1950s as Dave Garroway's *Today* show and a Walt Disney series. Thus, confronted with the reality of Sputnik, he blurted, "Vanguard will never make it. We have the hardware on the shelf. For God's sake, turn us loose and let us do something. We can put up a satellite in sixty days, Mr. McElroy." To this, Medaris retorted, "No, Wernher, ninety days."[39]

McElroy would not become secretary of defense until October 9, and even then he did not give von Braun and Medaris an answer right away. But although predictable in some degree, the beeping satellite was a deep blow to American pride. More important, it suggested to Democratic opponents of President Dwight D. Eisenhower's Republican administration that the Soviets now had the ability to launch missiles at the United States. Eisenhower had based his military strategy, called the New Look, on a limited military budget with spending focused on nuclear weapons to deter Soviet aggression. Although the administration said it was not surprised by the Soviets' satellite and had never thought it was "in a race with the Soviets," Sputnik 1 put it on the defensive. When Sputnik 2 joined its smaller companion in space on November 3, carrying a dog and weighing 1,121 pounds, Eisenhower and his cabinet needed some way to show that America too was capable of launching a satellite.[40]

Medaris told von Braun to prepare for a satellite launch, but Washington was slow to give him permission. On November 8 the Pentagon announced: "The Secretary of Defense today directed the Army to proceed with launching an Earth satellite using a modified Jupiter C." Realizing that this verbiage camouflaged the reality that "Jupiter C" (actually, Juno I) was merely a backup to Vanguard, Medaris pressed his superiors for a firm commitment to allow two attempts with his four-stage launch vehicle—Jupiter C plus one additional solid-propellant motor. Permission finally came after the first attempted launch of a Vanguard satellite failed on December 6, 1957, with a spectacular explosion on the launching platform at Cape Canaveral. Actually, the Vanguard launch was planned only as a test to check out the upper stages for the first time. But to the surprise of the Vanguard engineers, the president had announced, in the wake of Sputnik, that the launch would take the first American satellite into orbit.[41]

Meanwhile, the engineers in Huntsville and Pasadena had been making preparations for the launch of Juno I, which subsequently became the launch vehicle for other satellites. Besides the upper stages, JPL was responsible for the Explorer I satellite weighing some 18 pounds. The Juno I launch vehicle was substantially the same as Jupiter C with the addition of a single scaled-down Sergeant rocket as stage 4. This fit inside the three motors for the third stage, but it was slightly different from the other fourteen solid-rocket mo-

Figure 21. Juno I launch vehicle with Explorer I, the United States' first satellite, as its payload. The von Braun team developed the first stage of Juno I, while JPL provided the upper stages and the satellite. NASA Historical Reference Collection, Washington, D.C. Courtesy of NASA.

tors. It was a little shorter, and its nozzle was smaller to facilitate its separation from the middle of stage 3. It also contained a higher-performance solid propellant than the other motors. Indeed, the propellant later changed to a still higher-performance composition for the launching of the Explorer IV satellite.[42]

Juno I's guidance and control system, in order to launch satellites into orbit, had to be modified from the Redstone missile's. Located in a compartment above the first stage attached to a tub containing stages 2 through 4, it exercised control of that stage through the LEV-3 system (also used on Jupiter C because funding for both Jupiter C and Juno I was limited and neither mission required great precision; LEV-3 was much less expensive than the ST-80). A partially separate spatial attitude control system used four air-jet nozzles to align the upper stages before ignition of their motors for boost to orbital speed—about 18,000 mph at lower altitudes.[43]

On January 31, 1958, the Juno I launched vertically and pitched to an angle 40 degrees to the horizon by the time of engine cutoff. Instead of using the inertial cutoff mechanism of the Redstone missile, Juno I employed probes in the two propellant discharge lines. These energized after 149 seconds of flight, and when one of them sensed a lack of pressure, indicating depletion of a propellant, it sent a signal that closed the valves to both propellant tanks. On the Explorer I launch, this occurred at 157 seconds after launch rather than the expected 155 seconds.

Five seconds after cutoff, a timer activated explosive bolts to separate the upper stages from the Redstone booster. The delay was to allow the thrust of the Redstone to terminate fully so that the booster did not bump the upper stages and misalign them from their trajectory (although this happened anyway on one mission). The spatial attitude control system then used the same gyroscopes that had stabilized the booster by commanding the jet vanes. These gyros also controlled the air-jet nozzles in the back of the instrument compartment, which after separation of the booster was still attached to the upper stages. While the upper stages were coasting, the nozzles shifted the trajectory so it would be fully horizontal by the time the three upper stages, still attached together, reached their apex at about 404 seconds after launch.

Because Earth is curved and the second stage had to ignite parallel to the so-called "*local* horizon," the guidance team at ABMA used three different systems to help identify the correct moment to ignite the high-speed stages and have them propel the satellite in the right direction for achieving its intended orbit. First, they had radar to track the vehicle. Second, an accelerometer on the first stage provided data by telemetry to the control room. This indicated incremental velocity up to the time of first-stage separation. A computer on the ground took the cutoff velocity and calculated the moment when the vehicle should achieve the apex of its ascent. The third source of information was a Doppler tracking network.

For the first Juno I launch, Dr. Ernst Stuhlinger handled the data from these three sources. (Stuhlinger, who had done research on the guidance and control system of the V-2 at Peenemünde and had come to Fort Bliss in 1946, was now director of the Research Projects Laboratory at ABMA.) He entered the three sets of data, which he weighted according to his judgement about their quality, into an electromechanical analog computer that he had built largely in his garage at home and called an apex predictor. After this device calculated when the apex should be reached, he set a clock to send a radio signal initiating the sequence of upper-stage ignitions at a moment such that, by the time the fourth stage burned out, the vehicle would be on the correct trajectory. (In fact, he sent the signal manually, but the clock would have done so if he had not pressed the button himself.) He had to do all of this in the approximately four minutes between cutoff of the first stage and apex. Since each of the upper stages burned for 5–6 seconds and there was a slight delay between burnout of one stage and ignition of the next, the time between second-stage ignition and fourth-stage burnout was about 24 seconds.

For the second Juno satellite launch, Dr. Walter Haeussermann, then director of the Guidance and Control Laboratory at ABMA, replaced Stuhlinger to calculate the time of apex. Thereafter it was done by an automated system. For the first two launches, there had not been enough time to develop a computerized system to perform the tasks that Haeussermann and Stuhlinger carried out.[44]

Scheduled for January 29, 1958, the launch of the first Juno I with the Explorer I satellite as its payload was delayed for two days because of high winds. The launch finally occurred at 10:40 p.m. on January 31, with the launch vehicle and payload heading eastward across the Atlantic. For the launch, von Braun from ABMA and William Pickering from JPL were at the Pentagon with Army Secretary Wilbur Brucker. Informed of the launch time, von Braun calculated that the satellite should travel around the world and reach a tracking station in California by 12:41 a.m. Another tracking station at Antigua picked up signals from the satellite at exactly the time expected, but until Explorer I circled the globe and was picked up by the station on the West Coast, there could be no assurance that it had achieved its desired orbit.

Because the extra weight would have imposed limits on performance, there were no signals coming from the last two stages or the satellite back to the blockhouse at Cape Canaveral where General Medaris, Stuhlinger, and others were located. Tensions were high in both Washington and Florida; so was the consumption of coffee and cigarettes. Finally a careful study of Doppler figures showed that all four stages had fired and the correct velocity had been achieved. The question that remained was whether the horizontal attitude of the final stage matched the orbital trajectory once it was fired. Pickering was in contact with the West Coast tracking station, but there

Table 6.3. Satellite Launch Attempts by Juno I

Satellite	Date	Result	Corrective Action
Explorer I	1/31/58	Achieved orbit.	None
Explorer II	3/5/58	Failed to orbit. Fourth stage did not ignite because a structural support failed.	Strengthening of structural support for igniter. No succeeding missions suffered such failures.
Explorer III	3/26/58	Achieved orbit.	None
Explorer IV	7/26/58	Achieved orbit. Propellants with higher performance for stages 3 and 4 of Juno permitted heavier satellite.	None
Explorer V	8/24/58	Failed to orbit. After separation, first-stage booster struck instrument compartment.	Designed better separation devices.
Explorer attempt, unnumbered	10/22/58	Failed to orbit. Vibrations from spin of upper stages caused satellite to separate from rockets.	None

Sources: Stuhlinger, "Army Activities in Space," 68–69; Wolfe and Truscott, "Juno I," 72–79, NHRC.

was no signal at 12:41. Eight minutes later—an eternity to those waiting for confirmation—the satellite appeared. It was a bit high but in a stable orbit. It turned out that the angle of the fourth stage when it went into orbit had been 0.81 degrees off. Explorer I would have orbited even with an error as high as 4.0 degrees.[45]

Subsequent launches of Explorer satellites by the Juno I launch vehicles were not all successful, as table 6.3 shows.[46]

Mercury-Redstone

The need to provide a longer-range, more powerful tactical missile and to launch one or more satellites to help the United States recover from the blow to its technological pride dealt by Sputnik, improve its military capabilities, and advance space science were not the only cold war concerns that affected the Redstone program. Soon after Sputnik, the United States entered a competition with the Soviets to place a human being in orbit as part of an emerging space race between the two superpowers. The organization that started the program to launch an individual into space was the venerable National Advisory Committee for Aeronautics, created by Congress in 1915 "to supervise and direct the scientific study of the problems of flight with a view to their practical solution." As this mandate would suggest, the NACA had been concerned primarily with airplanes, but in June of 1958 the organization's director, Hugh L. Dryden, assembled a group of engineers from two NACA research laboratories under Langley Aeronautical Laboratory engineer Robert R. Gilruth to plan a program that would place a human in space.[47]

Gilruth's group came up with a plan that he presented to the Senate Committee on Astronautics and Space Exploration on August 1, 1958. The plan involved use of Air Force Atlas missiles to place a small capsule containing an astronaut into orbit as the main goal of what came to be Project Mercury. Early suborbital flights to test the capsule would be done with a Redstone missile.[48]

While Gilruth's group was planning for Project Mercury, Sputnik had already energized Congress and the Eisenhower administration to create a new civilian space agency with the NACA as its core. On July 29, 1958, President Eisenhower signed legislation creating NASA, the National Aeronautics and Space Administration, which began operating on October 1 of that year. Five days later, the new agency decided to continue with Project Mercury under Gilruth's leadership. This decision constituted the formal birth date of the project.[49]

At this point the Redstone was the only booster in the American inventory that could be trusted. The Atlas was still being tested, and the Jupiter and Thor intermediate-range missiles were as yet unreliable. Consequently, in October 1958 several people from Gilruth's organization flew to Huntsville, where representatives of the Army Ordnance Missile Command (established March 31, 1958, as the parent command for ABMA, JPL, and other organizations) agreed that Redstone missiles could be made available within twelve to fourteen months of their being requested. On October 6 the Army command agreed tentatively to provide ten of the missiles for Project Mercury.[50]

On November 3, 1958, NASA formally asked Army Ordnance Missile Command to provide eight Redstone missiles. The civilian agency transferred the requisite funding on January 8, 1959, and ABMA began planning to have Chrysler provide the modified missiles. Meanwhile, on November 5, 1958, Gilruth's group became the separate Space Task Group (STG), initially located at Langley Research Center (as Langley Laboratory became on October 1, 1958) but reporting to NASA Headquarters.[51]

Working with the STG to prepare the Mercury-Redstone launch vehicles was the Development Operations Division of ABMA, directed by von Braun. He established a Mercury-Redstone Project Office to aid in redesigning the Jupiter C version of the Redstone to satisfy the requirements of the Mercury project. Von Braun put Joachim P. Kuettner in charge of the effort to make the Redstone safe for launching humans, a process referred to as "man-rating" at a time when there were no female astronauts. Kuettner was himself a flight engineer and test pilot, having worked for Messerschmitt during the Nazi period in Germany.[52]

Kuettner's group recognized that the Redstone missile could not satisfy the mission requirements for Project Mercury. These necessitated a vehicle with sufficient performance and reliability to launch a two-ton payload with an astronaut aboard into a flight path reaching an apogee of 100 nautical miles (115 statute miles) during which the astronaut would experience at least five minutes of weightlessness. The Jupiter C with its elongated propellant tanks and lighter structure had the necessary performance but not the safety features for human flight. To add these features, Kuettner's group instituted changes in three phases: initial modifications that the engineers thought would satisfy the requirements, further changes following ground tests, and still further modifications to solve problems experienced in flight tests.[53]

Mercury-Redstone, as the initial Mercury launch vehicle was called, needed elongated tanks like those of Jupiter C to provide an engine burning period of 143.5 seconds. This required an auxiliary hydrogen peroxide tank to keep the turbopumps operating that long, plus a seventh high-pressure nitrogen tank to provide pressure for the larger fuel tank. During the period of Project Mercury, Rocketdyne was in the process of moving from the A-6 to the A-7 engine for the Redstone program, so Kuettner's group immediately opted for the newer version to preclude a shortage of hardware as the

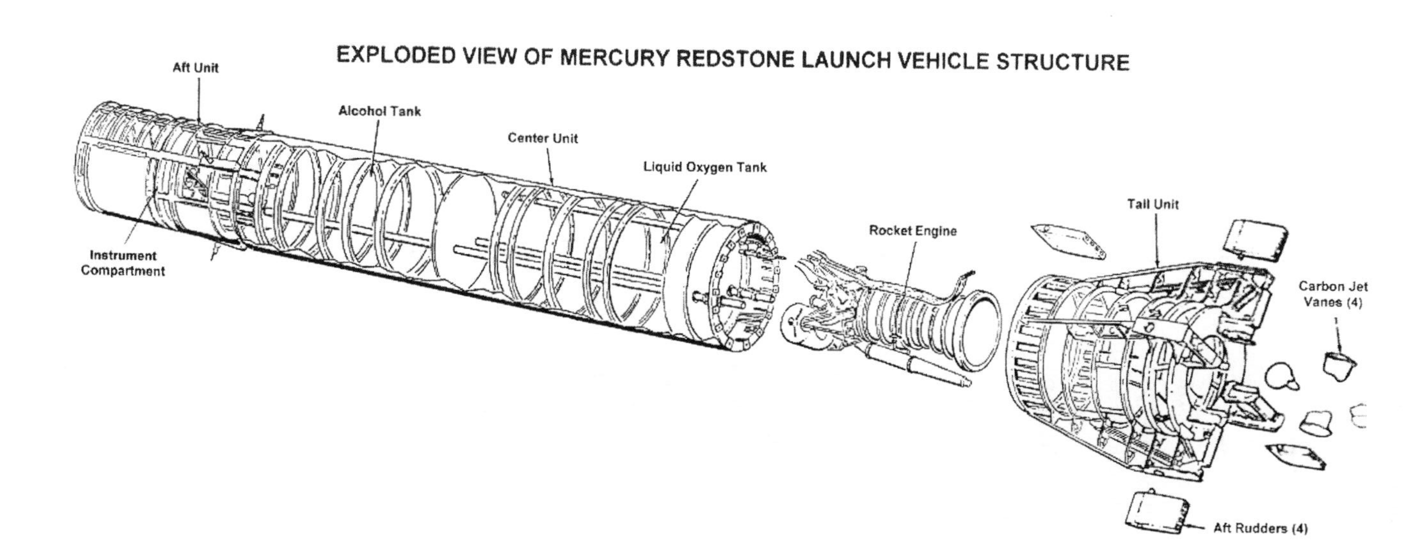

Figure 22. Technical drawing giving an exploded view of the Mercury-Redstone launch vehicle structure as of December 1964. Courtesy of NASA.

A-6 phased out. Another propulsion change was the reversion to alcohol as the fuel in place of the highly toxic Hydyne. This latter propellant contained unsymmetrical dimethylhydrazine, which—at least in animal experiments—resulted in convulsions and possible death from even short exposure. But the lower hazards of alcohol came at a cost in performance and required higher-quality jet vanes, as the extended burning time with alcohol for fuel caused extensive erosion of the original carbon vanes.[54]

Another significant modification of the tactical missile was the reversion from the ST-80 stabilized platform for guidance and control to the LEV-3 autopilot, which provided only control. The LEV-3 was simpler and more reliable, according to more than one source. Since it met the requirements of the Mercury mission, which did not require great precision, it became the choice of the von Braun group. In two areas, however, it was improved. Air bearings in the pitch gyroscope and the integrator gyroscope that initiated cutoff for the propulsion system increased the accuracy of the overall system. As with the early test flights of the tactical missile, Ford Instrument Company supplied the LEV-3s. Relatedly, Kuettner's group had Chrysler attach what had been the aft unit on the nose section of the missile to the center tank portion of Mercury-Redstone. This was the section containing the instrument compartment, including the LEV-3 and other electronic components. On the Redstone missile, this unit remained with the payload after engine cutoff and separation to provide terminal guidance. As a result of this change for Mercury-Redstone, some air rudders on the forward portion of the missile also disappeared as no longer necessary.[55]

An especially key feature of Mercury-Redstone was an automatic in-flight abort system. It monitored a number of different sources of potential catastrophic failure of the launch vehicle and could initiate engine cutoff, ignition of an escape tower's three solid-propellant rockets on the Mercury capsule containing the astronaut, the resultant spacecraft separation, and the deployment of drogue, main, and reserve parachutes to carry the astronaut safely to the land or water below. Triggers of this process included excessive deviations in the attitude of the launch vehicle, loss of electrical power, loss of thrust, and/or excessive turning rates. Since some of these failures could produce catastrophe very rapidly, the design called for the abort to occur automatically, but there was manual backup so that the astronaut, the blockhouse, the mission control center, or range safety could push a button and begin the abort.[56]

To ferret out potential sources of failure, Chrysler instituted a special test program to promote greater reliability. It subjected all major missile sections

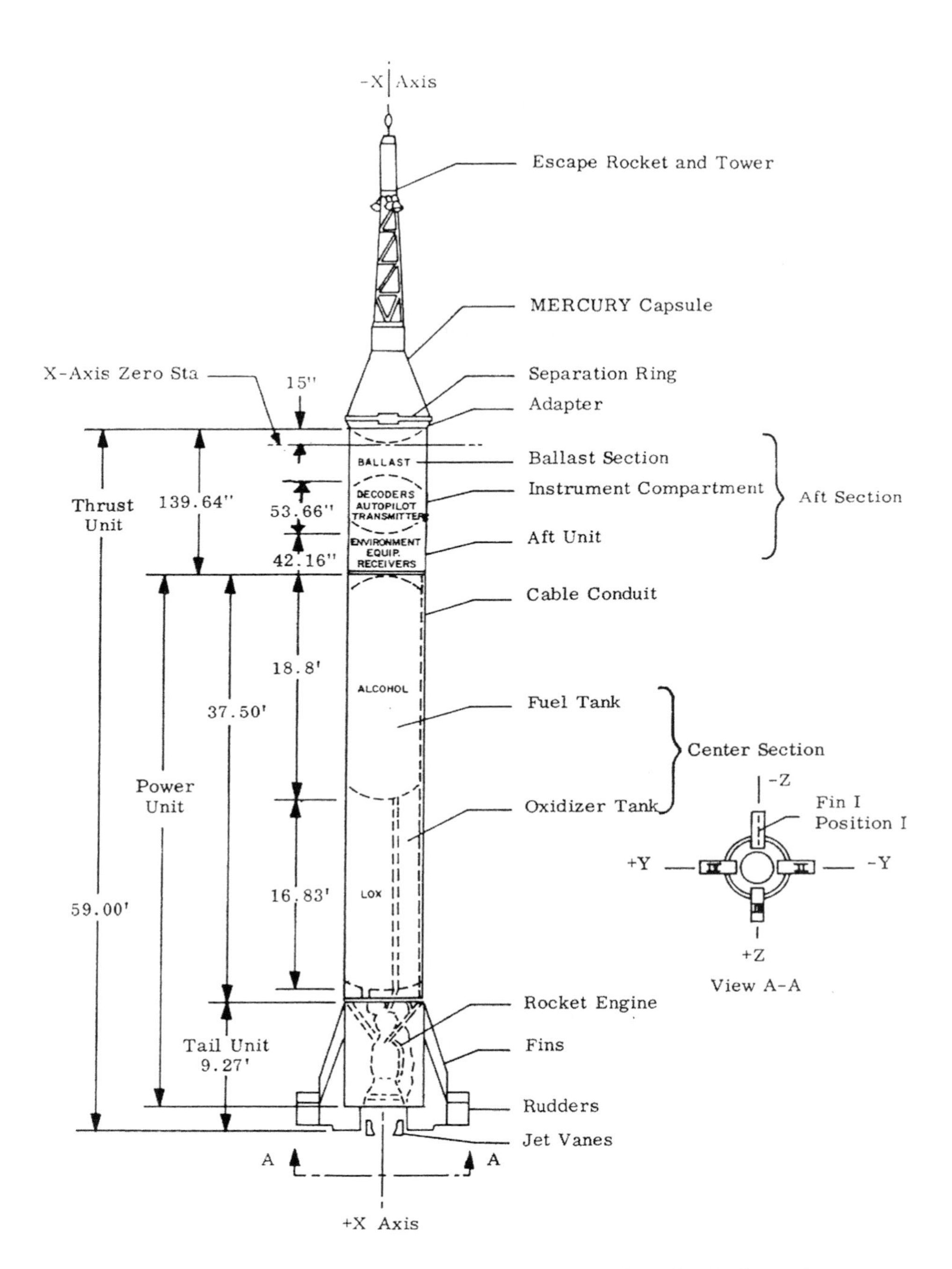

Figure 23. Technical diagram of the overall Mercury-Redstone launch vehicle as of May 1959. Notice the escape tower at the top of the drawing. Courtesy of NASA.

and the systems they contained to vibration tests under conditions of humidity and temperature that simulated expected transportation, prelaunch, and flight environments. It applied loads up to 150 percent of those anticipated in flight. When it found sources of trouble, it conducted additional systems tests. Von Braun's engineers followed these tests with noise and vibration tests at their test stands during static tests of the engine. Several components experienced damage or failure during the vibration test program and had to be modified. Because the A-7 engine was new, there were many such static tests. These revealed combustion instability at 500 cycles per second, which Rocketdyne corrected by enlarging the injector holes. Another low-frequency oscillation proved, on investigation, to result from the design of the static test stand, which had to be modified.[57]

In all, some eight hundred changes to the Redstone resulted from the combined efforts of the contractors and the von Braun team, some of them occurring after flight testing began. As Gilruth's engineers learned of the number of changes being made, they became concerned. As a group, the STG tended to favor underdesign, whereas von Braun's engineers tended in the direction of overdesign. Gilruth's people feared that the number of impending changes would essentially negate the record of safety the Redstone missile and the Jupiter C had demonstrated, the reason they were chosen for Project Mercury in the first place. When confronted with these concerns, von Braun's engineers stressed that their long experience dating back to Peenemünde had proved to them that their exhaustive testing really worked.[58]

Before even the test flights of Mercury-Redstone could take place, von Braun's Development Operations Division moved to NASA jurisdiction and became the George C. Marshall Space Flight Center. Some 4,670 people and 1,200 acres of Redstone Arsenal transferred to NASA on July 1, 1960, with von Braun becoming the center director.[59] Despite, or perhaps because of, all the testing and modification of the Redstone missile to become the Redstone-Mercury launch vehicle, the initial flight tests did not go well. But that is what flight tests are for—to find problems and correct them before actual launches occur. On November 21, 1960, Mercury-Redstone 1 with a "simulated man" in its capsule completed its countdown. Ignition occurred normally, the launch vehicle rose 3.8 inches from the pad, and then it settled back on its fins after tilting slightly. The parachutes deployed and fell to the launchpad, while the capsule remained in place on the booster. Intensely embarrassed, the Mercury-Redstone team conducted an investigation that revealed faulty grounding of electrical circuitry. This had allowed sufficient

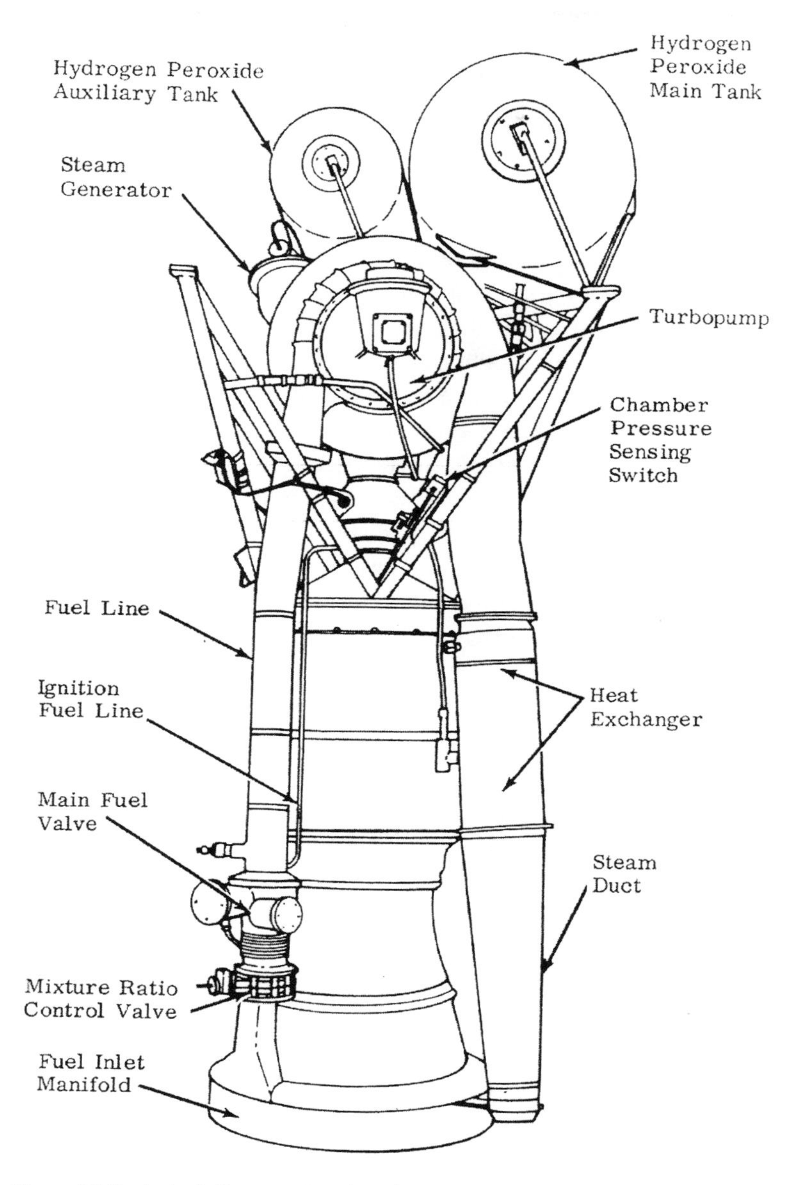

Figure 24. Technical diagram showing the components and shape of the Mercury-Redstone rocket engine as of December 1964. Courtesy of NASA.

electrical current to go through the cutoff relay that it caused the abort system to terminate thrust. Someone had replaced a specially shortened Mercury cable with a longer Redstone control cable, which caused the problem. To prevent this from recurring, the engineers added a special strap that maintained vehicle grounding in subsequent launches.[60]

Neither the booster nor the capsule was damaged beyond repair, so both would be reused, the capsule on the next mission, Mercury-Redstone 1A. This likewise simulated having a human in the capsule. On December 19, 1960, the vehicle launched successfully, but a malfunction of an accelerometer caused the velocity to exceed the target by 260 ft/sec. This boosted the capsule 6 miles higher than intended, making the reentry deceleration 0.4–1.0 Gs more than the permissible maximum of 11.0 Gs. (A G is a measurement of acceleration equal to the force of gravity at sea level.) The control center's flight surgeon thought this would have affected the astronaut's alertness if the mission had been a manned. The capsule also landed 20 miles beyond its designated spot in the ocean, but it was recovered. On the whole, the flight had been a success, demonstrating accurate performance of the abort system. The cause of the accelerometer malfunction lay in the wiring arrangement, which engineers adjusted for future flights.[61]

On January 31, 1961, Mercury-Redstone 2 launched with a chimpanzee named Ham in the capsule. The booster carried him 42 miles higher and 124 miles further downrange than planned. He went up to an altitude of 156 miles and landed some 415 miles from the launching pad. Failure of an oxidizer control valve plus excessive pressure in the turbopump had resulted in the feeding of too much oxygen to the engine and early shutdown. The excessive oxygen raised the cutoff velocity, while the early shutdown triggered the abort mode. The escape motor added to the velocity of the capsule, subjecting Ham to about 17 Gs going up and about 15 on the way down. Through it all, he performed the tasks he had been taught to do, enabling him to avoid electric shocks that penalized nonperformance. Despite the capsule's taking on water before Ham was rescued (by helicopter rather than ship), he accepted an apple and half an orange after being lowered to the deck of a naval vessel. This behavior gave confidence to the astronauts that they too could tolerate their launches.

Engineers made adjustments to the engine cutoff system in response to the anomalies on this flight. On both flights 1A and 2, moreover, the LEV-3 autopilot had experienced high vibration levels. Engineers had selected it for the Mercury-Redstone because it had worked well on the early test flights of the Redstone missile. However, the lengthening of the vehicle for

Figure 25. Launch of Mercury-Redstone 2 on January 31, 1961, carrying the chimpanzee Ham aboard in a Mercury capsule. Courtesy of NASA.

Mercury plus the heavier payload had caused the natural bending frequency of the booster to decline by a factor of four. This made it resonate with the frequency of the control system and caused vibrations that could have destroyed the system. The solution was a filter network that the engineers installed before the next flight.[62]

The Space Task Group was basically satisfied with progress at this point, and the original schedule called for the next Mercury-Redstone flight to carry an astronaut. But von Braun, Kuettner, and other Marshall engineers were concerned about the "chatter" in the control system and the distance missions 1A and 2 had both traveled. They decided the Mercury-Redstone needed one more flight without a passenger to test the solutions they had devised. This mission, designated Mercury-Redstone Booster Development or MR-BD, launched on March 24, 1961, and landed very close to its intended target area. George M. Low, future deputy administrator of NASA, wrote to Administrator James Webb that it showed "all major booster problems have been eliminated." This proved to be the case, as Mercury-Redstone 3 boosted astronaut Alan Shepard into a successful suborbital flight on May 5, 1961. Shepard reached an altitude of 116 miles, well above the 62-mile international standard for space flight. On July 21 of the same year, Mercury-Redstone 4 carried Virgil I. Grissom on the program's second suborbital space flight, to 118 miles maximum altitude, concluding the involvement of Redstone in Project Mercury and as a launch vehicle. The success of these two missions had led to the cancellation of any further Mercury-Redstone flights, although two more had been planned.[63]

Meanwhile, however, the Soviets had succeeded in placing cosmonaut Yuri A. Gagarin into a fully orbital flight on April 12, 1961. The USSR was still ahead in the space race.[64]

Summary and Conclusions

The development of the Redstone missile and its adaptation for use as a space booster marked the beginning of actual launch-vehicle technology in the United States. Use of three stages of scaled-down Sergeant motors in the Juno I complemented the Redstone achievements and showed that not just the von Braun team with its V-2 experience but also the expertise garnered at JPL was important for progress in American rocket technology. Significant contributions were also made by engineers at North American Rocketdyne. These engineers had learned a great deal from the V-2 and the Germans who developed it. But with the Redstone engine, developed from

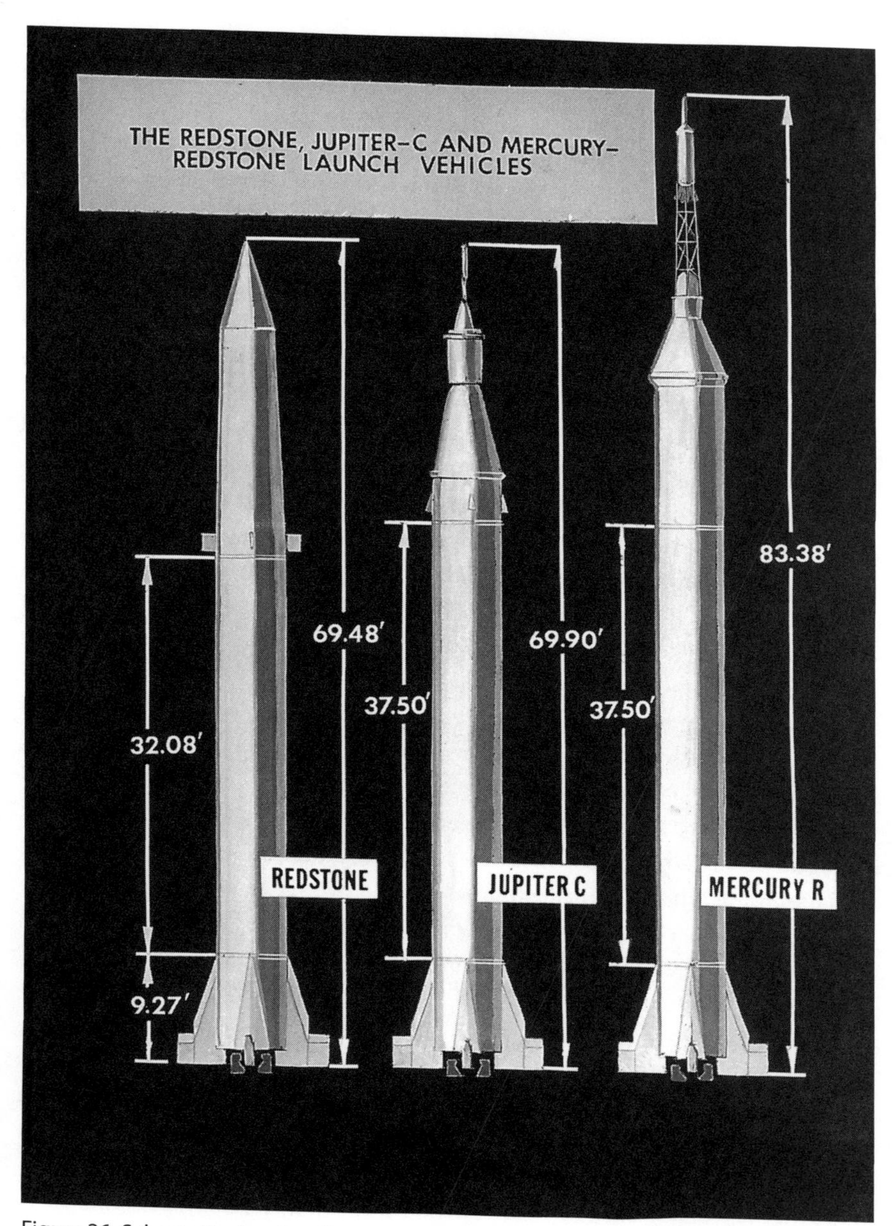

Figure 26. Schematic drawing illustrating the differences among the Redstone missile, the Jupiter-C, and the Mercury-Redstone launch vehicle. The Mercury-Redstone launched astronaut Alan Shepard in his Mercury spacecraft, *Freedom 7*, on May 5, 1961. Courtesy of NASA.

an offshoot of the Air Force's Navaho program, they advanced beyond V-2 technology to make original American contributions to rocketry, even if they were helped by Germans like von Braun and Huzel. Their work, moreover, launched Rocketdyne on a path that made it the leading rocket-engine manufacturing firm in the country by the 1960s, despite the head start that Reaction Motors, Aerojet, and Thiokol had enjoyed and despite the later competition of other firms.

Both the Rocketdyne engineers and their Army/NASA counterparts under von Braun struggled at times to effect modifications to engines and entire launch vehicles. Problems with combustion instability, resonant frequencies, and the like showed that despite a considerable database of information about how rocket and missile components and systems worked, and despite a lot of accumulated experience, developing reliable space-launch systems was still partly a trial-and-error process. Engineers used what they knew, a great deal of testing, and their best methodology to arrive at what they thought would be viable systems, only to have unexpected problems occur. When that happened, the fund of engineering experience and data could and did yield solutions. But it often took actual flight testing to show whether or not those solutions worked.

In addition to improved engine technology, the Redstone demonstrated significant developments in structures and in guidance and control. The lighter structure of the missile as compared with the V-2 and the improved inertial guidance system both showed advances over rockets and missiles that had preceded the Redstone except for the Air Force's MX-774B and the Navy's Viking. The structural changes carried over to the Jupiter C and Juno I, but the advanced guidance-and-control system with the air bearings did not. The Jupiter C, Juno I, and Mercury-Redstone all used the earlier and simpler LEV-3 instead of the ST-80.

Mercury-Redstone brought the first attempt at "man-rating" launch vehicles. This too was a trial-and-error process, but it would pave the way for Mercury-Atlas and other launch systems for astronauts on up through Saturn,[65] which would carry humans to the Moon, and the space shuttles, which would provide routine human access to space. America was still behind the Soviets in the space race (though probably not the missile race, despite fears to the contrary), but the United States was showing signs that it would eventually catch up and even surpass the other superpower in some aspects of the continuing space effort, which also would involve some cooperation with the Soviets.

7

The Atlas, Thor, and Jupiter Missiles, 1954–1959

The Atlas, Thor, and Jupiter missiles were all important in the development of launch vehicle technology. All brought further innovations, and all became the first stages of launch vehicles themselves, with the Air Force's Atlas and Thor being more significant in this role than the Army's Jupiter. Additionally, all three programs illustrated the roles of interservice and interagency rivalry and cooperation that were key features of rocket development in the United States. They also showed the continued use of both theory and empiricism in the complex engineering of rocket systems. Perhaps even more important than these matters, however, was the sheer size of the effort to produce these and other ballistic missiles rapidly to meet the threat of Soviet missile development, about which not much was known at the time. Because of concerns about the Soviets, technology "had to advance steadily on a very broad front. It was not one important 'breakthrough' that enabled this advance; rather, it was a thousand different refinements, a hundred thousand tests and design modifications, all aimed at the development of equipment of extraordinary power and reliability," according to Milton Rosen, writing in 1962. The new missiles "combined to create a new and complex industry" to provide the components and support they needed. "The space industry had to grow very quickly from a few hundred people to several hundred thousand."[1]

The scope of this change is best seen in a comparison of the numbers of people involved in the Vanguard and Atlas programs. Whereas 15 people at the Naval Research Laboratory and 180 contractor employees had sufficed for Vanguard,[2] already in 1957 Atlas required some 1,209 companies in 32 states to produce its components. The Air Force organization responsible for its development had 635 people overseeing the process in that year, assisted by 1,900 contractors responsible for systems engineering and technical direction. By 1960 there were 17 major contractors working on Atlas, 200 subcontractors, and about 80,000 workers.[3] Atlas obviously was a very much larger effort than Vanguard, and it began to create the infrastructure in talent, knowledge, data, and capability that were necessary for the maturation of missile and launch-vehicle technology in the decade of the 1960s.

Until the Air Force became serious about Atlas, it had lagged behind the Army and the Navy in the resources it devoted to the development of purely ballistic missiles. (The important Navaho project involved air-breathing ramjet technology but did contribute to ballistic missile technology, forming an important qualification to this generalization.) The reasons for this were the limited post–World War II research and development budgets the president and Congress provided, doubts that technical hurdles to effective ballistic missiles could be overcome quickly, and an Air Force culture that emphasized piloted aircraft and did not like the threat that unpiloted ballistic missiles posed to the role of humans in the cockpit.[4]

The series of developments that led the Air Force to shift from limited support to major emphasis on ballistic missiles was complex. A key factor was the development of thermonuclear warheads. On January 23, 1951, when the Air Force awarded Convair a contract for MX-1593—the project that soon became Atlas—the specifications called for study of a missile that could launch an 8,000-pound warhead and hit a target 5,000 nautical miles (5,750 statute miles) away with an accuracy of 1,500 feet circular error probable. In Convair's initial design, this required a missile 160 feet in length and 12 feet in diameter, with five or seven big engines.[5] The large size and then-impracticable accuracy at such a range lent credence to arguments against spending large sums of money developing ballistic missiles.[6]

But by 1953, development of thermonuclear weapons by both the United States and the Soviet Union promised warheads that were significantly lighter and smaller, as well as much more powerful, than earlier fission-type (atomic) warheads. This meant that Atlas and other ballistic missiles could be smaller and less accurate but still as effective as larger, more precise weapons with the older types of warheads. The new nuclear technology plus the increased threat from the Soviets provided one condition for greater Air Force support of Atlas and other purely ballistic missiles.[7]

These twin developments constituted a necessary but not a sufficient impetus for change in the Air Force's research-and-development priorities. It still took two heterogeneous engineers to nudge the Air Force and the Department of Defense (DoD) in a new direction. One of them was Trevor Gardner, assistant for research and development to Secretary of the Air Force Harold Talbott in the Eisenhower administration. Born in Cardiff, Wales, Gardner had become an American citizen in 1937 and earned a B.S. in engineering from the University of Southern California that same year. After getting a master's in business administration from USC two years later, he had risen to become general manager and executive vice president of Aerojet's parent company, General Tire and Rubber, in 1945. Three years later he became vice president and then president of a small electronics, photographic, and ordnance firm near Caltech named Hycon Engineering Company.

Someone who knew Gardner well, Herbert York—first director of the Lawrence Livermore Laboratory, operated by the University of California for the Department of Energy, and later chief scientist of the DoD's Advanced Research Projects Agency and then the first director of Defense Research and Engineering—described Gardner as "intelligent, vigorous, somewhat volatile, and impatient to make changes quickly." Short and stocky with wire-rimmed glasses, Talbott's assistant did not arouse such generally favorable assessments in everyone. Some described him as "sharp," "grating," "abrupt," "irascible," "cold," "unpleasant," and "a bastard." But regardless of the impression he made, he became an effective advocate for much greater support of ballistic missiles.[8]

The other key supporter of ballistic missiles was the "brilliant and affable" polymath Dr. John von Neumann, who was research professor of mathematics at Princeton's Institute for Advanced Study and also director of its electronic computer project. The son of a well-to-do Jewish banker, von Neumann was born in Budapest in 1903. He attended the elite Minta Gymnasium, run by Theodore von Kármán's father, whose graduates also included two other prominent influences on American nuclear-weapons development and policy, Leo Szilard and Edward Teller. At the Minta "open educational laboratory," scholars from the University of Budapest directed the "experimental learning" of the students.

Although John wanted to study mathematics, his father pressured him to pursue chemical engineering, where the elder von Neumann thought there was more money to be made. So Johnny earned a degree in chemical engineering from the Federal Institute of Technology in Zurich *and* a Ph.D. in mathematics from the University of Budapest, both in 1926. After teaching mathematics at the universities of Berlin and Hamburg, von Neumann, like the much older von Kármán (the aerodynamicist, not the pedagogue), left Germany for the United States. There he taught mathematics at Princeton University until 1933 and then moved to the Institute for Advanced Study, where he remained the rest of his life. He did important work in such diverse fields as theoretical and applied physics (including quantum theory), game theory, and computer architecture. In the process, he contributed to the development of both the atomic and thermonuclear bombs, in the latter case working out "the mathematical phases needed to decide whether the H-bomb would function."[9]

Like von Kármán, von Neumann enjoyed working with the military and was able to explain difficult concepts in clear terms. Such a stellar and genial individual was bound to have great influence. In 1953 he headed the Nuclear Weapons Panel of the Air Force's Scientific Advisory Board, which confirmed beliefs that in the next six to eight years the United States would have the capability of fielding a thermonuclear warhead weighing about 1,500 pounds and yielding 1 million tons of explosive force. This was fifty times the yield of the atomic warhead originally planned for the Atlas missile, and it fit in a much lighter package. With this information in hand and with Talbott's blessing, Gardner commissioned von Neumann to head a Strategic Missiles Evaluation Committee, known as the Teapot Committee.[10]

The committee's report, dated February 1, 1954, speculated (in the absence of good intelligence data) that the Soviets might be ahead of the United States in the development of intercontinental missiles, and suggested that the CEP specifications of America's three ongoing long-range missile projects—the Snark and Navaho cruise missiles and the Atlas ballistic missile—were out of date because of "very recent progress toward larger yield warheads." The committee urged revamping all three projects, with the CEP for Atlas relaxed from 1,500 feet to "probably" 3 nautical miles, and noted the urgency of developing an intercontinental ballistic missile capability. Gardner forwarded the report to the assistant secretary of defense for research and development, at that time Donald A. Quarles. He added the notation that he was in complete agreement with the report's general findings and that Secretary Talbott had directed a "mid-1958 preliminary intercontinental ballistic missile" operational capability, with "an extremely high priority . . . assigned to the project."[11]

Creation of the Western Development Division

As a result of the Teapot report and Gardner's efforts, in May 1954 the Air Force directed that the Atlas program go on an accelerated development schedule, assisted by the service's highest priority. This highly important decision set Atlas on the path to becoming both a key missile and a significant launch vehicle. The Air Research and Development Command within the air arm created a new organization in Inglewood, California, named the Western Development Division (WDD). Recommended by Gardner, Brig. Gen. Bernard A. Schriever became the first commander on August 2. The

Figure 27. Maj. Gen. Bernard A. Shriever in 1956, when he was Western Development Division commander. Official U.S. Air Force photo, courtesy of the 45 Space Wing History Office, Patrick AFB, Fla.

new organization gained "complete control and authority" over the Atlas program, including the uncommon option of communicating directly with other major commands, Air Force headquarters, and the secretary of the Air Force. Schriever also had unusual authority to select highly qualified Air Force officers to work on his staff. Among the first were Col. Charles H. Terhune Jr., who became Schriever's deputy director for technical operations; propulsion expert Lt. Col. Edward Hall; Lt. Col. Benjamin Paul Blasingame, knowledgeable in electronics and guidance and control; and personnel expert Lt. Col. John Hudson. Hudson was authorized to select officers in the top 4 percent of Air Force effectiveness ratings.[12]

Schriever had been born in Bremen, Germany, in 1910. His father, Adolf, an engineer with the North German Lloyd steamship line, was interned in America in 1916, and the family joined him there, moving to Texas when he was released. Adolf Schriever died in an industrial accident in 1918, and the family was hard pressed to support itself for a time. But Bennie Schriever, as he was called, graduated in 1931 from the Agricultural and Mechanical College of Texas (since 1964, Texas A&M University) with a degree in architectural engineering. Tall, slender, and handsome, the determined young man accepted a reserve commission in the Army and completed pilot training, eventually marrying the daughter of Brig. Gen. George Brett of the Army Air Corps in 1938.

Obtaining a regular commission in the Air Corps that year, Schriever became an engineering officer and test pilot, graduating from the Air Corps Engineering School and then earning an M.S. in aeronautical engineering at Stanford University before going to the Pacific theater in World War II, where he flew sixty-three combat missions in B-17 bombers. After the war he headed the Scientific Liaison Branch of the Army Air Forces (a successor of the Air Corps and the predecessor of the Air Force), working closely with von Kármán, who headed the Scientific Advisory Board. In 1951 Schriever served as assistant for development planning at the headquarters of the now-independent Air Force, where he devised a method that identified technological advances likely to be promising and applied them to future requirements of the new service. Clashing with Gen. Curtis LeMay, commander in chief of the Strategic Air Command, over LeMay's desire for an impractical nuclear-powered bomber, Schriever started a pattern of disagreements with the senior general that continued for many years but did not prevent the younger man from becoming a pivotal force in furthering ballistic missile and space technology.[13]

Schriever in essence inherited from Gardner and von Neumann the role of heterogeneous engineer, promoting and developing the Atlas and later missiles. Gardner left the federal government in 1956, and von Neumann died early the following year, both before Atlas was fully developed. Schriever had a dramatically different personality from Gardner's, being generally calm and persuasive rather than intense and abrasive. He managed to select highly competent people for his staff, many of his managers later becoming general officers themselves. An extremely hard worker, Schriever demanded much of his staff but was somewhat aloof to most of them and (apparently unintentionally) inconsiderate of their time, being frequently late for meetings without realizing it. One of Schriever's early staff members, who later became a lieutenant general, Otto J. Glasser, said that Schriever was "probably the keenest planner of anybody I ever met" but that he was "one of the lousiest managers." Good, as Glasser said, at planning and organizing, gifted with vision, he often overlooked matters that needed his attention—perhaps not surprising since he spent much time flying back and forth to Washington, D.C. His secretary and program managers had to watch carefully over key documents to ensure that he saw and responded to them.

Terhune, who also later became a lieutenant general, called Schriever a "superb front man" for the organization, "very convincing. . . . He had a lot of people working for him [who] were very good and did their jobs, but Schriever was the one who pulled it all together and represented them in Congress and other places." Glasser added, "He was just superb at getting over there [to Capitol Hill] and laying out the wisdom of his approach so that the Congress *wanted* to ladle out money to him." Glasser also said Schriever was good at building camaraderie among his staff.[14]

Before he left government service, Gardner, who became the Air Force's initial assistant secretary (instead of special assistant) for research and development on March 1, 1955, provided Schriever with one more mechanism, known as the Gillette Procedures, for developing Atlas and later ballistic missiles. In mid-September Gardner asked Hyde Gillette, a budget official in the office of the secretary of the Air Force, to study how to simplify procedures for managing what were coming to be called intercontinental ballistic missiles (ICBMs).[15] Schriever had complained that there were forty different offices and agencies he had to deal with. Approval of his annual development plans, he said, took months to steer through all of these bodies. The procedures that emerged from Gillette's committee and won DoD approval on November 8, 1955, greatly simplified the process by reducing the bureaucracy Schriever had to deal with essentially to two ballistic missile committees, one at the Office of the Secretary of Defense and the other at the Air Force level. Coupled with other arrangements, this gave Schriever unprecedented authority to develop missiles—authority he used to good advantage.[16]

Another key element in the management of the ballistic missile effort was the Ramo-Wooldridge Corporation. Simon Ramo and Dean Wooldridge were classmates at Caltech who had both earned their Ph.D.'s there at age twenty-three. After World War II they had presided over an electronics team that built fire control systems for the Air Force at Hughes Aircraft. In 1953 they set up their own corporation, with the Thompson Products firm, which specialized in electronic equipment, automobile parts, and pumps, buying 49 percent of the stock.

For a variety of reasons, including recommendations by the Teapot Committee and by another committee to advise about Atlas, also headed by von Neumann,[17] Schriever suggested that Ramo-Wooldridge become a systems engineering–technical direction contractor to advise his staff on the management of the Atlas program. The Air Force issued a formal contract on January 29, 1955, although the firm had begun working in May 1954 under letter contract on a study of how to redirect the Atlas program. This unique arrangement caused considerable concern in the industry, especially on the part of Convair, that Ramo-Wooldridge employees would be in a position to make unfair use of knowledge they gained to bid on other contracts, although the firm was not supposed to produce hardware for missiles. To ward off such criticism, the firm created a Guided Missile Research Division and kept it separate from other divisions of the firm. Louis Dunn from JPL, who had served on the Teapot Committee, became its director, bringing several JPL people with him. This arrangement did not put an end to controversy about Ramo-Wooldridge's role, so in 1957 the division became Space Technology Laboratories, an autonomous division of the firm with Ramo as president and Louis Dunn as executive vice president and general manager.[18]

The Ramo-Wooldridge staff outnumbered the Air Force staff at WDD, but the two groups worked together in selecting contractors for components of Atlas and later missiles and in overseeing their performance, testing components and whole missiles, and analyzing results. For such a large undertaking as Atlas, soon joined by other programs, there needed to be some system to inform managers and allow them to make decisions on problem areas. Thus WDD, which became the Air Force Ballistic Missile Division (BMD) on June 1, 1957, developed a management control system to collect information for planning and scheduling.

Schriever and his program directors collected all of this data in a program control room, located in a concrete vault and kept under guard all the time. At first, hundreds of charts and graphs were attached to the walls, but in the course of time, this system was supplemented by digital computers for tracking information. Although some staff members claimed Schriever used the control room only to impress important visitors, program managers benefited from preparing weekly and monthly reports of status, because they had to verify their accuracy and thereby keep abreast of events. Reports from a procurement office the Air Force Air Materiel Command assigned to WDD on August 15, 1954, provided Schriever with a separate check on information from his own managers. The thousands of milestones—Schriever called them "inchstones"—in the master schedule kept not only him but also his key managers advised of how development matched planning. All of the information came together on "Black Saturday" meetings once a month starting in 1955. Here program managers and department heads presented problem areas to Schriever, Ramo, and Brig. Gen. Ben I. Funk, commander of the procurement office. As problems arose, discussions sometimes resolved them. If not, a specific person or organization would be assigned to come up with a solution, and the staff of the program control room would track progress. Sometimes Ramo brought in outside experts from industry or academia to deal with particularly difficult challenges.[19]

Because the Soviet threat was perceived as great and the technology far from mature, Schriever and his team employed a practice called concurrency, which was not new but was not routinely practiced by the Air Force and the federal government. Used on the B-29 bomber, the Manhattan Project, and the development of nuclear vessels for the U.S. Navy, it involved developing not only all subsystems on overlapping schedules but also the facilities to test and manufacture them, the systems for operational control, and the training system—in this case, for the Strategic Air Command, which would assume responsibility for the missiles when they became operational.

This was different only in degree from practices on the Redstone and Vanguard. But Atlas was a larger missile with a greater number of components and contractors. Schriever claimed that implementing concurrency was equivalent to requiring that a car manufacturer not only build the automobile but also construct highways, bridges, and filling stations as well as teach driver's education. He argued that concurrency saved money, but this seems doubtful. As one example of the costs of concurrency, each model of the Atlas missile from A to F involved improvements, and the F-models were housed in silos. Every time the design for the F-model changed, the Army Corps of Engineers had to reconfigure the silo. There were 199 engineering change orders for the silos near Lincoln, Nebraska, and these raised the costs from $23 million to more than $50 million.[20] Concurrency was important in ensuring rapid development, however, and thus justified.

A final element in WDD's management portfolio was parallel development. To avoid being dependent on a single supplier for a system, Schriever required two contractors for many of them. Eventually, when Thor and Titan I came along, the testing program threatened to become overwhelming, and Glasser argued that Ramo-Wooldridge just ignored the problem. He presented his concerns to Schriever, who directed him to come up with a solution. Glasser decided which systems would go on Atlas, which on Titan, and which on Thor. This avoided duplicative suppliers within each program but still provided an alternate source for each type of system if one proved wanting. In the process, Glasser became the deputy for systems management and also the Atlas project manager.[21]

Developing Atlas

Convair had begun developing Atlas under the MX-774B project, which entailed more innovations than the swiveling engines that are discussed in chapter 1 of this book's sequel, *Viking to Space Shuttle*. The other two major innovations were (1) monocoque propellant tanks that were integral to the structure of the rockets and pressurized with nitrogen to provide structural strength with very little weight penalty, and (2) separable nose cones so that the missile itself did not have to travel with a warhead all the way to the target and thus didn't need to survive the aerodynamic heating from reentering the atmosphere. The second innovation made the first possible by allowing the structure to be extremely light, improving the mass fraction. Moreover, the separate nose cone produced less drag than the entire missile, adding to the distance it would travel. Commissioned by the Army Air Forces' organization at Wright Field, Ohio, in April 1946 and flight tested in 1948, the MX-774B antedated the Redstone missile by several years and may have contributed to similar features on the Redstone, even though MX-774B never went into production.

The engineer generally credited with these innovations was Charlie Bossart, who had graduated from the University of Brussels in 1925 with a degree in mining engineering. He then got a scholarship through the Belgian-American Education Foundation to MIT, where he studied aeronautics and structures. He stayed in America and worked as a structural engineer for fifteen years, gaining some experience with airplanes and also with missiles on the Lark tactical missile project at Convair's Downey plant, where he had become chief of structures. (Convair was short for Consolidated Vultee Aircraft Corporation, formed by the 1941–43 merger of Vultee Aircraft with Consolidated Aircraft, which had been established in Buffalo in 1923 and moved to San Diego in 1935.) Intrigued by the idea of a large missile, Bossart persuaded Consolidated Vultee's chief engineer to let him head the effort on project MX-774. Although he started from the design of the V-2, the innovations mentioned above plus the use of aluminum for the skin yielded a much better mass fraction.[22]

No one has successfully explained how Bossart conceived these innovations. Atlas historian John Chapman quoted an unnamed colleague of Bossart as saying Charlie was "one of those rare types who can make practical engineering out of very complex theories." Chapman also says Bossart borrowed the idea of integral propellant tanks from his experience with the airplane industry, which did use integral tanks, but not in the same way as on the MX-774 and Atlas. Richard Martin, a colleague of Bossart since 1951, added, "He could quickly understand all the requirements of a subsystem and then conceptualize a design that would perform all the critical functions most efficiently." Obviously his being a structural engineer rather than a chemist, an aerodynamicist, or a practitioner of some other discipline helped with two of the innovations, especially the integral tanks, but not the swiveling engines. Whatever the explanation, Bossart ultimately earned the U.S. Exceptional Civilian Service Award in 1958 and the 1959 Collier Trophy (for the Air Force and General Dynamics, which acquired Convair in 1953) for his contributions to Atlas.[23]

Following the cancellation of the MX-774 project in 1947 and after some static and then flight testing of its three test rockets in 1948, Bossart reverted to his job as chief of structures, and his team of engineers went to other jobs. But Convair allowed him to continue some modest work on rocketry, with team members taking time from their current jobs on occasion to help him. One issue they had not resolved was staging for a large missile. They had experienced ignition problems in static and then flight tests of the three rockets, fired at White Sands, so they did not want a two-stage vehicle with the second stage igniting at altitude. Bossart and another engineer named Lloyd Standley were considering arrangements for tanks and engines that would avoid two stages by clustering engines and tanks. Standley, who later headed the mechanical design effort for the Atlas missile, suggested that having a single tank for more than one engine would reduce complication, although if some of the engines were dropped off along the trajectory, the single tank would be heavier than multiple tanks after some of them were discarded with their engines. With light construction, Bossart decided, it was a small price to pay. In this way was born the idea of a stage-and-a-half missile, employed for Atlas, with one engine staying with the tank as the sustainer while other, booster engines were discarded after they had imparted an initial velocity to the missile.

Another suggestion during this period came from electrical engineer James W. Crooks Jr. Crooks had joined Convair in 1944, left in 1945 to earn an electrical engineering degree at Kansas State University, and returned to the company in 1946. There he developed the Azusa tracking system that found widespread use in the world of missiles and rockets. Given the stringent requirements for missile accuracy, it seemed barely possible to arrange engine cutoff with enough precision to meet the demands of a flight control system. Crooks suggested two small rockets near the back of the propellant tank for more exact control after the sustainer engine cut off. To make maximum use of their small thrust to counterbalance their weight, these vernier engines would fire at launch and continue to burn after sustainer cutoff for a brief period, during which they would adjust the missile's attitude and velocity.[24]

With the outbreak of the Korean War in June 1950 came increased funding and, on January 23, 1951, the contract to Convair for project MX-1593, a two-phase study to determine whether a ballistic or a glide rocket was a better option for a long-range missile. (Instead of following a ballistic path, like that of a bullet, a glide rocket would use wings or other air-bearing surfaces to extend the distance it would fly.) Immediately Bossart began to reassemble his team from project MX-774. He suggested calling the missile Atlas, a designation that the Air Force approved on August 6, 1951. Sources differ on whether the name derived from the Greek mythological figure who bore the weight of the world on his shoulders or from the fact that Convair was then owned by the Atlas Corporation. One version holds that Bossart based it on the mythological figure and someone said that was a good choice because the firm would not object.

A glide missile would encounter aerodynamic heating problems upon reentry and its necessary heat-resistant structure would obviate the advantages of the light-tank arrangement Bossart had devised with its separable nose cone. So he argued for a ballistic missile. The Air Force agreed, cancelling the glide study in September 1951. While the Air Force and the DoD debated about support for ballistic missiles in the next three years, Atlas received $26.2 million in funding, $18.8 of it in fiscal year 1954 funds, and Convair continued to design a missile according to the original specifications with a heavy atomic warhead. During this period Bossart and a Convair employee named Bill Patterson also served as heterogeneous engineers, helping sell the Air Force and Congress on funding the Atlas program.[25]

The Steel-Balloon Structure

The initial design for the MX-1593 version of Atlas is now of only antiquarian interest, including the size of the integral tanks, but work on the tank concept contributed directly to the final version of the Atlas missile. Since the MX-774 test vehicle had used an aluminum design with success, Bossart's team tested aluminum for a large, riveted tank. The engineers found no sealant that would make the riveted seams leakproof, so they sought a weldable substance and arrived at stainless steel, which could tolerate the aerodynamic heating during the missile's ascent without insulation and with little reduction in strength.

One of the early members of Bossart's Atlas team after its 1951 resurrection was Richard E. Martin, who had earned a master's degree and studied structural analysis, vibration, and flutter at the University of Illinois. "Charlie asked me to calculate vibration modes for a simple, thin-walled, pressurized, monocoque cylinder," Martin recalled. He then had to "develop a test plan to prove that vibration frequencies and mode shapes could be accurately calculated for this type of structure." Martin said he "was a good candidate" for the task because he "had no experience and [thus] no reason to believe the concept would not work." He soon learned that "many technically trained people" thought it wouldn't.

Martin's procedure is illustrative of the way rocket engineering was frequently done. He started with "ordinary beam theory" and was able to derive "an exact solution to the differential equations" by assuming that the beam was uniform and neglecting "transverse shear stiffness and rotary inertia." He tested small tanks when they were empty and full of water, finding that the test results were in close agreement with theory. Next, Convair issued a subcontract to the nearby Solar Aircraft Corporation to build 12-foot-diameter tanks, the size that would be needed for the missile then (1954) being contemplated for the heavy fission-type warhead. The engineers tested the tanks at a Point Loma test site on the Pacific overlooking San Diego Bay. Again, the results accorded well with existing theory. One primitive but convincing test was to drop the "flimsy" tank filled with water. Although the drop was only a few inches, the loading on the tank was high. Despite skepticism, the tank held up, but "even later, just before the first Atlas flight, a Rocketdyne executive at the launch site" thought his firm's powerful engines would "tear right through your Reynolds Wrap tanks."[26]

Even as Convair was pursuing this research and testing, there was some doubt that the Air Force would retain the firm as the airframe and assembly contractor for Atlas. By October 25, 1954, however, Schriever recommended that arrangements with Convair continue, and the company began redesigning Atlas to the three-engine configuration with a 10-foot-diameter tank that became the production version of the missile. On January 6, 1955, the Air Force awarded a contract to the Convair Division for the development and production of the Atlas airframe, the integration of other subsystems with the airframe and one another, and their assembly and testing.[27]

In the final configuration, the pressurized tanks provided the structure for the missile except for a forward ring to which the nose cone or other payload would be mounted, an aft ring for mounting the two booster engines, a rear cone and stringers to support the sustaining engine, and a bulkhead between the fuel and oxidizer tanks. The skin thickness varied from 0.010 to 0.051 inches of corrosion-resistant stainless steel, with the average being 0.020 inches, roughly half as thick as a dime. Most equipment that was usually placed between the stages of more conventional designs was attached to the tanks, which derived their strength from the pressure of the propellants, maintained by a system using chilled helium, a nonflammable and lightweight gas. Typically attached to rails welded to the tanks, where the vibrations were less severe than for interstage adapters, the equipment was easily accessible to mechanics and technicians through doors hooked to brackets that themselves were welded to the tanks. This occasioned problems only when design changes occurred. During the early part of the development cycle, the location for rate gyros changed several times, prompting Convair's general manager, James Dempsey, to suggest that the rails be run the whole length of the vehicle so that the locations could change right up until launch day. This probably was a good idea, but Dempsey didn't insist, and it wasn't implemented.

A concern in 1955 was the extent to which the thin steel balloons, as they were called, would be affected by airflow over a wide range of speeds and high dynamic pressures as the missile ascended through the atmosphere on its way to space in its ballistic trajectory. The project consulted leading experts at Caltech, MIT, UCLA, and Columbia University, who provided equations to quantify the problems. Unfortunately, no computers yet available could solve the equations. UCLA aerodynamicist John Miles offered some simplifying assumptions. Calculations using these assumptions indicated the steel balloons would bear the loads and buffeting they would encounter. The subsequent flight history of the missile and space booster proved this to be correct. During extensive static testing of the missile with engines firing at Convair's Sycamore Canyon test site in the hilly region northeast of San Diego and at a rocket test site the Air Force set up on Edwards Air Force Base near Boron in the Mojave Desert, the tank structure withstood strains and vibrations later determined to be at higher levels than during flight. Although doubters remained, Bossart's conception seemed to be verified by all of the calculations and testing.[28]

Atlas Engines

Although Convair had hoped to produce most of the components of the Atlas, it chose in 1951 not to get into engine development and subcontracted that task to both Aerojet and North American Aviation. NAA was already working on engines for the Navaho. By June 1954 they included a

120,000-pound-thrust engine that had been modified to burn liquid oxygen and kerosene as part of the rocket engine advancement program (discussed further in chapter 1 of *Viking to Space Shuttle*). Atlas borrowed heavily from the innovations that NAA had introduced for this engine,[29] while Aerojet ultimately became the contractor for the engines of the Titan missiles and launch vehicles instead of Atlas.

In particular, the innovations in the Navaho 120,000-pound engine at NAA included conversion of the combustion chamber from welded sheet metal to lighter-weight brazed tubes for regenerative cooling along the lines first developed by Ed Neu at Reaction Motors (which NAA engineers may have heard about and adopted); increased chamber pressure and nozzle area ratio; powering of the turbopumps with a gas generator burning the same propellants as the combustion chamber instead of a separate hydrogen-peroxide system; and smaller, more efficient turbopumps. The result was a lighter, more efficient engine. Even though Navaho was cancelled in 1958, through these innovations it contributed to the next generation of NAA engines at what became the Rocketdyne Division.[30]

On October 28, 1954, WDD and the Special Aircraft Projects (procurement) Office that the Air Materiel Command had located next to it issued a letter contract to NAA to continue research and development of liquid-oxygen and kerosene (RP-1) engines for Atlas. The cooperating Air Force organizations followed this with a contract to North American for twelve pairs of rocket engines for the series A flights of Atlas, which tested only two outside booster engines and not the centrally located sustainer engine. The Rocketdyne Division, formed in 1955 to handle the requirements of Navaho, Atlas, and Redstone, also developed the sustainer engine, which differed from the two boosters in that its nozzle had a higher expansion ratio for optimum performance at higher altitudes once the boosters were discarded.[31]

Using knowledge gained from the Navaho and Redstone engine development programs, NAA engineers began developing the MA-1 Atlas engine system for Atlases A, B, and C in 1954. (Atlas B added the sustainer engine to the two boosters. Atlas C had the same engines but an improved guidance system and thinner skin on the propellant tanks. Both were test vehicles only.) The MA-1 system, like its successors the MA-2 and MA-3, was gimballed and used the brazed "spaghetti" tubes forming the inner and outer walls of the regeneratively cooled combustion chamber, developed at NAA in 1951. NAA/Rocketdyne began static "hot fire" tests of the booster engines in 1955 and of all three MA-1 engines in 1956 at Santa Susana. The two booster engines, designated XLR-43-NA-3, had a specific impulse of

245 lbf-sec/lbm and a total thrust of 300,000 pounds. The sustainer engine, XLR-43-NA-5, had a lower specific impulse of 210 lbf-sec/lbm and a total thrust of only 54,000 pounds.[32]

Produced in 1957 and 1958, these engines ran into failures of systems and components in flight testing that also plagued the Thor and Jupiter engines, which were under simultaneous development and shared many component designs with the Atlas. They used high-pressure turbopumps that transmitted power from the turbines via a high-speed gear train to the propellant pumps. Both Atlas and Thor used the MK-3 turbopump, which failed at high altitude on several flights of both missiles, shutting down the propulsion system. Investigation showed that lubrication was marginal. Rocketdyne engineers redesigned the lubrication system and a roller bearing. They also strengthened the gear case and related parts. Turbine blades experienced cracking, attributed to fatigue from vibration and flutter, so engineers tapered the blade's profile to yield a different natural frequency and added devices called shroud tips. These extended from one blade to the next, restricting the amount of flutter. One test flight suffered explosion of the sustainer engine, caused by rubbing in the oxygen side of the turbopump and solved by increasing clearances in the pump and installing a liner.

Another problem encountered on the MA-1 was a high-frequency acoustic form of combustion instability resulting in vibration and increased transfer of heat that could destroy the engine in less than a hundredth of a second. The solution proved to be rectangular pieces of metal called baffles, attached to a ring near the center of the injector face and extending to the chamber walls. Fuel flowed through the baffles and ring for cooling. The baffles and ring served to contain the transverse oscillations in much the way that the eighteen compartments on the V-2 had done, but without the cumbersome plumbing. Together with a change in the injection pattern, this innovation reduced the instability to manageable proportions. These improvements came between the flight testing of the MA-1 system and the completion of the MA-3 engine system (1958–63).[33]

The MA-2 "was an uprated and simplified version of the MA-1," used on the Atlas D, which was the first operational Atlas ICBM and was later used as a launch vehicle under Project Mercury. Both MA-1 and MA-2 systems used a common turbopump feed system in which the two turbopumps (for fuel and oxidizer) operated from a single gas generator and provided propellants to booster and sustainer engines. For the MA-2, the two boosters delivered a slightly higher specific impulse of 248 lbf-sec/lbm, with the sustainer's increased to 213 lbf-sec/lbm. The overall thrust rose to 309,000 pounds for

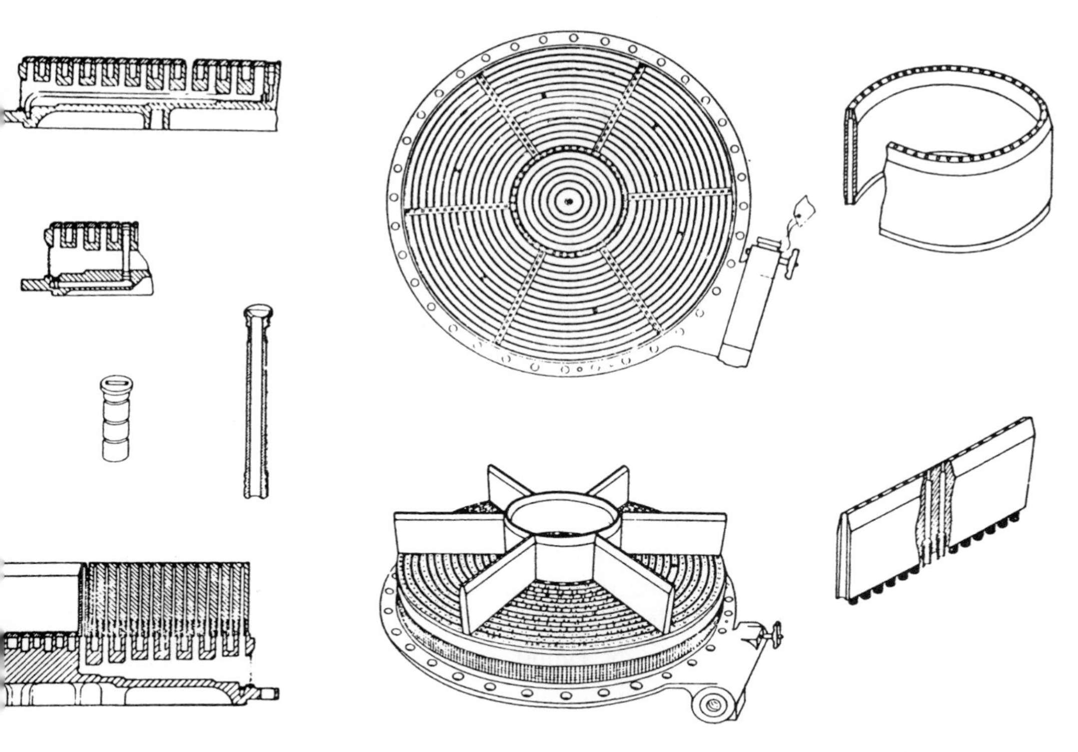

Figure 28. Technical drawing of a baffled injector similar to the one used on the Atlas MA-1 engine to prevent combustion instability by containing lateral oscillations in the combustion chamber. From Huzel and Huang, *Design*, 122. Courtesy of NASA.

the boosters and 57,000 pounds for the sustainer. An MA-5 engine system was initially identical to the MA-2 but was used on space launch vehicles rather than missiles. In development from 1961 to 1973, the booster subsystem went through several upratings, leading to an ultimate total thrust of 378,000 pounds, compared to 363,000 for the MA-2. Most of the thrust increase came from increased pressure in the combustion chamber.

In the MA-3 engine system, the two identical boosters and the sustainer each had their own turbopump and gas generator. The system was also simplified, and starting reliability improved, by hypergolic thrust chamber ignition. A single electrical signal caused solid-propellant initiators and gas-generator igniters to begin the start sequence. Fuel flow through an igniter fuel valve burst a diaphragm holding a hypergolic cartridge and pushed it into the thrust chambers. Oxygen flow occurred slightly ahead of the fuel, and the cartridge with its triethyl aluminum and triethyl boron reacted with the oxygen in the thrust chamber and began combustion. Hot gases from combustion operated the turbopump, a much more efficient system than previous turbopumps operated by hydrogen peroxide in such rockets as the V-2 and Redstone.

The MA-3 sustainer engine's specific impulse at sea level was only slightly higher than the MA-2 sustainer's, at 214 lbf-sec/lbm, and the thrust of 57,000 pounds was the same. The total thrust of the boosters, however, went up to 330,000 pounds, with a climb in specific impulse to 250 lbf-sec/lbm at sea level. At altitude the specific impulse rose to 309 lbf-sec/lbm for the sustainer and 288 for the boosters. The higher value for the sustainer engine at altitude was attributable to nozzles that were designed for the lower pressure outside the atmosphere. The MA-3 appeared on the Atlas E and F missiles, with production running from 1961 to 1964.[34]

Most of the changes from the MA-1 to the MA-3 followed a decision in 1957 by Rocketdyne management to create an Experimental Engines Group under the leadership of Paul Castenholz, a design and development engineer who had worked on combustion devices, injectors, and thrust chambers. Castenholz "enjoyed a reputation at Rocketdyne as a very innovative thinker, a guy who had a lot of energy, a good leader." The group consisted of about twenty-five mostly young people, including Dick Schwarz, fresh out of college and later president of Rocketdyne. Bill Ezell, who was the development supervisor, had come to NAA in 1953 and was by 1957 considered an "old-timer" in the company at age twenty-seven. Castenholz was about thirty. Before starting the experimental program, Ezell had just come back from Cape Canaveral, where there had been constant electrical problems on attempted Thor launches. The Atlas and Thor contracts with the Air Force each had a clause calling for product improvement, which was undefined, but one such improvement the group sought was to reduce the number of valves and electrical wires and connections that had to operate in a precise sequence for the missile to function.

The experimental engineers wanted a system with one wire to start the engine and one to stop it. Buildup of pressure from the turbopump would cause all of the valves to "open automatically by using the . . . propellant as the actuating fluid." This one-wire start mechanism became the solid-propellant initiator for the MA- 3, but the engineers under Castenholz first used it on an X-1 experimental engine on which Cliff Hauenstein, Jim Bates, and Dick Schwarz took out a patent. They used the Thor engine as the starting point and redesigned it to become the X-1. Their approach was mostly empirical, which was different from the way rocket development evolved by the

1980s, when the emphasis shifted to more analysis on paper by a computer, with simulation coming before actual hardware development. In the period from 1957 through the early 1960s when Castenholz's group was operating, they had ideas, built the hardware, and tried it out, learning from their mistakes, including the explosion of engines.

A further difference from the 1980s was noted by Stan Bell, another engineer in Castenholz's group. "We were allowed to take risks and to fail and to stumble and to recover from it and go on," he said. "Now everything has got to be constantly successful." Jim Bates added that there weren't any "mathematical models of rocket engine combustion processes" in the late 1950s and early 1960s. "There weren't even any computers that could handle them." But, he said, "we had our experience and hindsight."

The move to a hypergolic igniter came because existing pyrotechnic devices posed a delicate problem. It was difficult to get one that had sufficient power for assured ignition without going to the point of a hard start that could damage hardware. This led the engineers to the hypergolic cartridge used on the MA-3. In the process of developing it, however, they discovered that a little water in the propellant line ahead of the hypergolic cartridge, or "slug," produced combustion in the line but not in the chamber, where the propellants built up and caused a detonation that blew "hell out of an engine," as Bill Ezell put it. They learned from that experience to be more careful. Still, Ezell said, "there's probably no degree of analysis that could have prevented that from happening." There were simply lots of instances in rocket engine development where, unless the experimenters could "make the right guess or assumption," there was "no way to analyze it. So you've got to get out and get the hard experience." Ezell also opined that "without the Experimental Engine program . . . in my opinion there never would have been a Saturn I."[35]

That remark, suggesting a line of evolution from their work to engines used on the Apollo program, goes beyond the subject of this chapter. But the experiences and comments of Castenholz's group shine a spotlight into the often dimly lit processes of early rocket engineering. Without the product-improvement clauses in the Atlas and Thor contracts, a common practice of the Non-Rotating Engine Branch of the Power Plant Laboratory at Wright-Patterson Air Force Base, the innovations made by this group probably would not have occurred. They thus would not have benefited Thor and Atlas as well as later projects such as Saturn I. Even with these clauses, not every company would have put two dozen bright young engineers to work on pure experiment or continued their efforts after the first engine explosion. That Rocketdyne did both probably suggests why it became the preeminent rocket-engine firm in the country.

However, using the MA-3 engines that resulted from the experimental group's work did not resolve all of the problems with the Atlas E and F configurations. The Atlas lifted off with all three engines plus the two verniers firing. Once the missile—or, later, launch vehicle—reached a predetermined velocity and altitude, it jettisoned the booster engines and structure, with the sustainer engine (and verniers) propelling the remaining part of the rocket to its destination. The separation of the booster sections occurred at disconnect valves that closed to prevent loss of propellant from the feed lines. This system worked through the Atlas D but became a major problem on the E and F models, with their independent pumps for each engine rather than a common turbopump for all. Unlike the earlier models, the E and F did not use water in the regenerative cooling tubes that formed the combustion chamber because of the hypergolic slug's reaction with water. With previous igniters, the water had ensured a gentle start. With the hypergolic device, testing of the engines by Convair at its Sycamore Canyon site had produced some structural damage in the rear of the missile. Design fixes were instituted, but no realized there was a large pressure pulse created when the new models ignited.

On June 7, 1961, the first Atlas E lifted off from Vandenberg Air Force Base on the California coast at an operational launch site that used a dry flame bucket rather than water to receive the missile's thrust. The Atlas flew for some 40 seconds before the propulsion system failed and the missile crashed. After its parts were recovered from the ground, Rocketdyne specialists analyzed the hardware and data, concluding that a pressure pulse had caused a sudden upward force from the dry flame bucket back onto fire resistant blankets, called "boots," that stretched from the engines' throats to the missile's firewall to form a protective seal around the gimballed engines. The blowback caused one boot to catch on a drain valve at the bottom of a pressurized oil tank that provided lubrication for the turbopump gearbox. The tank drained and the gearbox, without lubrication, ceased to operate. To solve this problem engineers placed a new liquid in the cooling tubes ahead of the propellants to soften the ignition process.

Repeated failures of different kinds also occurred during the flight test program of the E and F models at Cape Canaveral. Control instrumentation showed a small, short-lived pitch upward of the vehicle during launch. Edward J. Hujsak, General Dynamics' assistant chief engineer for mechanical and propulsion systems, pondered the evidence and spoke with Gene Armstrong, the firm's director of engineering. Hujsak believed that the problem lay with a change in the geometry of the propellant lines for the E and F models that allowed RP-1 and liquid oxygen to mix and escape. Engineers "did not really know what could happen behind the missile's traveling shock front" as it ascended, but if the escaped propellants were being contained in such a way as to produce an explosion, it would explain the various failures.

The solution entailed additional shutoff valves in the feed lines on the booster side of the feed system, preventing expulsion of propellants. These had to be retrofitted in the operational missiles. However, the Air Force decided that, since there could be no explosion if only one of the propellants were escaping, the shutoff valves would be installed only in the oxygen lines. A subsequent failure on a test flight moved the service to approve installation in the fuel lines as well. This solved the problem.[36] Here was a further example of the extent to which engineers did not always fully understand how changes in a design could affect the operation of a rocket. Only failures in flight testing and subsequent analysis pinpointed problems and suggested solutions.

Atlas Guidance and Control

The picture for Atlas's guidance and control systems was more complicated than for the structure or the engines because more contractors were involved. Convair had awarded the Instrumentation Laboratory at MIT a subcontract in early 1954 to work on inertial guidance for Atlas. The first contract that WDD and its contracting partner issued in this field was also with MIT to research and develop a fully inertial guidance system. There were additional arrangements for the AC Spark Plug Division of General Motors to work with MIT to produce and test the completed system. As it turned out, this team produced the guidance and control systems for Thor and Titan, not for Atlas. But the MIT laboratory began its guidance system work on Atlas.[37]

Headed by MIT professor Charles Stark Draper—and named for him after 1973 when the Charles Stark Draper Laboratory separated from MIT—the Instrumentation Laboratory had its beginnings in Draper's development of fire control equipment for the Army Air Forces and the Navy during World War II. After the war, he and the laboratory that he founded developed inertial guidance systems for submarines, aircraft, and missiles. Both the inertial guidance systems and the fire control systems used gyroscopes, and already in his wartime work Draper had successfully "floated" them in a liquid to reduce friction and thus increase their accuracy, not an easy feat. The floated gyroscope became a hallmark of Draper's work. Minneapolis-Honeywell credited him with the successful development of the technology, which Honeywell developed into a production gyroscope used first for aircraft and then for the Vanguard launch vehicle. Draper, however, had worked originally with Sperry Gyroscope and then AC Spark Plug to produce the equipment he designed,[38] with Minneapolis-Honeywell apparently working on its own in its development of basically the same technology.

Initially, only Richard H. Battin and his boss, J. Halcombe Laning Jr., at the Instrumentation Laboratory worked on analysis for the Atlas guidance system. Battin had earned an undergraduate degree in electrical engineering in 1945 and a Ph.D. in applied mathematics in 1951, both at MIT. He began work at the lab in 1951. Since there was "no vast literature to search on 'standard' methods of guiding ballistic missiles," Battin said he and Laning "invented" a method called "Delta guidance," in which, according to Donald MacKenzie, "the missile's position and velocity at the end of powered flight are preplanned, but there is still flexibility in how those conditions are achieved," as distinguished from what MacKenzie termed "fly-the-wire" guidance in which "the requisite trajectory is calculated in advance, on the ground, and the guidance and control system is then simply required to return the missile to the preplanned trajectory when it appears to deviate."

As Battin pointed out, "The electronic digital computer industry was in its infancy then [the mid-1950s], so that an on-board guidance system could be mechanized only with analog components." And "though simple in concept, the Delta guidance method . . . is not easy to mechanize especially with analog hardware." But he and Laning were determined to make it work "despite all its deficiencies" until Battin traveled to Convair in summer 1955. He met with Walter Schwidetzky, who headed Convair's guidance group after having worked at Peenemünde during World War II and come to the United States in the first group with Wernher von Braun. Several ideas that Battin brought back from his conversations in San Diego led Laning to abandon delta guidance and take up a concept called Q-guidance, used on the Thor guidance and control system. Battin continued working on delta guidance for a few more weeks, then joined Laning in developing the more sophisticated Q-guidance system. But it was "some form of Delta guidance" that ultimately came to be used on Atlas, Battin said.[39]

Meanwhile, Lieutenant Colonel Blasingame was the individual at WDD who was responsible for overseeing development of the guidance and control system for Atlas. He was an MIT graduate, having earned his Ph.D.

working in conjunction with the Instrumentation Laboratory, but he was initially unable to put an inertial guidance system on the Atlas because of the weight of the systems then available and the influence of people on the von Neumann committee and at Ramo-Wooldridge who favored radio guidance. He had managed to arrange the contract with MIT and AC Spark Plug, but on February 24, 1955, there was a separate contract with General Electric to design, develop, and produce three ground-based systems to do the tracking and issue commands to the Atlas missile using radio guidance.

The system that evolved through Atlas D was a radio-inertial system with an onboard autopilot that transmitted data about missile velocity and attitude to ground stations, but because of reflections from the ground, the stations could not track the missile until it had ascended to sufficient altitude for these reflections to dissipate. During the first part of the ascent, the gyro-stabilized autopilot provided guidance and control. Later in the flight, when the ground equipment received signals and tracking data from the missile, it fed them to a Burroughs computer that processed the data and then sent commands to the autopilot. Through the autopilot, the commands went to the gimballing system used on Atlas for changes in the trajectory. The guidance and control system also commanded a slow pitch-over in the direction of the target after the missile reached 15,000 feet. About two minutes into the launch, the booster engines would cease burning and drop away. The sustainer engine would burn for another 146 seconds. Then the control system would effect cutoff, the two 1,000-pound vernier engines would make final course corrections, and the reentry vehicle would separate to travel alone toward the target.

Arguably, the radio-inertial system was more accurate than the later inertial system installed on the Atlas Es and Fs. According to one source, in 1963 the Atlas D missiles with radio-inertial systems were achieving accuracies of within a nautical mile of their target on 80 percent of their test flights, whereas Atlas Es and Fs with inertial systems achieved only a 1.5-nautical-mile accuracy on the same percentage of flights. However, missiles with radio-inertial guidance had to be located so close to their ground guidance equipment that one nuclear warhead or a bad accident could destroy the whole complex. Jamming of the radio signals was another worry. And the guidance system limited the rate of fire to one missile every fifteen minutes. So WDD decided in 1958 to adopt an all-inertial system for the Atlas Es and Fs.[40]

The company that built the inertial guidance-and-control system for Atlas was the Arma Division of American Bosch Arma Corporation, located at Roosevelt Field on Long Island near New York City. On April 12, 1955, WDD and its procurement arm had awarded this organization a contract to design, develop, construct, and test a complete inertial system for "a ballistic missile." It served as a backup for the efforts of the MIT Instrumentation Lab and AC Spark Plug, and the initial plan was for the Arma guidance and control system to operate on the Titan I missile. Arma had begun its existence in 1918 as an engineering company to make searchlights for Navy boats. It did this so well that after World War I, the Navy asked the company to develop a shipboard compass, leading to the development of the Arma gyroscopic compass. During World War II, Arma became a leading manufacturer of fire control systems. In 1948 it merged with American Bosch and became the Arma Division, which soon got Air Force contracts for fire control systems, leading to the commission to work on the missile guidance system for WDD.

Arma had introduced a "fluid suspended gyro" after World War II and had begun to study use of digital computers in late 1952 under Navy contracts, despite what Battin at MIT said about their lack of maturity. By 1954 the Long Island firm had reduced the weight of its fluid-suspended gyro to less than two pounds. It was floated in flurolube, like the Minneapolis-Honeywell hermetic integrating gyros used on Vanguard, but unlike them, Arma's gyros operated in two degrees of freedom so that two gyros could stabilize a platform in space on a three-axis gimbal system similar to the LEV-3 used on the V-2 and some of the Redstone missiles and launch vehicles. These gyros were of course more sophisticated than those used on the V-2, and they had a feature that apparently was unique—the use of four thin wires to steady and center them. Thus, while the Arma system may have borrowed from many sources, it appears to have been a separate innovation that combined borrowings with ideas it developed on its own.

Following award of its Air Force contract for what turned out to be the Atlas missile, Arma engineers began to design accelerometers of high precision. The three accelerometers they developed for the Atlas guidance and control system were small single-axis devices that sensed changes in their axis and transmitted the information to the digital computer, which sent signals to an autopilot. This in turn "commanded" the engine gimbals to rotate in the appropriate direction to correct the trajectory. For the computer, Arma used miniaturized transistors, diodes, transformers, resistors, capacitors, and printed circuitry to achieve compactness. By the spring and summer of 1957, Arma had components of its system going through vibration and acceleration tests on the supersonic sled track at the Naval Ordnance Test Station (NOTS) at Inyokern, California. The company also had its own temperature test chambers, and facilities for testing the effects of altitude and humidity, salt spray, rain, shock and vibration, radio interference, and fungus. In addition, NASA's Langley Research Center performed acoustic noise tests on the system.[41]

The tests at NOTS, in particular, marked a new departure in evaluating components for ballistic missiles and rockets. Inertial guidance systems had to be somewhat delicate to sense accelerations and trajectory changes with the degree of accuracy demanded of them. Yet like the rest of the components on rockets, they had to undergo heavy forces and vibrations during the launch process. Rocket-propelled sled tracks were uniquely qualified to simulate these launch conditions. In 1949 the Navy's Bureau of Ordnance had funded studies for a Supersonic Naval Ordnance Research Track (SNORT) at NOTS for testing rockets and missiles at transonic and supersonic speeds without destroying the test articles. After study, NOTS arranged for construction beginning in January 1952 of a 4.1-mile track. Ready in March 1954, the track proved its effectiveness in a public demonstration. It provided a test that was more realistic in many respects than a wind tunnel's and much cheaper than an actual flight test.[42]

At the time that WDD arranged for its testing at NOTS, the Arma inertial system was still intended for use on Titan I, and almost simultaneously Schriever's organization arranged for testing AC Spark Plug's inertial system to be used on the Thor intermediate range ballistic missile (IRBM). NOTS personnel discussed the upcoming testing with Ramo-Wooldridge. Using an Aerojet liquid-propellant engine to power the sled, the operators conducted a number of test runs in the fall of 1956 to demonstrate the sled's safety and reliability for this type of testing, determine what vibrations would be present, and prove the reliability of instrumentation to be used.[43] The tests on the actual Arma guidance and control systems were highly successful.[44] They and the acoustical tests at Langley show the range of organizations involved in developing an Air Force missile and the high degree of interagency cooperation.

On March 8, 1960, an Atlas D subjected the Arma guidance and control system to its first test in flight. "The 110-ton Atlas made a 90 degree twist as it rose from the pad," according to a March 10 news release. "The spin oriented a missile guidance system from a vertical to a horizontal plane, in anticipation of turning the Atlas down range. The maneuver did not alter the planned course of the missile which was aimed at an ocean target 5000 miles away." This flight showed the accuracy of the Bosch Arma system as well as several innovations that might have won the firm future contracts in the missile and rocket field, but Arma developed problems in organization and management that kept it from winning large contracts beyond its Atlas work.[45]

Nose Cones

Although it had little application to launch vehicles, the issue of how to design a nose cone for the warhead of ballistic missiles was a difficult one for Atlas engineers. The problem of aerodynamic heating upon reentry was severe, and initial thinking favored a slender, pointed nose. However, the speeds at which an ICBM would reenter the atmosphere made the problem of cooling a slender shape seem insuperable. Aerodynamicist Harvey Julian Allen at the NACA's Ames Aeronautical Laboratory on the southern part of the San Francisco Bay determined that a blunt nose would be a better shape because the bow shock wave it produced would help to dissipate the heat. Allen conceived the theory of blunt-body reentry in 1951, but it did not come to significant Air Force attention until 1956.[46]

Following its usual practice, WDD and its contracting organization awarded contracts on May 25 and June 5, 1956, to General Electric and to AVCO to develop nose cones for Atlas, with AVCO being the backup contractor. The two firms proceeded independently, but available research tools included a supersonic wind tunnel at JPL and a hypersonic wind tunnel at the Naval Ordnance Laboratory at White Oak, Maryland. AVCO's research facilities included shock tubes, and WDD had already contracted on January 24, 1955, with Lockheed Aircraft for an X-17 reentry test rocket. This had two stages to lift a nose cone out of the atmosphere and another to propel it back downward at a speed approximating that of reentry.

Originally, the plan was to use a blunt heat-shield nose cone made from copper that would absorb any heat that the shock wave did not dissipate. But the weight penalty was immense—1,115 pounds for GE's Mark 2 copper nose cone. Nevertheless, the B and C models of Atlas used the Mark 2, the A model not having included a nose cone in its testing. Gradually the Air Force contractors came to the conclusion, already successfully tested by the Army in August 1957, that an ablative nose cone, which reduced heating by vaporization, was better than a heat sink. After the Army test, the Air Force had argued that the results for an IRBM with its much shorter range did not necessarily carry over to an ICBM, which had a heating rate four times as great because of the higher speed of the ICBM nose cone upon reentry. Consequently, the Air Force used a Thor-Able launch vehicle from April 1958 through May 1959 to test an ablative nose cone. The initial mission failed, and during two flights in July 1958 the nose cone could not be recovered. But telemetry indicated that ablation could dissipate the aerodynamic heating from reentry at Atlas speeds (Mach 22, versus Mach 15 for Titan). In mid-1958 GE began full design and production of its Mark 3 ablative nose cone, which was lighter and produced less drag than the Mark 2. The Atlas D converted to the Mark 3, while the Atlas Es and Fs employed AVCO's Mark 4, with a different ablative design using quartz ceramic and metal honeycomb for the front cone, which heated, softened, and flowed backward before vaporizing and dissipating the heat.[47] Thus the Air Force's research did lead to different materials for ablation than the Army used but ended up confirming the general conclusions of the von Braun group about ablation's superiority to the heat-sink technique. This is not in any way to suggest that the Army program was superior to that of the Air Force. In some degree, the Army was lucky to discover ablation through unsuccessful testing of a substance for jet vanes. Rather, what this example does suggest is the sharing of information between services despite interservice rivalry—a sharing that went in both directions.

Testing and Deployment of the Atlas

The first flight test of the Atlas A with just its booster engines occurred on June 11, 1957. The missile lifted off and, after one engine ceased firing, performed several violent maneuvers including a loop. The test did confirm that the gimballing (in which the engines rotated for steering) worked, although the concept had already been proved by Viking and Vanguard. More important, it showed that the steel balloon tanks were not only viable but extremely resilient. The third launch was the first success, on December 17, 1957, but it was one of only three successes out of eight tries for the Atlas A. The Atlas B with a sustainer engine and a separable nose cone had a better record, with five successes against only three failures. It failed on its first launch, on July 19, 1958, before the nose cone could be separated. But on August 2 the missile flew through separation of the boosters, the cutoff of the sustainer, and the firing of the retro-rockets that separated the nose cone.

Atlas C with an improved guidance system and thinner skin had its first launch on December 23, 1958. It was successful, as were two others of the model's six launches. The first launch of the Atlas D, the model that was first deployed, was a failure on April 14, 1959, but of a total of 117 launches, the D model had 90 successes. It incorporated the MA-2 engines, and on May 20, 1960, an Atlas D with an operational Mark 3 nose cone plus some 1,000 pounds of test instruments flew 9,040 statute miles. In the midst of all the failed launches, meanwhile, evaluation of the causes had led to numerous

Figure 29. Erection of an Atlas intercontinental ballistic missile at Cape Canaveral in 1957. Official U.S. Air Force photo, courtesy of the 45 Space Wing History Office, Patrick AFB, Fla.

Figure 30. Erection and mating of an Atlas missile's nose cone on Pad 11 at Cape Canaveral in August 1962. Official U.S. Air Force photo, courtesy of the 45 Space Wing History Office, Patrick AFB, Fla.

modifications to hundreds of the roughly 40,000 parts on the developmental missiles. Atlas E experienced 19 successful launches out of 35, while the F model was 38-for-52, making a total of 158 successful launches for all models against 71 failures, for a success rate of almost 69 percent.[48]

The Atlas D became operational in September 1959, with the first E and F models following in 1961. All three remained operational until 1965, when they were phased out of the missile inventory. Only the F models were housed in silos. Research and development costs for the Atlas missile program came to $2.23 billion, and procurement costs for the entire weapon system were $2.85 billion, for a total of $5.08 billion. This worked out to $14.81 million per missile.[49]

The Thor-Jupiter Controversy

In the midst of Atlas development, President Eisenhower became concerned about recent developments in thermonuclear weapon technology and asked his Scientific Advisory Committee to have a group of eminent scientists evaluate the offensive and defensive capabilities of the United States and the Soviet Union. Formed in the fall of 1954, a Technology Capabilities Panel headed by James R. Killian Jr., president of MIT, issued its report in mid-February 1955. After considering recent intelligence, the committee indicated that the Soviets might threaten the strategic balance early in the next decade unless the United States took extraordinary measures. Many experts doubted that an ICBM could be deployed before 1962, so the committee recommended the development of both sea-based and land-based IRBMs. In November 1955 the Joint Chiefs of Staff recommended in turn that the Air Force develop the land-based version while the Army and Navy collaborate on an IRBM that could be both land- and sea-based. Thus were born the Air Force's Thor and the Army's Jupiter, with the Navy eventually developing the solid-propellant Polaris after initially trying to adapt the liquid-propellant Jupiter to shipboard use.[50]

The JCS recommendation gave rise to the so-called Thor-Jupiter Controversy, which the House of Representatives' Committee on Government Operations called a "case study in interservice rivalry," but which also featured some degree of cooperation. The cooperation occurred not only between the Army and the Navy initially but even between the Army and the Air Force, although they fought to outdo one another and cooperated only to the degree dictated by the needs of national defense and the exigencies of the situation. The principal protagonists in the rivalry and forced cooperation were the "quiet and reserved" Gen. Bernard Schriever for the Air Force and the "forceful and dramatic" Gen. John Bruce Medaris for the Army.[51]

Thor

The Air Force did not wait for endorsement of an IRBM from the Pentagon. Already on May 25, 1955, it assigned WDD the responsibility for developing what it then called a tactical ballistic missile. About this time, Cdr. Robert C. Truax had tired of trying to interest the Navy in ballistic missiles, in which he personally was interested. So he went to see Trevor Gardner, whom he knew from his tour at Aerojet when Gardner had been an executive with the parent company, General Tire and Rubber. Truax arranged with Gardner and the Navy to be assigned to WDD. There, Colonel Terhune assigned him to head a team with Ramo-Wooldridge to develop a plan for the tactical ballistic missile, soon redesignated an IRBM. Working with him was Ramo-Wooldridge's Adolf K. Thiel, who had helped develop the V-2 from the Institute for Practical Mathematics at the Technical Institute of Darmstadt. Thiel had come to the United States with the von Braun group, worked on the Redstone, and left Huntsville in March 1955 to join Ramo-Wooldridge.[52]

Assisting Truax and Thiel were an Air Force major and experts in guidance and structural design from Ramo-Wooldridge. "Our team concluded," Truax wrote, "that the quickest road to a 1500-mile-range capability was to take the AVCO reentry vehicle, one barrel of the Rocketdyne Atlas 'booster' engine, the AC [spark plug] guidance platform, and incorporate these into a new airframe designed from scratch." After selling this concept to the Air Force, Truax moved on to a job in the Ballistic Missile Division working on a reconnaissance satellite, and Col. Edward Hall became head of the Thor program, later succeeded by Lt. Col. Richard K. Jacobson.[53]

On November 8, 1955, Secretary of Defense Charles E. Wilson formally directed the Air Force to proceed with developing IRBM number 1, which became the Thor. He told the Army to develop an alternate IRBM that became the Jupiter, with the Army and Navy collaborating on a sea-launched version of that missile. In an accelerated source-selection process—given further impetus by President Eisenhower's decision on December 1 to assign both the ICBM and IRBM the highest national priority—WDD and its procurement arm chose the Douglas Aircraft Company as the airframe and assembly contractor for the Thor IRBM (XSM-75), with a contract signed just after Christmas. The speed was necessary because the Air Force wanted 120 missiles to be deployed by January 1960, only four years away.[54]

Thor did not have a steel-balloon structure like Atlas's but an aluminum airframe with integral tanks and stringers, plus frames in sections around and between the tanks. The engine section required three thrust beams as well to carry the thrust from the engines—one main and two verniers—to the airframe. Although a fact sheet gives the length of the Thor as 65 feet, the technical manual says the missile was 63 feet long and 8 feet in diameter.[55]

The guidance and control system for Thor, developed by the AC Spark Plug Division of General Motors in Milwaukee, used liquid-floated gyros and the Q-guidance system developed by the Instrumentation Laboratory at

MIT. The 1,500-nautical-mile range of Thor made it feasible in Blasingame's mind to proceed with an all-inertial system, which was more difficult to develop for Atlas's 5,500-mile range. But it still required a long talk with MIT's Draper to convince Blasingame that it would work even for the Thor. The Q-guidance system permitted much of the computation for guidance (the Q-matrix) to be performed long before the missile was fired, leaving only a small amount of calculation to be done by the analog computer on the missile. Battin and Laning at the MIT lab had a difficult time coming up with the Q system. In their final report, they had to create a fourteen-page appendix just to describe their equations.[56]

AC Spark Plug was producing similar inertial systems for the Air Force's Mace and the Navy's Regulus II cruise missiles, so there was a pooling of technological information between tactical and strategic missiles as well as between the two services' projects. The basic system used a three-gyro-stabilized platform that remained fixed in its spatial orientation and provided a reference for all of the movements of the missile around the pitch, roll, and yaw axes. Three pendulous integrating gyroscopes sensed deviations from the planned trajectory and fed this information to an analog computer that calculated the effects and sent corrective signals to the missile's autopilot. The autopilot, in turn, controlled the gimballing of the main and vernier engines to correct the trajectory. The system had to correct for deviations not only in pitch and yaw but also in roll. If the missile rolled, the pitch and yaw axes would be interchanged, causing a reversal of control and resultant guidance-and-control failure. Douglas had built Thor with fins, despite Vanguard's successful launching of a satellite without them in March 1958, but on June 4 of that year Thor 115 launched with an intentionally violent pitch in two directions to simulate the effect of violent winds as it ascended through the atmosphere and to see how the guidance and control system would react. Two of the fins fell off under the heavy aerodynamic loads, but the missile flew a perfect trajectory, showing that fins were unnecessary. Because fins added weight, future production missiles discarded them. The violent pitching also demonstrated the basic integrity of the airframe,[57] apart from the fins.

The Thor guidance and control system showed influences of V-2 technology as well as the more direct involvement of the Instrumentation Laboratory at MIT. Draper, indeed, credited the Peenemünde group with significant contributions to inertial guidance, including stable platforms and integrating accelerometers.[58] But most of the credit belongs to Battin, Laning, Draper himself, and AC spark plug, as well as Blasingame for the Air Force. The division of GM developed the system in eighteen months and successfully put it through laboratory tests, followed by the sled tests at NOTS. Manufacturing it involved about 3,200 subcontractors and required extreme precision. To ensure accuracy, the plant in Milwaukee employed 1,400 engineers and electronics specialists, 70 percent of them women. Female operators examined every part under a 30-power microscope, using dental tools, steel wool, and other equipment to eliminate imperfections such as burrs on metal pieces. The overall system as of 1958 weighed 650 to 700 pounds (75 percent lighter than a design in 1956), using vacuum tubes and magnetic amplifiers instead of transistors because transistors weren't yet considered a fully proven technology.[59]

The main engine for the Thor was essentially one-half of the Atlas booster system already discussed. Because Thor's main engine was the sole power plant except for two small verniers, it had to burn longer than the Atlas boosters. This ultimately necessitated a reduction in thrust from 165,000 pounds (half of the 330,000 delivered by the twin MA-3 boosters) to 150,000 pounds, though this later improved to 170,000 pounds for the operational missile. Rocketdyne developed the early versions of the Thor main engine quite rapidly, and in June 1956 the firm delivered the first experimental engine (only 135,000 pounds of thrust at this point) to Douglas Aircraft for tests of compatibility with the airframe. Static testing of the engine took place at three locations: Rocketdyne's Santa Susana test site, a Douglas facility in Sacramento, and the rocket test site on Leuhman Ridge overlooking the dry-lakebed landing site for experimental aircraft at Edwards Air Force Base in the Mojave Desert.[60]

The growing facility at Edwards came to play an important part in the history of rocketry. After World War II, the Air Materiel Command's Power Plant Laboratory at Wright Field (later Wright-Patterson Air Force Base) was responsible for all propulsion research and development in the Army Air Forces and then the Air Force. The laboratory's Non-Rotating Engine Branch had started a rocket propulsion program in 1939, and by 1947 the branch decided to move the testing and evaluation of rockets to the remote Leuhman Ridge site on Muroc Army Airfield (later Edwards Air Force Base). It awarded a contract in April 1947 to Aerojet Engineering Corporation for construction of technical facilities at the site, with construction beginning in November 1949. By February 26, 1953, test stand 1-5 was ready, and static testing began on the 30,000-pound-thrust Bomarc surface-to-air missile engine, followed in March by testing of a Navaho booster engine on test stand 1-3. Other test stands followed, including 1-A designed for a million pounds of thrust. And tests of the Atlas began on test stand 1-3B, followed by Thor on test stand 1-5 from July to December 1956. These tests on Thor involved extra-heavy tanks and served to train Douglas personnel on the engine. Nearby, on test stand 1-3B, Rocketdyne personnel carried out Thor engine-development tests. The people and facilities who had come to Edwards became the Rocket Branch within the Air Force Flight Test Center in 1950, but their name changed frequently in ensuing years. On July 1, 1959, the former Power Plant Laboratory, now called the Propulsion Laboratory, transferred to Leuhman Ridge. It too underwent several name changes, with its reporting unit changing in the early 1960s from the Flight Test Center to Space Systems Division and then to a division of Systems Command at Bolling Air Force Base in Washington, D.C. The unit itself remained at Leuhman Ridge, however, and was redesignated the Air Force Rocket Propulsion Laboratory on October 25, 1963.[61]

Meanwhile, flight testing for the Thor missile had begun in three phases. Phase 1 tested the compatibility of the airframe, engine, and autopilot. Phase 2 added the all-inertial guidance system, and phase 3 tested the Mark 2 heat-sink reentry vehicle, developed by GE. On January 25, 1957, less than thirteen months after the contract with Douglas was signed, the first Thor test flight ended almost before it began when the missile exploded during liftoff from pad 17B at Cape Canaveral. Investigation revealed the cause to have been contamination of the liquid oxygen fill-and-check valve. The Thor team instituted more careful procedures for handling the volatile cryogenic fluid and waited for the launchpad to be repaired for another try. This second launch, twelve weeks later on April 19, might have been successful if not for a wiring problem in the console of the range safety officer. The faulty wiring made the missile appear to be heading inland instead of out over the ocean, so the officer commanded its destruction.

There followed a malfunction of a main fuel valve on May 21, causing the missile to blow up on the pad. And on August 30 an electrical signal to the mechanism controlling yaw did not work as designed, resulting in violent maneuvers that tore the missile apart some 92 seconds after liftoff. Only on September 20 did the Thor team have a completely successful launch, with the missile reaching its operational speed of 10,000 mph and a range of 1,100 miles. Without the heavy instrumentation needed for test flights (but not operational launches), the missile might have reached its target distance of 1,500 miles.[62]

Failures continued to outnumber full successes, with twelve failures or partial successes against only six full successes out of the first eighteen launches in 1957 and 1958. However, on October 24, 1957, the date of the eighth launch, a Thor missile achieved a distance of 2,700 miles, exceeding specifications. On December 7 and 9 two missiles tested the guidance

Figure 31. Wreckage from a Thor explosion during liftoff on January 25, 1957. Official U.S. Air Force photo, courtesy of the 45 Space Wing History Office, Patrick AFB, Fla.

and control system. The first of these witnessed a failure of the main-engine cutoff signal from the guidance system. This caused it to lose stability when the valves remained open after propellant depletion. The follow-on mission two days later went without a hitch, and a nose-cone test on February 28, 1958, showed that that system worked even though the main engine prematurely ceased functioning. Subsequent nose-cone launches were not uniformly successful, but at the end of the eighteen research-and-development launches, the Air Force decided in September 1958 that Thor was ready for operational deployment.[63]

Before that time, problems with the turbopumps, common to the Atlas, Thor, and Jupiter, and differences of approach to these problems had led to considerable disagreement between the Thor and Jupiter teams. The solutions to the turbopump problems have already been described in conjunction with Atlas development, but the problems themselves first became evident in the testing of Thor and Jupiter. Indeed, they formed a core element in the Thor-Jupiter controversy. There had been friction between the Army and the Air Force even before the turbopump issue simply because of interservice rivalry and the use of a common supplier of engines. Personalities and personal backgrounds heightened the level of disagreement. Thiel, for example, had come to Ramo-Wooldridge from the von Braun group to work on Thor in support of the Air Force. That required that he revisit Huntsville numerous times for coordination because of the "shared engines." He recalled encountering "a lot of hostility." Some people even called him a traitor.[64]

The first inkling of problems with the turbopumps came in mid-1957 when Rocketdyne engineers were tearing down Thor engines for inspection of parts following static flight-rating tests. They found some evidence of "bearing-walking," the dysfunctional movement of bearings within the turbopump. Ballistic Missile Division and Ramo-Wooldridge engineers asked the firm to look into the problem and seek a solution. The next hint came on October 11, when a Thor test flight experienced premature shutdown of its engine 152 seconds into the mission. As this was near the end of the planned engine "burn," it was attributed to a minor problem and not investigated thoroughly. Only later did engineers carefully screen the telemetered data from the mission and discover that it indicated a turbopump problem. In the interim, a Jupiter flight had failed in November 1957. Suspecting a turbopump problem, engineers from the Army Ballistic Missile Agency inserted extra instrumentation in the Jupiter launched in December. This indicated problems in the turbopump gear case. The ABMA engineers tested similar engines in a vacuum chamber and decided that "bearing-walking" at high altitudes was the culprit. At their insistence, Rocketdyne implemented an interim fix involving a bearing retainer, which worked.[65]

Air Force and Army representatives discussed the matter at a meeting at ABMA on February 24, 1958, requested by von Braun to pool information about common equipment. The Air Force still blamed the problems on abnormal bearing loads from flight stresses, not on a design deficiency. The Army, on the other hand, stopped the flight testing of Jupiter missiles to begin retrofitting all propulsion systems with the bearing retainer Rocketdyne had designed for ABMA. Thiel recalled that the Huntsville engineers understood the problem earlier than did his own bosses, who refused to accept the Army's arguments. He said that at a number of meetings with von Braun's engineers, he had to side with his former rather than his present colleagues, making for "a bitter, bitter fight." Once the Jupiters resumed test flights, they experienced no further turbopump problems. Meanwhile, failures of an Atlas and a Thor missile in April 1958 "aroused increased concern that a design deficiency existed in the Thor and Atlas engine and turbopump installations," as Schriever admitted later.[66]

Figure 32. A Thor, an early Intermediate Range Ballistic Missile, being erected in the Missile and Space Gallery at the National Museum of the United States Air Force, Wright-Patterson AFB, Ohio. Courtesy of the museum.

The Thor failure occurred on April 23, with telemetry pointing to the turbopump gearbox as the culprit. Engineers from the Ballistic Missile Division, Space Technology Laboratories, and Rocketdyne again looked at the flight of the previous October and determined that it too had resulted from a gearbox failure. Rocketdyne was already working on a solution for Thor and Atlas as well as Jupiter, but the plan was to install the improved turbopumps on the next batch of engines, not retrofit them as the Army had done. The question for the Thor program was whether to suspend flights and retrofit the existing Thor missiles with improved equipment. One group argued that it would be politically damaging to have another failure due to a turbopump. Another pointed out the urgency of getting the flight results from the scheduled missions so Thor could be deployed on time. Thiel, interestingly, was in the latter group. There had been fourteen launches up to that time, only two of which had failed because of turbopump problems. That statistic yielded presumptive 6-to-1 odds of *not* having a turbopump fail. Consequently, the Thor engineers urged Schriever to continue testing so they could get data on the guidance-and-control and nose-cone systems sooner. Schriever courageously agreed to take the chance of a public-relations disaster. Thor and Atlas failures in August and September were again due to turbopump problems, but the program continued even as these mishaps fueled the Thor-Jupiter controversy further.[67]

With the turbopump redesign by Rocketdyne, however, these were the last of three Thor and three Atlas failures due to pump deficiencies. The first Thor squadron went on operational alert with the Royal Air Force in Great Britain in June 1959, three others followed by April 1960. With the operational readiness of Atlas and Titan ICBMs in 1960, the last Thors could be removed from operational status in 1963, making them available for space launch activities.[68]

Jupiter

While WDD/BMD and its contractors were developing Thor, ABMA and its contractors were busily at work on Jupiter. They went about their business without a clear indication whether the Army or the Air Force would eventually deploy the missile or even whether they would be able to complete its development and production. The Army, and particularly von Braun's group at Redstone Arsenal, had been advocating a longer-range battlefield missile than the Redstone for some time. In the wake of the Technology Capabilities Committee report in early 1955, von Braun exercised his proclivities as a heterogeneous engineer and met with Secretary of Defense Wilson to point out that experience on the Redstone prepared his team for development of a missile with a 1,500-nautical-mile range. Following Wilson's decision to let the Army work with the Navy to develop an IRBM, the land service quickly designated General Medaris as commander of an Army Ballistic Missile Agency with powers somewhat analogous to Schriever's.[69]

In November 1956, Secretary Wilson decided to limit the Army to jurisdiction over missiles with no more than a 200-mile range, for use on a tactical battlefield. Under this ruling, operational control over all IRBMs would go to the Air Force. By April 1957 there were hints that one of the two IRBMs would be discontinued, but following Sputnik, President Eisenhower approved both programs for rapid development. Meanwhile, on December 8, 1956, the Navy had dropped out of the Jupiter program to develop its own solid-propellant Polaris missile, but not before it had had some effect on Jupiter's length and diameter. Originally the Army had planned a missile of 90 feet in length, later reduced to 65 feet. However, to fit it aboard a submarine, the Navy wanted a 55-foot missile. The two services compromised on a length of 58 feet and a diameter of 8.75 feet as necessary to achieve a range of 1,500 nautical miles, whereas the Army's initial plan called for a diameter of 7.9 feet. Ultimately the missile was 60 feet long but still 8.75 feet in diameter.[70]

As Redstone Arsenal began to prepare, late in 1955, for development of the Jupiter (so baptized in April 1956), the Guided Missile Development Division that formed the core of the new ABMA consisted of some 1,600 people, including about 500 scientists and engineers, of whom about 100 had worked on the V-2. Recruiting began to increase that number to a planned 2,349, but with growth in the numbers of ballistic missiles being produced, the Army had to compete with industry and the Air Force for its technical staff. By June 30, 1956, ABMA's quota of civilian positions had risen to 3,301, of which 2,702 had been filled. The military staff at that time was 513 personnel.[71]

As with the Redstone, the Army planned to design, develop, and test prototypes of the Jupiter before turning them over to the prime contractor for production, following the usual practices of the arsenal system. Chrysler remained the prime contractor, as it had been on the Redstone. A key issue early in the development process was which contractor would produce the engine. Although NAA had provided the engine for Redstone, ABMA engineers feared that the Rocketdyne engine being used for Thor and Atlas would have only marginal performance for Jupiter. The Army therefore requested a separate contractor for its engines, but the Ballistic Missile Committee at the Secretary of Defense level turned this down. Eventually ABMA settled on the same basic engine as the Air Force was using, but Medaris and Schriever

Figure 33. Jupiter missile engine. Built by North American Aviation, this was basically the same as the engines used for the Atlas and Thor missiles, and represented much technological advance over previous designs. Courtesy of NASA.

disagreed on whether the Army and its prime contractor had to obtain them through the Air Force. Initially Schriever generously provided ABMA with four of eight engines available at one point in 1956, and in the end he agreed to let Chrysler contract for the engines with Rocketdyne through the Army's Los Angeles Ordnance District. As a result, while Rocketdyne developed the Thor engines from the MB-1 to the more advanced MB-3 configuration used on the operational missiles, it developed the Jupiter engine separately, based upon an original experimental configuration.[72]

The first of these gimballed engines arrived at ABMA in July 1956, and the testing began smoothly. Within a few months, however, four engines had burned through, and the Army engineers requested modifications to strengthen the spaghetti-type regeneratively cooled combustion chamber. When two more engines arrived, in September and November, the ABMA engineers found that they delivered only 135,000 pounds of thrust instead of their rated level of 150,000. These two down-rated engines flew on the first launches in 1957, but ABMA requested engine modifications that Rocketdyne implemented in late 1956 and into 1957. With one of the modified engines, a test showed a thrust of 195,000 pounds without damage to the engine or the turbopump. In its eventual configuration, used on the operational missile, the engine had a thrust rating of 150,000 pounds with a specific impulse of 247.5 lbf-sec/lbm, as compared with 170,000 pounds and 252 lbf-sec/lbm for the operational Thor engine.[73]

Another major difference between Thor and Jupiter involved the vernier engines. The Thor used the same vernier engines that the Atlas employed. These burned the same propellants as the main engines and delivered roughly 2,000 pounds total thrust. By contrast, the Jupiter initially used a hydrogen peroxide propellant for its single vernier but shifted in August 1958 to a solid-propellant motor with about 500 pounds of thrust. Whereas the Thor vernier engines burned while the main engine was operating and beyond, the vernier on the Jupiter was located on a so-called aft unit between the thrust unit and the warhead unit. Following cutoff of the main engine, the Thor's aft unit separated from the thrust unit. Two seconds later, the vernier ignited and propelled the aft unit plus the warhead with ablative nose cone to the desired final velocity, whereupon squibs effected thrust termination through separation of the vernier's nozzle from the motor.[74]

An even more significant difference between the Thor and the Jupiter lay in their guidance and control systems. Although both were inertial, Jupiter's system was a continued development by ABMA engineers from the one used on the Redstone. Ford Instrument Company then built the system.

At ABMA, Fritz Mueller was able to reduce the size of the air-bearing gyros from the Redstone system. He also increased their accuracy. Three of them kept the ST-90 stabilized platform fixed in space as a reference for deviations from the intended trajectory. Before launch, two air-bearing gyros on pendulums sensed the local horizontal and aligned the platform with it, as corrected for Earth's rotation. In flight, three additional pendulous air-bearing gyro-accelerometers sensed accelerations of the missile in the yaw, pitch, and roll planes and sent signals to altitude, range, and cross-range computers, which transmitted corrections not only to the servo devices to gimbal the main engine but also to a turbine exhaust nozzle that provided roll control. The missile launched vertically and was tilted gradually by the system into its ballistic trajectory. The range computer solved an equation that determined the cutoff point of the main engine, followed by the vernier, so that the missile would land as close as possible to its target. Once the upper portion of the missile had passed out of the atmosphere and separated from the thrust unit, eight jet nozzles, operated by bottled nitrogen under pressure, provided spatial attitude control. At 250–300 pounds, this guidance and control system's weight was less than half the 650–700 pounds of

Thor's, but it provided a circular error probable of 0.8 nautical miles compared to Thor's 2.0 or slightly less.[75]

Once the combined aft and warhead units achieved the correct attitude and velocity, two spin rockets built by Thiokol activated to rotate the missile body at 60 rpm, whereupon prima cord explosive detonated, separating the warhead unit from the aft unit. The warhead then followed a ballistic reentry path, protected by its ablative coating. This, of course, differed substantially from the Mark 2 heat-sink nose cone developed by GE for the Thor.[76]

The first actual Jupiter missile—as distinguished from the Jupiter A and Jupiter C, which were really Redstones—launched on March 1, 1957, at Cape Canaveral. Designated missile 1A because it preceded the first scheduled mission, it rose to 48,000 feet. However, after 74 seconds it went out of control and broke apart. Analysis of transmitted data revealed that flame from the exhaust had entered the rear portion of the missile and burned wires in the control system. Jupiter engineers solved this by providing a fiberglass heat shield to the area around the engine.

Jupiter 1B followed on April 26, 1957, and achieved an altitude of 60,000 feet and a flight time of 93 seconds. But then sloshing of the propellants in their tanks created uneven forces on the sides of the missile, sending it so far off course that the control system was powerless to overcome them. This flight was considered partly successful, but obviously the sloshing, revealed by flight instrumentation, had to be fixed. ABMA engineers placed the thrust section of a Jupiter missile on a railroad flatcar and simulated the conditions of flight by moving the vehicle back and forth, testing various types of baffles in the propellant tanks to reduce the sloshing of the liquids. Baffles shaped like truncated cones worked best in the fuel (RP-1) tank, while accordion shapes successfully reduced the motion of the liquid oxygen.[77]

Medaris blamed the sloshing on the diameter of the missile, which had been more slender until the Navy requested a shorter, hence thicker shape. Although the Jupiter was only 9 inches wider than the Thor, there seems to have been no sloshing problem on the Air Force's IRBM, so perhaps that small difference was critical.

In any event, the baffles worked, and the originally scheduled Jupiter missile number 1 launched successfully on May 31, 1957. The missile landed 253 nautical miles short of its estimated 1,400-nautical-mile impact point, but the rocket engine and control system functioned adequately. Since this flight preceded a successful launch of the Thor, it prompted Col. Edward Hall in Schriever's organization to say about the launch, "I think it is a tragic thing for the country!"[78] Hall's comment testifies to the bitterness aroused by the Thor-Jupiter rivalry.

Figure 34. Jupiter A missile, designed and developed by the Army Ballistic Missile Agency and launched at Cape Canaveral on March 1, 1957. Note the lack of fins at the rear. Courtesy of NASA.

In any event, both programs continued their flights. On October 22, 1957, Jupiter missile 3 successfully tested the ablative nose cone and the ST-90 stabilized platform with the latter's circuits partially closed. Turbopump problems caused the failure of missiles 3A and 4 on November 26 and December 18. On May 18, 1958, following retrofitting of modified turbopumps, Jupiter number 5 tested separation of the main power unit of the missile from the "aft" section and nose cone, with the nose cone being recovered from the ocean. On July 17, 1958, Jupiter missile 6B delivered a nose cone 1 nautical mile short and 1.5 nautical miles to the right of the predicted impact point on the first flight test of the complete inertial guidance system. In the total program of 29 research-and-development flight tests of the Jupiter, the Army rated 22 successful, 5 partially successful, and 2 as failures. Of the

29, 19 were tactical prototypes and 16 of them landed within 0.81 nautical miles of the target, the eventual circular error probable of the production missile.[79]

The Army arranged delivery of the first production Jupiter missile to the Air Force in August 1958. In May 1959 the missile achieved initial operational capability, with an Air Force crew successfully launching a Jupiter from Cape Canaveral under simulated tactical conditions on October 20, 1960. The first Jupiters actually became operational in July 1960 at a launch position in Italy, and two squadrons of the missile were fully operational in that country by June 20, 1961. A third squadron in Turkey was not operational until slightly later, with all of the missiles taken out of service in April of 1963 after enough Atlas and Titan ICBMs were operational.[80]

As table 7.1 shows, the Thor and Jupiter IRBMs had roughly comparable characteristics. The Thor had greater average thrust, especially with the two vernier engines included, since these burned throughout the launch and beyond main-engine cutoff, whereas the less powerful vernier motor on the Jupiter added its thrust only after cutoff. This may have contributed to the Thor's becoming a standard first-stage launch vehicle, whereas Jupiter served in that capacity to only a limited degree. Another factor may have been that there were 160 production Thors to only 60 Jupiter missiles. These differing quantities no doubt contributed to the difference in cost between the two missiles.[81]

Conclusions

The Thor, Jupiter, and Atlas vehicles constituted the first strategic missiles in the U.S. arsenal. Developed with comparatively great rapidity to meet a perceived Soviet threat, they necessitated new organizations for missile development in the Air Force and the Army—and, incidentally, the Navy. And they brought in their train a series of new management procedures. They also greatly expanded both the management structures for missile development and the industrial infrastructure available for missile and then launch-vehicle development and production.

Defense necessity became the mother of invention in a variety of fields. These included a missile structure with the steel-balloon tanks of Convair's Charlie Bossart; many innovations in engine development at Rocketdyne (encouraged by the product improvement clause in Air Force contracts); digital computing at Bosch Arma and inertial guidance at Arma, MIT's Instrumentation Laboratory, AC Spark Plug, and ABMA; nose cone development at Cornell University, ABMA, AVCO, and GE; and testing appara-

Table 7.1. Comparison of Thor and Jupiter Missiles

	Thor	Jupiter
Length (feet)	63	60
Diameter (feet)	8	8.75
Launch weight (pounds)	108,804	109,800
Range (nautical miles)		
maximum	1,500	1,500
minimum	300	300
Accuracy/CEP (nautical miles)	1.7	0.8
Guidance	Inertial	Inertial
Reentry vehicle	Mark 2, heat-sink	Mark 3, ablative
Warhead (megatons)	1.44, nuclear	1.44, nuclear
Propulsion	Rocketdyne MB-3	Rocketdyne S-3D
Thrust (pounds)		
main engine	170,000	150,000
vernier	1,000 (each of 2)	500
Speed		
Mach number	15.0	15.1[a]
mph	~10,000	~10,000
Cost per missile system	$6,696,000	$8,469,000

Sources: Lonnquest and Winkler, *To Defend and Deter*, 259, 261, 268; Nicholas and Rossi, *Missile Data Book*, 3–5, 3–7, 3–8, 3-10; T.O. 21-SM75-1, 1-1, UP; Grimwood and Strowd, "Jupiter," 5-1; Standard Missile Characteristics, SM-78 Jupiter (see note 73 to chapter 7); "Thor Fact Sheet," NASM; North American Rockwell, "Jupiter Propulsion System," NHRC.

a. From Nicholas and Rossi. Lonnquest and Rossi give a figure of Mach 15.4 on p. 261.

tuses ranging from shock tubes to rocket sleds and railroad cars. Rocketdyne showed that a willingness to take risks could pay off in engine improvements that would continue into the future. But some of those very improvements caused problems in flight testing, which remained the ultimate arbiter of whether systems engineering on a complicated rocket had been successful or not.

Through both ground and flight tests, engineers discovered the limitations inherent in their use of theory and accumulated knowledge for the development of rockets. Combustion instability, burn-throughs, turbopump failures, propellant sloshing, and other problems revealed the uncertainties that remained a part of rocket engineering. Cut-and-try methods continued to be necessary to overcome unexpected results. Ingenuity played an important role in ascertaining the exact nature of problems and then solving them.

One other feature in the development of Thor, Jupiter, and Atlas was cooperation among sometimes competing agencies. While much has rightly been made of the intense interservice rivalry between the Army and the Air Force in the Thor-Jupiter controversy, even those two programs cooperated to a considerable extent. They exchanged more information than the heat of the animosity between the two services would have suggested. Although Medaris complained to Congress about the lack of information he received from the Air Force, and although Schriever's Air Force organization was not altogether forthcoming in complying with the Army's requests, Schriever claimed that his Ballistic Missile Division had transmitted to ABMA, at its request, a total of 4,476 documents between 1954 and February 1959. By his count, BMD had withheld only 28 documents for a variety of reasons including contractors' proprietary information.[82]

There also were meetings between ABMA and BMD engineers and managers. In the cases of ablative nose cones and turbopump failures, the Army engineers reached early solutions that were later adopted by the Air Force (if only in its own way). But both Jupiter and Thor reached operational status with remarkable speed, in significant part because Jupiter was preceded by the V-2, Navaho, and Redstone, while Thor borrowed technology originally designed for the Navaho and Atlas. But a number of instances showed how interagency cooperation was becoming an increasingly important feature of rocket development. Examples included a Naval Ordnance Test Station rocket sled testing Air Force and Army inertial guidance systems, Langley Research Center's assistance for Air Force inertial guidance, and borrowings of technology from Navaho, Mace, and Regulus II. Interservice and interagency rivalry helped to encourage competing engineers to excel. But without the sharing of information and technology, rocketry might have advanced much less quickly than it did. In retrospect, it appears that this technology transfer was more significant than the competition between services. Without the sharing of information, in this case especially between the Air Force and the Army, neither service would have been as successful as in fact it was.

This would become even more evident in the solid-propellant breakthrough that resulted in the Polaris and Minuteman missiles (see chapter 9). But it was also a factor in the use of missiles for space launch vehicles. Furthermore, there were instances of technology transfer from space launch vehicles to missiles, as the Viking and Vanguard's pioneering use of gimballing would illustrate.[83] Jupiter, Thor, and Atlas marked a huge step forward in the maturation of U.S. rocketry, but they could not have advanced so far without the many pioneering developments, from a variety of sources, that preceded them.

8

The Titan I and II, 1955–1966

Development of the Titan family of missiles and launch vehicles proceeded in building-block fashion to a greater degree than most other rocket families. This was not intended in the beginning. But although Titan II used very different (storable) propellants than Titan I, its engines bore the same basic designations as those for the earlier missile. Thus engineers built upon the original design for at least the propulsion system of the Titan II, which they developed further for Titans III and IV.[1] In addition, Titans III and IV added upper stages using the same propellants as the Titan II engines, plus strap-on solid rocket motors derived from those for Polaris and Minuteman. This was not altogether different from what happened with the Atlas series of launch vehicles or the Thor, which became the core first stage for the Delta family of launch vehicles. But the building-block concept was more a central feature of the Titans than of the Atlas- and Thor-derived space launchers.

Titan I began essentially as insurance for Atlas in case the earlier missile's technology proved unworkable. Accordingly, the major new feature of the first of the Titans was the capability to start a large second-stage engine at high altitude.[2] The WAC Corporal had proved the viability of the basic technology involved, and Vanguard would develop it further after Titan I was started. But in 1955, to use a full second stage on a ballistic missile, and to ignite it only after the first-stage engines had exhausted their propellants, still appeared to pose substantial risk.

Titan II became the first (and last) large ballistic missile to use storable hypergolic propellants, and the capability such propellants offered for rapid launch became important for the Titan III launch-vehicle concept. Titan III's proponents foresaw a military need for launching some payloads into space on short notice, which was not possible with cryogenic propellants. Huge strap-on boosters added significantly to the thrust Titans III and IV could generate, hence to their capabilities to launch large satellites into high geosynchronous orbits and to propel spacecraft into trajectories that would allow them to escape Earth's gravity. NASA's space shuttles essentially borrowed the strap-on concept and technology from the Air Force's successful

Figure 35. Left to right, the Jupiter Intermediate Range Ballistic Missile and the Titan II and Titan I Intercontinental Ballistic Missiles in the Missile and Space Gallery of the National Museum of the United States Air Force, Wright-Patterson AFB, Ohio. Courtesy of the museum.

Titan III, and the Titans themselves continued to operate as launch vehicles into the twenty-first century.

Although most of the technologies used in the Titan rockets were not radically new, problems arose during the development and testing phases that required empirical solutions. The growing body of knowledge about the operations of missiles and launch vehicles—what was called rocket science in popular parlance—still did not permit designs guaranteed to withstand the rigors of testing on the ground and in the air. Engineers had to rely on trial and error as well as their own knowledge to overcome problems that designers could not, or at least did not, foresee.

Titan I

While the principal motivation for developing Titan I was to serve as a backup for Atlas,[3] there were other factors involved. One was the possibility of producing a better missile. Another was the perceived desirability of competition. In October 1954, Schriever wrote that he thought it "wise to sponsor an alternate configuration and staging approach [to those of Atlas] with a second source. . . . It is possible that such an approach might provide a . . . substantially superior" design. It might also stimulate Convair to do better work on the Atlas, Schriever thought. Meanwhile, Ramo-Wooldridge had been studying various missile designs through contracts with Convair, the Glenn L. Martin Company, and Lockheed. The last of these contractors had recommended a two-stage missile, and Louis Dunn of Ramo-Wooldridge had forwarded the suggestion to von Neumann's Intercontinental Ballistic Missile Scientific Advisory Committee, an offshoot of the Strategic Missile Evaluation Committee that von Neumann had also chaired. The ICBM committee agreed that it was a good idea, and Schriever requested permission to develop it. His Air Force superiors approved such a program on May 2, 1955.[4]

Meanwhile, the Western Development Division had awarded a contract on January 14, 1955, to Aerojet to develop engines burning liquid oxygen and a hydrocarbon fuel for possible use on Atlas. These soon evolved into engines for Titan. On May 6 the Air Materiel Command invited five contractors to submit proposals for the structure of the alternate ICBM, and on October 27 a letter contract authorized Martin to design, test, and develop the airframe for a two-stage missile the Air Force designated XSM-68 (later named Titan I after the powerful sons of Uranus and Gaea in Greek mythology). The contract also called for Martin to plan the overall development of the missile, although Ramo-Wooldridge remained responsible for systems engineering and technical direction, overseeing Martin's planning and development.[5]

Titan I Engines

Aerojet developed a pair of identical stage-1 engines and a smaller second-stage engine for Titan I, all burning liquid oxygen and RP-1. The gimbal-mounted stage-1 engines bore the initial designation LR87-AJ-1, while the stage 2 engine, also gimballed, was called LR91-AJ-1. All three were regeneratively and film cooled, with turbopumps to feed the propellants. Connecting the pumps for the fuel and oxidizer to the gas turbine was a gear train. Initially the Aerojet engineers planned to use regenerative cooling for both the LR91 and its large nozzle skirt, required by the larger expansion ratio the second-stage engine needed for operation at higher altitudes. This arrangement made the cooling jacket too large for proper phasing of propellants into the combustion chamber, so Aerojet adopted an ablative skirt for stage 2, reducing the area to be cooled regeneratively. Engineers first tried high-silica fibers, but asbestos-reinforced phenolics provided more satisfactory ablative cooling.[6]

Although the basic designs of the engines held up well under testing, there were many problems that required comparatively minor adjustments. One involved the arrangements for starting the stage-1 engines. The original design used tanks outside the missile to feed RP-1 and gaseous oxygen to a gas generator where their burning would initiate the delivery of propellants from tanks in stage 1 to the combustion chambers. This system was cumbersome, requiring an excessive number of supply lines to the missile, and the oxygen ducts proved difficult to purge. So Aerojet engineers introduced an internal gaseous nitrogen feeder system to provide pressure on the propellants, which were bled from lines below the turbopumps and fed into the gas generator. There, once ignited, they produced gas to operate the turbopumps, feeding the propellants into the thrust chambers under high pressure (587 psi).[7]

For stage 2, pressurized gaseous helium from a rubber-lined fiberglass bottle inside the missile initiated operation of an auxiliary turbopump that forced the propellants into a gas generator to operate the main turbopump. The first stage-2 engines used in flight tests beginning on February 2, 1960, used this system, but for the LR91-AJ-3 engines that began flying in May 1960, a titanium bottle for the helium replaced the fiberglass version, which leaked. Another change between versions -1 and -3 of the stage-2 engine was occasioned by coke deposits on the turbopump assembly stemming from fuel-rich gases in the gas generator. During tests, the deposits caused a 20 percent loss in the effective area of the turbine nozzles. Engineers first tried modifying the turbine manifold so it forced the particles of coke into baffles or traps, but this failed to restore proper functioning. They then tried a turbulence-generating device called a swirler at the entrance to the manifold. This mixed the gases effectively, increasing combustion efficiency, hence exhaust velocity. In the process, by lowering the percentage of raw fuel from which the deposits had formed, it solved the coking problem.[8]

One other example illustrates what can happen with technology transfer. It occurred after Titan II engine development had begun in 1960, so presumably it involved a version -3 rather than a -1 engine. The problem lay with a gearbox for the first-stage engine (presumably for the turbopump, but the source does not specify). In ground tests for flight qualification, the component failed two hours into a twelve-hour test. This was a major issue, since no further flight testing could occur until the gearbox was modified to eliminate the cause of failure, then installed in all of the affected engines, roughly one hundred of which had already been delivered. Aerojet's president, Dan Kimball, called Chandler Ross, vice president for engineering, and assigned him to review the troubleshooting process. Engineers had spent about two weeks checking out the gearbox and trying to discover what went wrong. Additional testing had brought recurrence of the problem, which involved self-destruction of the pinion gear (a small wheel with few teeth that meshed with a larger gear or gears). There was no obvious cause, and the materials all satisfied the specifications. By now about half the engineers in Aerojet's Liquid Rocket Plant were wrestling with the issue.

Ross says he was "ready to give up when we stumbled on the answer." Aerojet had subcontracted the gearboxes from Western Gear Corporation. The subcontractor was behind in its deliveries because it had so little assembly space in a clean-room environment, which was required for the component. So Aerojet arranged for parts to be shipped to Sacramento, where the rocket firm's employees began assembling the gearboxes. Faced with the failures, Ross asked Western Gear's chief engineer to describe his company's assembly process in detail. This man mentioned treatment of the gears with light oil and an abrasive. Aerojet's assemblers, unaware of this step, had omitted it. All of the failures, it turned out, involved boxes assembled in Sacramento. Since only a few units had been put together there, it was relatively easy to install properly treated gearboxes in those engines.[9] This apparently solved the problem. It also illustrated the care that was needed with rocket parts and subassemblies, the attention to small details that rocket engineering demanded, and the ease with which technologies and procedures transferred from one firm to another could inadvertently be changed.

These sorts of "minor design changes" were numerous enough to warrant redesignating both the LR87 and LR91 engines to LR87-AJ-3 and LR91-AJ-3. In the process of solving the problems, there had been some 6,900 Titan I engine tests, 5,440 tests of the thrust-chamber assembly, 3,930 of the turbopump assembly, and 9,900 others, for a total of more than 26,000. This was close to twice the number of such tests that would later be needed on the Titan II engines, not surprising given "the strong similarity of this later design."

The two LR87-AJ-3 first-stage engines together yielded 300,000 pounds of thrust at sea level (344,400 in a vacuum) with a sea-level specific impulse of roughly 250 lbf-sec/lbm (about 290 in a vacuum). The LR91-AJ-3 second-stage engine yielded 80,000 pounds of thrust in the vacuum of space, where it operated. Its vacuum specific impulse was about 310 lbf-sec/lbm using the same propellants as stage 1. No doubt because of the follow-on contracts Aerojet won for engines on Titans II, III, and IV, Ross pronounced the January 14, 1955, contract for what turned out to be Titan I engines "the most important in the company's history."[10]

Titan I Airframe

The Martin Company constructed Titan I's airframe from an alloy of aluminum and copper called 2014 aluminum. Though high in strength, the alloy was considered by the American Welding Society not to be weldable. But technicians at Martin's Baltimore plant had developed a process called tungsten inert gas welding that worked, and they trained the workforce at the new Denver plant, where Titans were tested and assembled, in the intricacies of the process. Martin also arranged with the University of Denver to carry out a comprehensive research program testing the material at high and low temperatures to determine their effects on welds. In the construction process, the welders joined twelve longitudinal panels to form a cylinder for each propellant tank. They then welded domes onto each cylinder to create a self-supporting integral-tank structure that, unlike Atlas's steel balloons, needed no internal pressure for strength but followed the essential design principles of Viking and Vanguard. The stage-1 tanks did need internal frames, while the stage-2 tanks, being shorter, required no additional stiffening. Where structural loads were less great, the tanks were lightened by chemical milling, using sodium hydroxide to remove thickness from areas not protected by an asphaltlike substance. The parts of the airframe on either end of the tanks consisted of 2014 aluminum supported by stringers and frames.[11]

Titan I Guidance and Control System

On October 18, 1955, the Air Force contracted with the Western Electric Company for its Bell Telephone Laboratories to develop a radio-inertial guidance system for the Titan I. The system that Bell Labs developed had evolved from a shipboard radar used in World War II. The inertial portion of the system included a platform oriented in space by two gyroscopes. This provided inputs to an autopilot that signaled the pitch-over maneuver and provided roll commands during the initial part of the missile's flight, when the first stage's exhaust inhibited communications between a ground antenna and one on the Titan I. The missile also apparently used three strapped-down rate sensors (integrating gyroscopes) and an analog computer that helped stabilize the vehicle by sending control signals to the gimballed engines.[12]

Once the missile had reached a position where the radio of the launch complex could communicate with the vehicle unimpeded by the rockets' blast, commands from the ground guidance system modified signals from the inertial guidance system. The system on the ground consisted of two 20-foot-tall antennas, a huge digital computer built by Remington Rand Univac, and a radio link to the receiver/transmitter on the missile. When the Titan I took off from a launch complex, one of the two antennas would rise out of its protective silo and track the missile, remaining exposed to potential enemy countermeasures for some twenty-five minutes as it did so. (The other antenna was a backup.) Univac's Athena missile guidance computer included ten cabinets plus two remotely located motor generator sets. The program for each trajectory that could be flown by a given Titan I was on a 10-inch reel of Mylar tape. At the launch complex, the guidance officer could take one of the reels from a safe and feed its contents into the drum memory of the Athena computer. This would guide the missile once it was loaded with propellants, raised from its silo, and launched. Each computer controlled three missiles. The computer itself was located in the control center, where it was protected against nuclear blasts up to overpressures of 100 psi. There were ten tapes with different target data, and a new one could be read into the memory in roughly a quarter hour if needed.[13]

After the missile launched and either of the first-stage propellants was depleted, liquid-level sensors in the tanks signaled cutoff of the two engines. Following a short period of tail-off for the thrust, explosive bolts separated the second stage from the first. Small solid rockets on the second stage ignited, propelling the second stage forward and inertially positioning the propellants in the tanks for engine start, initiated by a staging timer. Since the second stage had only one engine, its gimbals could provide control in only pitch and yaw, with roll control coming from four vernier thrusters. The guidance system commanded thrust termination of the second-stage engine just before the missile achieved terminal velocity, with the verniers providing final adjustments of the trajectory. When the vehicle reached terminal velocity, two ordnance devices released the reentry vehicle from stage 2, and two small retro-rockets backed the spent stage away from the nose cone.[14]

The ground equipment for the radio-inertial guidance system included vacuum tubes as well as transistors. There were few problems with the transistorized portions of the computer itself, but the vacuum tubes in equipment such as the radar receiver often went bad in multiples, requiring isolation of the faulty devices so they could be replaced. Sometimes such troubleshooting took days of effort by a guidance technician flown by helicopter to the launch complex. Thus, although the Titan I guidance and control system was accurate to within 0.80 nautical miles, it was not without its problems.[15]

Titan I Reentry Vehicle

The development of the Titan I missile occurred late enough in the sequence of ICBM and IRBM projects at Western Development Division that it could benefit from the work both AVCO and General Electric had done on reentry vehicle design. The contractor for the Titan I reentry vehicle was AVCO, and that company built several different versions, first for testing and then for the production model. The RVX-3 was 72 percent of the size of the production model and was tested in the third lot of flight tests. The RVX-4 was supposed to be a full-sized test article, but in March 1959 the Lawrence Livermore Laboratory increased the diameter of the warhead for Titan I, making the RVX-4 a 94-percent-scale version. The final, production reentry vehicle for Titan I was the Mark 4, also used on the Atlas E and F. It had ablative material called Avcoite bonded to a metal structure beneath it for protection against reentry heating.[16]

Titan I Flight Testing and Deployment

The flight testing of Titan I missiles occurred in a series of letter-designated lots. Lot A, consisting of six simplified but live first stages with dummy second stages, produced four successful launches at Cape Canaveral between February 6 and May 4, 1959. The first two lot-A missiles were damaged by contractor personnel during handling and had to be returned to the Martin plant in Denver. Apparently they never were launched. The A-6 missile launch included a successful separation test of the dummy second stage, which was filled with water. These four launches without a failure seemed to show that the first stage was capable of performing its mission. They also demonstrated successful cutoff of the first-stage engine and satisfactory operation of the pitch programmer. The launches provided data suggesting that the missile's aerodynamic drag was lower than expected and that measures taken to protect against aerodynamic heating and vibration had been successful.[17]

Unfortunately, the lot-B firings were not so successful. Involving complete first and second stages but with the duration of the second-stage burn reduced, they were intended to show the compatibility of the airframe, propulsion units, and flight controls during the boost phase of the launch as well as to demonstrate the missile's staging and the overall effectiveness of the flight

Figure 36. First Titan I launching from Pad 15 at Cape Canaveral on February 6, 1959. Official U.S. Air Force photo, courtesy of the 45 Space Wing History Office, Patrick AFB, Fla.

controls. Both stages were powered by prototype engines, XLR87-AJ-1 and XLR91-AJ-1. Before missile B-4 even got to Cape Canaveral, a faulty liquid oxygen pump caused it to explode during static tests at Denver in May 1959. This was followed in July by a test crew cracking the casing of the second stage of the C-1 missile, also in Denver. On July 22, a test crew repeated this performance on a first-stage engine. Two days later, tests on Titan B-6 had to be halted for a faulty fuel pump, followed on July 31 by fuel problems that forced postponement of the scheduled launch of Titan B-5. Soon thereafter, the second stages of missiles B-7 and C-2 also suffered damage.

Culminating this series of mishaps was the attempted launch of Titan B-5 on August 14, 1959. The first-stage engine ignited as expected, but the hold-down bolts, intended to keep the missile on the launchpad until there was sufficient thrust for liftoff, released prematurely. The Titan I rose but did not have enough thrust to produce stable flight. It ascended about 10–12 feet, pulling an umbilical lanyard with it, whereupon the control system automatically shut off the engines. The vehicle fell back to the launching pad, emitting flame and smoke. The ensuing explosion caused heavy damage to the service tower.[18]

Despite Martin's experience with the Matador tactical missile and the Viking sounding rocket, as early as March 1957 the Western Development Division's Titan project manager, Col. Benjamin P. Blasingame, had expressed doubts about the qualifications of some of the personnel assigned to the Titan project. By July 1958 the firm had recruited additional engineers and generally been responsive to Air Force concerns. But the record of test failures in 1959, before and after the August 14 explosion, resurrected Air Force misgivings. In late October, Martin's chairman of the board, George M. Bunker, and company vice president William B. Bergen met repeatedly with Schriever, by this time commander of Air Research and Development Command; Maj. Gen. Osmond J. Ritland, who succeeded Schriever as commander of the Ballistic Missile Division; and Maj. Gen. Ben I. Funk, who commanded the procurement organization associated with BMD. The immediate consequence of these meetings was the assignment of Bergen to take over the Martin operations in Denver.[19]

This arrangement did not last long. On December 12, 1959, the first of the lot-C missiles was on the launching pad at Cape Canaveral. Similar to lot B in that it included complete first and second stages, with stage 2 designed to burn only a limited time, lot C also sought to demonstrate the compatibility of all the major subsystems, including separation of a scaled-down reentry vehicle. Designated C-3, the missile that was launched on December 12 achieved none of those goals. It did demonstrate full thrust, with the hold-down bolts releasing properly. But as the test vehicle started to rise, it exploded. As the *New York Times* reported, the blast "ripped the Titan apart in a mushrooming ball of orange flame and clouds of black smoke. Burning debris was hurled for hundreds of feet into the air." Investigation revealed that "chatter" in an electrical relay, caused by shocks from the liftoff plus the explosions that released the hold-down bolts, had set off the safety

Figure 37. Titan I missile exploding during launch at Cape Canaveral on August 14, 1959. Launch failures were common during this early period of missile and launch vehicle development. Official U.S. Air Force photo, courtesy of the 45 Space Wing History Office, Patrick AFB, Fla.

mechanism designed to destroy the missile upon command from the control tower.[20]

To prevent this obviously unforeseen sequence from recurring, Martin engineers redesigned the command destruction system. They relocated "all destruct relays and initiators . . . from the booster and sustainer engine compartments to the between-the-tank sections of Titan Missile B-7A and subsequent flight missiles." Here, they believed, the effects of the launch and explosions of hold-down bolts would be less severe. Engineers from Space Technology Laboratories were preparing on Christmas Eve to examine this redesign for their certification,[21] giving some indication of the urgency in the development of Titan I, even if it started as a backup to Atlas.

Meanwhile, Ritland had ordered a complete investigation and technical review of the Titan program. The resultant report, finished in December with its essential findings communicated to Martin managers on January 5, 1960, urged consolidation of all the company's Titan I operations under a single manager and implementation of improved procedures. Bunker did not dispute the findings and personally took over control of the Denver plant. He believed this move would reassure the Air Force that the firm was taking action to fix the problem. At the same time, it made all the employees in the Titan organization aware that the problem was important to the boss and had better be important to them as well. Bunker also communicated to at least one of his managers another point that illustrates his integrity and bona fides. This manager had briefed Bunker on a solution to an unspecified Titan problem that he believed was good for Titan but injurious to Martin's bottom line. Bunker told him always to do what was good for the program because, in the long run, that would also be good for the firm.[22]

Bunker's decision to take over Denver operations apparently did reassure the Air Force. On January 7 Schriever wrote to Gen. Thomas D. White, Air Force chief of staff, that he and Ritland believed the Titan program was "fundamentally sound from a technical viewpoint," a view that was "confirmed by an independent analysis conducted by an Ad Hoc Group of the Scientific Advisory Committee" headed by Clark Millikan of Caltech. Schriever added that "the Martin Company has made a good start towards solving its management problems."[23]

This assessment seemed to be confirmed on February 2, 1960, when composite missile B-7A (consisting of the first stage of B-7 and the second stage of B-6) launched successfully, flew 2,026 nautical miles, and demonstrated ignition of the second stage at altitude plus effective operation of the guidance and control system. This mission included the first use of the roll programmer to achieve the planned flight azimuth. The flight evidently also showed that Martin's fix of the problem on the C-3 launch was effective, and it satisfied all of the objectives of the lot-B flight program in one flight.[24]

While one of the two lot-B missiles was successful in its flight test, only one of five lot-C missiles was successful—all of them flying with prototype (version -1) engines. On February 5, 1960, missile C-4 launched successfully and pitched over on command, but after 51 seconds of flight the reentry vehicle unintentionally separated from stage 2. The remainder of the Titan I pitched and yawed violently, then exploded. The cause of the malfunctions was pressure on the guidance-compartment skin resulting from two deficiencies. One of these was due to an incorrect assumption that the internal pressure in the compartment would be equal to the external (ambient) pressure. Compounding this error, "improper structural verification testing techniques" had led to a structure for the guidance bay that was insufficient to withstand the design loads. Martin engineers quickly strengthened the guidance compartment structure before the C-1 flight on March 8, 1960.[25]

On that date, the missile completed stage-1 flight successfully, followed by the separation of stage 2. The second- stage engine then failed to ignite because of a malfunction in a gas-generator valve. Exactly a month later, on April 8, missile C-5 flew successfully through both stage-1 and stage-2 separation and ignition. Some 43 seconds into stage-2 operation, however, its liquid-oxygen pump or its drive-gear assembly failed, resulting in an explosion in the combustion chamber. Finally, on April 28, 1960, missile C-6 flew a fully successful mission, traveling 3,244 miles downrange and landing in the ocean some 3 nautical miles short and 0.8 nautical miles to the right of the aim point. In addition to redesign of the destruction system in response to the explosion of missile C-3 and the strengthening of the guidance compartment structure following the failure of C-4, the lot-C problems had led to reinforcement of the transition section between stage 2 and the reentry body plus a redesign of the gas generator and turbine pump assemblies.[26]

Meanwhile, the lot-G missiles had begun their flight tests. These were missiles with proto-operational subsystems, including first and second stages capable of producing 300,000 and 80,000 pounds of thrust respectively. G-4 launched on February 24, 1960, and completed a successful flight of nearly 5,000 nautical miles, including separation of the Mark 4 nose cone. Five other lot-G missiles flew successful missions between March 22 and June 24, with G-8 being the only partial success in the bunch on September 28, 1960. Intended to fly 8,700 nautical miles, it reached only about 6,000 because of early cutoff of the stage-1 engine. All of the other lot-G missiles had shown good accuracy, and by June 1960 the program had met the basic objectives of the initial test program.[27]

There were still many research and development tests. Twenty-two lot-J flight tests included six failures or partial successes. These were tests using the final versions LR87-AJ-3 and LR91-AJ-3 production engines, and they brought several minor corrections of problems such as a hydraulic power failure and a tendency of the second-stage engines to cut off prematurely or fail to ignite. Titan VS-1 with an operational stage 1 (of Titan I) and a dummy stage 2 was modified for launch from a silo in preparation for Titan II. On May 3, 1961, it had a successful launch at Vandenberg Air Force Base. The lot-M missiles were like the ones from lot J except that they employed inertial guidance systems for tests that also supported Titan II. These occurred between June of 1961 and the end of January 1962, with two failures and five successes. Overall, among 57 purely research-and-development launches of Titan I missiles (including 7 operational missiles launched from Vandenberg), 38 were successes, while 19 failed or were only partly successful, for a success rate of roughly 67 percent.[28]

The Air Force deployed six squadrons of Titan I's, at nine per squadron, in the course of 1962. Each squadron had three launch control centers, with at least 19 nautical miles between them and three missiles clustered around each center. The initial operational capability of the missile occurred in April 1962. All the missiles were deactivated by February 1965 once Minuteman I and Titan II missiles were in place, giving the Titan I's an active life of less than three years.[29] Their principal role, it seems, was to provide a basis for the development of the longer-lived Titan II, although Titan I did serve as an interim source of strategic defense, especially until later in 1962 when the first Minuteman I missiles began to be deployed.[30]

Titan II

The history of the transition from Titan I to Titan II is complicated. One major factor stimulating the change was the quarter hour it took to raise Titan I from its subterranean silo, load the propellants, and launch it. A second was the difficulty inside a missile silo of handling the extremely cold liquid oxygen used in Titan I. The twin problems could have been solved by conversion to solid propellants like those used in Polaris and Minuteman, but another possibility was storable propellants. Under a Navy contract in 1951, Aerojet had begun studying hydrazine as a rocket propellant, while many firms and government laboratories had been involved in developing unsymmetrical dimethyl hydrazine (UDMH). This was a candidate as a storable fuel, but it had a lower specific impulse than the Titan needed. Hydrazine had better performance but could detonate if used as a regenerative coolant. Aerojet was the first firm to come up with a good compromise: an equal mixture of hydrazine and UDMH, which it called Aerozine 50. This combination ignited hypergolically with nitrogen tetroxide. Neither was cryogenic. And both could be stored in tanks for extended periods, offering a much quicker response time than Titan I. Aerojet urged a switch to the new propellants, but it was apparently Robert Demaret, chief designer of the Titan, and others from Martin who successfully proposed the idea to BMD in early 1958.[31]

The quicker response time afforded by storable propellants reduced the missile's vulnerability to enemy attack, but if the Titan could actually be launched from its silo, as was already planned for Minuteman, rather than being brought to ground level before firing, this would further lower its vulnerability. At least as early as November 1958, Schriever became aware of studies in Great Britain for in-silo launch of the British Blue Streak IRBM. Col. William E. Leonhard, BMD's deputy commander for installations, went to England and reviewed the British design. Apparently his impression of its prospects was favorable, because on January 19, 1959, Schriever approved in-silo launch for future Titan facilities, leading to the test launch of Titan I from a silo at Vandenberg in 1961.[32]

Another vulnerability concern was the clustering of three Titan I's around each launch control center (LCC). An enemy could knock out nine missiles by destroying three LCCs. It made much more sense to have an LCC for each missile, forcing the enemy to destroy them one at a time.[33]

A final point of vulnerability was the radio-inertial guidance system, with antennas that were exposed to attack when deployed and were subject to jamming. An all-inertial system would not have either weakness.

In July 1958 the Air Force commissioned Martin to study these issues as well as cost. Along with presenting solutions already mentioned, the study revealed the need for a new reentry design, as uneven ablation on the Mark 4 AVCO warhead was causing inaccuracies in its flight as it approached its target. Mindful of these considerations, in November 1959 the Department of Defense authorized the Air Force to develop Titan II, limiting deployment of Titan I to six squadrons. Titan II (the new designations becoming official in April 1960) would use storable propellants, in-silo launch, and an all-inertial guidance system.[34]

On April 30, 1960, the BMD development plan for Titan II called for a length of 103 feet (compared to 97.4 feet for Titan I), a uniform diameter of

10 feet (Titan I's second stage being only 8 feet across), and greater thrust than its predecessor. Titan I had 300,000 pounds of thrust in stage 1, while Titan II was to have 430,000, with stage 2's thrust climbing from 80,000 to 100,000 pounds. This higher performance would increase the range with the Mark 4 reentry vehicle from about 5,500 nautical miles for Titan I to 8,400, while with the new Mark 6 reentry vehicle, which had about twice the weight and more than twice the yield of the Mark 4, the range would remain at 5,500 nautical miles. Because of the larger nuclear warhead it could carry, the Titan II served a different and complementary function to Minuteman I's in the strategy of the Air Force, persuading Congress to fund them both. It was a credible counterforce weapon, whereas Minuteman I served primarily as a countercity missile, offering deterrence instead of the ability to destroy enemy weapons in silos.[35]

In May 1960 the Air Force signed a letter contract with the Martin Company to develop, produce, and test the Titan II. General Electric won the contract to design the Mark 6 reentry vehicle. And in April 1959, AC Spark Plug had contracted to build an inertial guidance system for a Titan missile, although it was not clear then that this would be the Titan II.[36]

Titan II Propulsion System

On October 9, 1959, Aerojet had already won approval to convert the Titan I engines to burn storable propellants. Research and development to that end began in January 1960. The Aerojet engineers also worked to achieve the improved performance called for in the April 30, 1960, plan. Although the Titan II's twin stage-1 engines (called XLR87-AJ-5) and sole stage-2 engine (XLR91-AJ-5) were based on those for Titan I, the new propellants and the requirements of the April 30 plan necessitated considerable redesign. Because the new designs did not always work as anticipated, the engineers had to resort to cut-and-try methods to find combinations that worked correctly and provided the necessary performance. Since the propellants were hypergolic, the Titan II engines needed no igniter. The injectors for the Titan I engines had used alternating fuel and oxidizer passages, with oxidizer impinging on oxidizer and fuel on fuel (called like-on-like), to effect mixing of the two propellants. For the Titan II, Aerojet engineers tried fuel-on-oxidizer impingement. This evidently mixed the droplets of propellants better, and higher performance resulted. But so did erosion of the injector face, necessitating a return to like-on-like.

This older arrangement caused combustion instability in the stage-2 engines, and engineers tried several kinds of baffles before they came up with one that worked. One candidate, an uncooled stainless-steel baffle, did not last through a full-duration engine test. Copper baffles with both propellants running through them for cooling also failed, as the nitrogen tetroxide corroded the copper. An eight-bladed configuration cooled by the oxidizer, and evidently using another type of metal, yielded poor performance. The final configuration had six blades radiating outward from a central hub like a wagon wheel, again cooled by the oxidizer. This solved the problem, at least at for the time being.[37]

The turbopumps for Titan I and Titan II were similar, but differences in the densities of the propellants led to greater power and lower shaft speed for the Titan II pumps. As this increased the propellant flow rates for the Titan II, Aerojet engineers had to make the turbopump gears wider so they could withstand greater "tooth pressures." From two blades in the inducer for Titan I, the design went to three blades. Engineers also had to redesign the impeller and housing passages to accept the higher flow rates and find new materials that were not degraded by the storable propellants.[38]

In a significant innovation, Titan II engineers replaced the pressurized gases that initiated propellant flow (nitrogen in stage 1, helium in stage 2) with an autogenous, or self-generating, system. Solid-propellant start cartridges initiated the process by spinning the turbines, whereupon gas generators kept the turbines spinning to pressurize the fuel tanks in both stages and to pump the Aerozine 50 into the thrust chamber. The second-stage oxidizer tank did not require pressurization because acceleration was sufficient to keep the nitrogen tetroxide flowing. And in the first-stage propulsion system, oxidizer from the pump discharge served to pressurize the tank. The result was a simplified system saving the weight of the pressurized-gas storage tanks in Titan I and requiring no potentially unreliable pressure regulators. A similar increase in reliability and saving in weight came from using the exhaust stream from the turbopump in stage 2 to provide roll control in place of Titan I's auxiliary power-drive assembly for the vernier thrusters. Gimbals continued to provide pitch, roll, and yaw control in the first stage plus pitch and yaw control in the second stage.[39]

The Titan II propulsion system had significantly fewer parts than its Titan I predecessor. The number of active control components dropped from 125 to 30, while valves and regulators declined from 91 to 16. The engines had higher thrust and higher performance, as planned. The Titan II first-stage engines had a combined thrust of 430,000 pounds at sea level, compared to 300,000 for Titan I. The second-stage thrust rose from 80,000 pounds for Titan I to 100,850 pounds for Titan II. Specific impulses rose less dramatically, from slightly above 250 to almost 260 lbf-sec/lbm at sea level for stage 1 and remained slightly over 310 at altitude (vacuum) for stage 2.[40]

A 1965 Aerojet news release on Titan II propulsion credited it to "the efforts of hundreds of men and women" but singled out four of them as leaders of the effort. Their biographies illustrate in three of the cases the way that engineers in the aerospace universe migrated from one firm to another or from government work to the private sector, carrying their knowledge of various technologies with them. Robert B. Young was the overall manager of the design, development, and production effort for both Titan I and Titan II. He was a chemical engineering graduate of Caltech, where Theodore von Kármán had encouraged him to devote his knowledge to rocketry. In the midst of the Titan effort, he had worked for a year as director of industrial liaison on the Saturn program at NASA's Marshall Space Flight Center, and he had risen within Aerojet's own structure from a project engineer to a vice president and manager of the Sacramento plant. Another leader was Ray C. Stiff Jr., who had discovered aniline as a hypergolic propellant while working under Robert Truax at the Navy Engineering Experiment Station in Annapolis. His role in the use of self-igniting propellants in JATOs to assist a PBY into the air while carrying heavy loads proved a forerunner of "the storable, self-igniting propellants used to such advantage in the Titan II engine systems." Stiff was Aerojet's manager of Liquid Rocket Operations near Sacramento and had also become a vice president.

A third manager Aerojet mentioned in the release was A. L. Feldman, a rocket engineer educated at Cornell University. While working at Convair, he had been "in on the initial design and development effort for the Atlas engines." After transferring to Aerojet, he served as manager for both the Titan I and Titan II engines and became assistant manager of Aerojet's Liquid Rocket Operations. A fourth key manager was L. D. "Lou" Wilson, who earned a degree in engineering at Kansas State. "At 28, he managed the painstaking design, development, test and production of the 'space start' second-stage engine for Titan I." Wilson then managed the entire propulsion system for the Titan II.[41] Alone among the four managers, he appears not to have worked for another firm or government rocket effort.

Titan II Guidance and Control System

As on Thor, AC Spark Plug developed a guidance and control system for Titan II from a design of Draper's MIT Instrumentation Laboratory. Engineers placed the system in stage 2 between the two propellant tanks. A gyroscopically stabilized inertial platform provided accurate measurements of the missile's attitude, while three gyros sent signals on vehicle accelerations to a digital computer designed and built by IBM under subcontract to AC Spark Plug. The computer processed the signals from the sensors and sent the resultant information to the autopilot that directed operation of the engine gimbals on stages 1 and 2 plus the verniers on stage 2.[42]

The Titan II inertial guidance-and-control system yielded a CEP of roughly 0.8 nautical miles, comparable to the accuracy of Titan I with its radio-inertial system. Like the Titan I system, it was not without its problems. During two successful research-and-development flights on February 6 and April 27, 1963, air bubbles appeared in the flotation fluid of the gyros. A short-term solution was to remove the gas from the fluid and improve gasket seals. By May 15, 1965, a longer-term solution was to add an enclosure equipped with bellows to maintain atmospheric pressure around the gyros.[43]

Titan II Reentry Vehicle

General Electric's Mark 6 reentry vehicle for Titan II used a nose cap and heat shield that both employed ablative materials. The nose cap consisted of a nylon cloth soaked with phenolic resin. Technicians cut it into half-inch squares and molded them to the nose-cone shape under pressure. The heat shield consisted of a GE plastic molded from an epoxy and other ingredients and heated in an oven. Technicians bonded two conical sections of this plastic plus the nose cap to the aluminum reentry airframe using neoprene rubber to insulate the aluminum. As the nose cap heated up from reentry, it sequentially formed porous layers of char 1–2 millimeters thick. Aerodynamic forces caused the char to slough off, whereupon a new layer formed and peeled off, creating pyrolic gases that helped to inhibit heat transfer from the boundary layer to the ablative surface. In this way the heat shield protected the 6,200-pound nuclear warhead, which had the largest yield of any used with a strategic missile. The entire Mark 6 weighed 8,380 pounds and was equipped with decoys as well as the warhead itself.[44]

Titan II Airframe

Since engineers had to design Titan II to launch from a silo, Space Technology Laboratories contracted with Aerojet in Azusa, California, for the latter firm to construct a one-sixth-scale model of the anticipated silo for the missile. Test firings of a Titan II model began on June 6, 1959, with thirty-six "launches" by the time of the Titan I full-scale silo launch on May 3, 1961. These tests provided data on acoustic, thermal, and aerodynamic effects in the silo, making it clear that the Titan I airframe would require modification to launch from a silo. Equipped with the data on the acoustical environment in the scale-model silo plus the full-scale test of Titan VS-1, Martin engineers used sirens to simulate the noise level so they could evaluate the airframe structure for Titan II. They learned that they had to increase the thickness of the tank skin or add internal ring frames to withstand the acoustical loads.[45]

Martin engineers and technicians used essentially the same materials for Titan II as for Titan I and even employed the same welding fixtures and jigs. But in place of the twelve extruded panels that formed the propellant tank walls on Titan I, they welded only four mechanically machined and integrally stiffened panels to form the 10-foot-diameter cylinder. This cut the welding task by roughly two-thirds. Chemical milling again lowered the weight of the airframe, this time by 900–1,000 pounds, but less of this technique was necessary because of an increased amount of machine milling. The skin thickness and number of ring frames were greater for Titan II than its predecessor to accommodate both in-silo launch and increased loads from denser propellants. On Titan I, longerons bolted onto the fuel tank had resulted in leaking, so welding replaced bolting on Titan II. Instead of using solid-propellant staging rockets to propel stage 2 onward following stage-1 separation, as on Titan I, Titan II featured ignition of the second-stage engine while the two stages were still attached, with the exhaust escaping through vent holes. A layer of ablative material on the forward tank dome of stage 1 prevented explosion from the heat. Explosive bolts again separated the stages. The acceleration from the stage-2 engine kept the propellants against the turbopump inlets through inertia for continuous feed to the combustion chamber.[46]

Martin engineers had designed the skirts—the sections of airframe that connected the propellant-tank domes—to sustain loads 1.25 times as heavy as expected in flight. But ground tests, in which stage 2 was loaded with lead shot and hydraulic jacks were used to simulate conditions of flight, suggested the possibility of collapse under some circumstances. When test engineers made this discovery, the first nine airframes were already built. On these, aluminum "belly bands" were riveted in six problem areas, while subsequent research-and-development and production missiles featured internal strengthening. Then, during early deployment of Titan II missiles in silos, nitrogen tetroxide began leaking through holes in oxidizer-tank welds too small to be detected by the original quality-control procedures. When the chemical mixed with water vapor, it created nitric acid, whose corrosive properties expanded the holes still further. This problem caused 17 of 60 missiles in the production cycle to be recalled, examined, and rewelded where necessary. It also required much more sensitive test equipment, which used helium gas to pressurize the tanks and highly sensitive mass spectrometers to detect any leaks. Under this new regimen, quality control required rewelding in only three Titan IIs constructed after October 1963.[47]

Flight Testing and Implications for Project Gemini

Between March 16, 1962, and April 9, 1963, there were 33 research-and-development test flights of Titan II missiles, 23 of them from Cape Canaveral and 10 from Vandenberg. Depending on who was counting, there were variously 8, 9, or 10 failures or partial successes (none of which occurred during the last 13 flights), for a success rate of anywhere from 70 to 76 percent. The problems varied widely. On the otherwise-successful initial Titan II silo launch, from Vandenberg on February 16, 1963, electrical umbilicals failed to disconnect properly, pulling missile guidance cabling with them and causing an uncontrollable roll. Other launches suffered premature engine shutdown, an oxidizer leak, a fuel-valve failure, a leak in a fuel pump, and gas-generator failure. But the most serious problem was also the most baffling. On the launch of the first Titan II on March 16, 1962, some 90 seconds after liftoff from Cape Canaveral, longitudinal oscillations began occurring in the first-stage combustion chambers. They occurred about eleven times per second for about 30 seconds. They did not prevent the missile, designated N-2, from traveling 5,000 nautical miles and impacting in the target area, but they were disquieting.[48]

The reason for concern was that in late 1961 NASA had reached an agreement with the Department of Defense to acquire Titan IIs for launching astronauts into space as part of what soon became Project Gemini. Its mission, as a follow-on to Project Mercury and a predecessor to Project Apollo's Moon flights, was to determine whether one spacecraft could rendezvous and dock with another, whether astronauts could work outside a spacecraft in the near weightlessness of space, and what physiological effects humans would experience during extended spaceflight. The longitudinal oscillations—called "pogo" because they resembled the motions of a then-popular plaything, the pogo stick—were not a particularly significant problem for the missile, but for an astronaut already experiencing acceleration of about 2.5 Gs from the launch vehicle, they well might be. The pogo effect on the first Titan II launch added another ±2.5 Gs, which could perhaps render the pilot of the spacecraft incapable of responding to an emergency.

Fixing the problem for Project Gemini, however, was complicated by a reorganization that had occurred within the Air Force on April 1, 1961, in which Air Force Systems Command (AFSC) replaced the Air Research and Development Command. On the same date, within AFSC, the Ballistic Mis-

Figure 38. First Titan II missile launching from Cape Canaveral's Pad 16 on March 16, 1962. Official U.S. Air Force photo, courtesy of the 45 Space Wing History Office, Patrick AFB, Fla.

sile Division split into a Ballistic Systems Division (BSD) that would retain responsibility for ballistic missiles (and would soon move to Norton Air Force Base east of Los Angeles near San Bernardino) and a Space Systems Division (SSD) that moved to El Segundo, much closer to Los Angeles, and was given responsibility for military space systems and boosters. SSD became NASA's agent for developing the Titan II as a launch vehicle, along with Atlas-Agena, which would provide the other half of Project Gemini. Gemini involved the launching of an Agena upper stage with which a manned spacecraft would rendezvous and dock. The problem for NASA lay in the fact that while BSD was intent on developing Titan II as a missile, its SSD sister organization would be responsible for simultaneously adapting Titan II as a launch vehicle for the astronauts.[49] A conflict between the two Air Force interests would soon develop and be adjudicated by General Schriever, the AFSC commander.

Meanwhile, there were further organizational complications for Project Gemini. SSD assigned development of the Gemini-Titan II launch vehicle to the Martin plant in Baltimore, whereas the Denver plant was working on the Titan II missile. Assisting SSD in managing its responsibilities was the new nonprofit Aerospace Corporation, also located in El Segundo. This nonprofit R&D center had come into existence on June 4, 1960, to answer concerns over the Air Force's contracting with Space Technology Laboratories for systems engineering and technical direction while STL was part of Thompson Ramo Wooldridge. STL continued operations for programs then in existence, but many of its personnel transferred to the Aerospace Corporation for systems engineering and technical direction of new programs. For Gemini, Aerospace assigned James A. Marsh as manager of its efforts to develop the Titan II as a launch vehicle. BSD established its own committee to investigate the pogo oscillations, headed by Abner Rasumoff of STL. For the missile, it found a solution in higher pressure for the stage-1 fuel tank, which reduced the oscillations and resultant gravitational forces by half on the fourth launch on July 25, 1962, without the engineers' understanding why it did so. Martin engineers, on the other hand, correctly thought the problem might lie in pressure oscillations in the propellant feed lines. They suggested installation of standpipes to suppress surges in the oxidizer lines of future test missiles. NASA's Manned Spacecraft Center, which was managing Gemini, and BSD agreed with this solution.[50]

Although standpipes later proved to be part of the solution to the pogo problem, initially they seemed to make the oscillations worse. Installed on missile N-11, which flew on December 6, 1962, the modification failed to suppress severe oscillations that raised the gravitational effect from pogo alone to ±5 Gs. This lowered the chamber pressure to the point that instrumentation shut down the first-stage engine prematurely. The following mission, N-13 on December 19, did not include the standpipe but did have increased pressure in the fuel tank, which had seemed to be effective against the pogo effect on earlier flights. Another new feature was aluminum oxidizer feed lines in place of steel ones used previously. For reasons not fully understood, the pogo level dropped and the flight was successful in other ways. Missile N-15, evidently with the same configuration as N-13, launched on January 10, 1963. Although problems with the gas generator in stage 2 severely restricted the vehicle's range, the pogo level fell to a new low, ±0.6 Gs. This was still not low enough for NASA, which wanted it down to ±0.25 Gs, but it satisfied BSD, which "froze" the missile's design with regard to oscillation reduction. Higher pressure in the first-stage fuel tanks plus use of aluminum oxidizer lines had lowered the pogo effect below specifications for the missile, and BSD felt it could not afford the risks and costs of further experimentation to achieve the level NASA wanted.[51]

NASA essentially appealed to Schriever, whose command included BSD. NASA's Brainerd Holmes, deputy associate administrator for manned space flight, contended that since no one understood what caused the pogo oscillations or the unstable combustion that also afflicted Titan II, the missile could not be "man-rated" as a launch vehicle. The result, on April 1, 1963, was the formation of a coordinating committee to address both problems, headed by BSD's Titan II director and including people from both the Aerospace Corporation and STL. Engineers from Aerojet, Martin, STL, and Aerospace were studying the problem. Eventually Sheldon Rubin of Aerospace looked at data from static tests and concluded that during pumping of the fuel, a partial vacuum formed in the fuel lines, causing resonance. This explained why putting standpipes in the oxidizer lines had, by itself, failed to suppress the pogo effect.

The solution was to restore the standpipe, keep the aluminum oxidizer feed lines and the increase in fuel tank pressure, and add a fuel surge chamber (also called a piston accumulator) to the fuel lines. Nitrogen gas pressurized the standpipe after the nitrogen tetroxide had filled the oxidizer feed lines. An entrapped gas bubble at the end of the standpipe absorbed pressure pulsations in the oxidizer lines. The surge chamber included a spring-loaded piston. Installed perpendicular to the fuel feedline, it operated like the standpipe to absorb pressure pulses. Finally, on November 1, 1963, missile N-25 carried both of these devices. The successful flight recorded pogo levels of only ±0.11 Gs, well below NASA's maximum of ±0.25. Tests on December 12, 1963, and January 15, 1964, likewise included the suppression devices and met NASA standards, the January 15 mission doing so even with lower pressures in the fuel tank. This seemed to confirm that the two devices had fixed the pogo problem, whose mystery Rubin had solved.[52]

Meanwhile, the combustion instability Brainerd Holmes had mentioned along with the pogo problem turned out to be another, and totally separate, major issue. It never occurred in flight, but it appeared in a severe form during static testing of second-stage engines. Several engines experienced such "hard starts" that the combustion chambers fell from the injector domes as if they'd been cut away with a laser beam. Engineers examining the test data concluded that combustion instability at a frequency of 25,000 cycles per second had sliced through the upper combustion chamber near the face of the injector with a force of ±200 psi. This occurred on only 2 percent of the ground tests of second-stage engines, but for Gemini, which required "man-rating," even this was too high.

Aerojet instituted the Gemini Stability Improvement Program (GEMSIP) in September 1963 to tackle the combustion instability. Apparently the problem was occurring only when second-stage engines were tested in an Aerojet facility that simulated the air pressure at 70,000 feet. There was no combustion instability in the first-stage engines at this point in their development because air pressure at sea level slowed the flow of propellants through the injectors. Aerojet engineers could solve the second-stage problem by filling the regenerative cooling tubes that constituted the wall of the combustion chamber with a very expensive fluid, which provided resistance to the rapid flow of propellants similar to that provided by air pressure at sea level. However, they and the Air Force finally agreed on a more satisfactory solution. This involved a change in injector design with fewer but larger orifices and a modification of the baffles that Aerojet thought had ended the combustion instability. In place of the six-bladed baffles radiating from a hub in the Titan II missile, Aerojet engineers increased the number of blades to seven and eliminated the hub for the Gemini configuration. Testing in the altitude chamber at the Air Force's Arnold Engineering Development Center in Tennessee proved that the new arrangement worked.

The GEMSIP program cost the Air Force about $13 million and lasted eighteen months. It cost NASA another $1.45 million to implement the changes on the last six Gemini launch vehicles. Ironically, NASA flew the first six Gemini missions with the older injectors used on the missiles, in seeming contradiction of the agency's arguments about man-rating that had inaugurated GEMSIP in the first place. In fact, there was a good reason to accept the missile engines and avoid delays. Early in GEMSIP, Aerojet engineers found that keeping the solid-propellant start cartridges for the stage-2 engines above a critical temperature effectively reduced the combustion instability. This persuaded NASA to fly the first six Gemini missions with the missile engines, although only the injectors produced under GEMSIP provided the dynamic stability NASA really wanted for human spaceflight.[53]

A final issue that arose in flight testing and required reengineering involved the gas generator in the stage-2 engine. This was a matter of concern not just to SSD and NASA for Project Gemini but to the Titan II Program Office at BSD. Engineers and managers first noticed a problem on the second Titan II test flight, on June 7, 1962, when telemetry showed that the second- stage engine had achieved only half of its normal thrust soon after engine start. When the tracking system lost its signal from the vehicle, the range safety officer sent a manual signal to cut off the fuel flow, causing the reentry vehicle to splash down well short of its target area. There was inadequate telemetry data from this flight for the engineers to diagnose the problem, and it took two more occurrences of gas-generator problems to provide enough data to understand what was happening.

It appeared that particles were partially clogging the small openings in the gas generator's injectors, reducing propellant flow and thus thrust. Technicians very thoroughly removed all foreign matter from generator components in a clean room before assembly, separated the generators from the engines in transport to Cape Canaveral, and subjected the assemblies to blowdown by nitrogen before each test flight. These measures did not solve the problem but did narrow the list of sources for the particles. It became apparent that there was no problem with stage-1 gas generators because sea-level air pressure did not allow particles from the solid-propellant start cartridges to reach the injector plates. For the stage-2 system, Aerojet had conducted successful ground tests in its altitude chamber, which could simulate 70,000 feet of altitude. Engineers assumed this was high enough, but it turned out that even at 70,000 feet there was enough atmosphere to cushion the injectors from the particles. At 250,000 feet, where the stage-2 engine ignited, the atmosphere was so thin that particles from the start cartridge flowed into the gas generator, sometimes in sufficient quantity to cause clogging. To rectify the difficulty, engineers added a rupture disk to the exhaust of the gas generator. This kept enough pressure in the system to cushion the flow of particles until ignition, when the disk ruptured. Gas generator failure was not a problem after implementation of this design change.[54]

Table 8.1. Comparison of Titan I and Titan II

	Titan I	Titan II
Length (feet)	98	104
Diameter (feet)		
stage 1	10	10
stage 2	8	10
Weight, fueled (pounds)	220,000	330,000
Range (nautical miles)	5,500	5,500
Thrust (pounds)		
stage 1 (sea level)	300,000	430,000
stage 2 (altitude)	80,000	100,850
Speed (Mach number)	21.0	21.0
Guidance	radio-inertial	inertial
Reentry vehicle (ablative)	AVCO Mark 4	GE Mark 6
Number produced		
research & development	47	33
production	108	108
Average cost	$21,484,000	$17,087,000

Sources: Stumpf, *Titan II*, 36, 49, 281; Nicholas and Rossi, *Missile Data Book*, 3–7, 3–10; Lonnquest and Winkler, *To Defend and Deter*, 230–31; Gordon et al., *Aerojet*, 4, 8, 20, 22.

Titan II Deployment

Even before the conclusion of the last thirteen fully successful research-and-development flights for the missile, some of which were launched from Vandenberg, the Air Force was confident enough in the reliability of the Titan II to declare it fully operational on the final day of calendar year 1963, with initial operational capability dating from June of that year. Between October and December 1963, the Strategic Air Command deployed six squadrons of nine Titan IIs apiece. They remained a part of the strategic defense of the United States until deactivated between 1984 and 1987. By that time, fleet ballistic missiles and smaller land-based solid-propellant ballistic missiles could deliver (admittedly smaller) warheads much more accurately. Also, the hardening of the Titan II silos at 300 psi provided much less protection against a nuclear blast than did Minuteman III silos, hardened at 2,000 psi. The decommissioning of Titan II left the nation's largest ballistic missiles available for refurbishment as space launch vehicles.[55]

Table 8.1 sets forth the principal characteristics of Titan I and Titan II.[56]

Gemini Launch Vehicles

The Atlas launch vehicle not only would be used as a booster for the Agena upper stage during the rendezvous-and-docking missions of Project Gemini; it had also been the primary launch vehicle for most of the Mercury flights.

In this role, the Atlas was a precursor for development of the Titan II as a launch vehicle for Gemini and needs to be discussed here. During Project Mercury, Atlas obviously had to be man-rated in similar fashion to the Redstone. Atlas's counterpart to Joachim Kuettner, who had headed the organization responsible for reliability and astronaut safety in Mercury-Redstone, was another former German test pilot, Bernard A. Hohmann, who had been the project engineer on two versions of the Messerschmitt Me-163, an early rocket-powered aircraft. By 1959, Hohmann was working for STL. Under his overall technical direction, the detailed design and development of Atlas's emergency detection equipment, called the Abort Sensing and Implementation System, or ASIS, was the responsibility of General Dynamics' Astronautics group under Philip E. Culbertson, who would later work for NASA. Engineers identified six functions requiring dual sensing for ASIS, including the pressure in the liquid-oxygen tank, changes in the vehicle's attitude rates for all three axes (pitch, roll, and yaw), and primary electrical power. Out-of-parameter indications in any of these areas or failure of any system would automatically initiate separation of the Mercury capsule from the launch vehicle. Although the astronaut, test conductor, flight director, or range safety officer could do this manually as well, automatic initiation was the primary mode.

Besides ASIS, there had to be numerous modifications of Atlas to convert it to the Mercury-Atlas configuration. The Mercury-Atlas vehicle was about 20 feet taller than the Atlas D from which it derived. This required moving the rate gyros for the autopilot higher on the vehicle so they would sense more precisely the movement of the nose. Also, the Mercury capsule's separation rockets might damage the thin "steel balloon" skin of Atlas's liquid-oxygen tank, so General Dynamics engineers had to add a fiberglass layer covering its dome. Because Mercury-Atlas would not require operation of the vernier rockets for attitude correction after cutoff of the sustainer engine, this part of the vernier operation was deleted, reducing the weight and also the level of complexity of the control system. Such changes, plus increased quality control, caused the Mercury-Atlas launch vehicle to cost 40 percent more than the Atlas missile.[57]

The first flight of a Mercury-Atlas launch vehicle, labeled MA-1, occurred on July 29, 1960, at Cape Canaveral. Its goals were to test the adequacy of the unoccupied Mercury capsule's structure for reentry into the atmosphere after an abort and to evaluate the abort-sensing instrumentation, although this was not connected to the capsule and could not actually initiate separation. The day was overcast, so no observation or filming of the flight was possible once MA-1 entered the clouds. For most of a minute after launch, the Atlas appeared to be rising steadily, but suddenly telemetry from the rocket itself (though not the capsule) ceased. Under the circumstances, the range safety officer terminated the flight without any of its objectives being met. While the cause of the failure was unclear, it occurred at the time of maximum dynamic pressure on the vehicle, and telemetry from the capsule indicated violent motions after the launch vehicle's transmissions ceased, suggesting that the Atlas either exploded or experienced structural failure.[58]

Study of the flight data from MA-1, and debate over why it failed and how to fix it, delayed Project Mercury about six months. During that time, engineers improved the structure linking the booster and spacecraft, which they presumed was the problem. In December 1960 a special technical committee including representatives from all organizations involved looked at several recent failures of Atlas vehicles supporting NASA, including MA-1. The committee concluded that for space launch missions, Atlas's 0.1-inch-thick stainless steel skin needed to be doubled in thickness in its upper sections to support a spacecraft, which evidently imposed heavier loads than did the reentry vehicle. Since the modification could not be completed in time for Mercury-Atlas 2, the committee decided, over some opposition from Aerospace Corporation and STL members, to install an 8-inch-wide steel corset called a "belly band" or "horse collar" to reinforce the area between the top of the Atlas and the spacecraft. Other areas were also strengthened. MA-2 had the same goals as MA-1, except that the ASIS was operational. The mission, flown on February 21, 1961, was a success, showing that the "horse collar" had been strong enough to tolerate the bending forces operating on the vehicle as it pitched over and moved through the region of maximum dynamic pressure. It also showed that the capsule could withstand reentry.[59]

Coming shortly after Soviet cosmonaut Yuri Gagarin's orbital flight, Mercury-Atlas 3 with the redesigned, thicker skin was slated only to carry a simulated astronaut that could "breathe" in the capsule as it attempted to orbit the Earth—a modest goal compared with the Soviet feat. Nevertheless, when MA-3 lifted off on April 25, 1961, it failed to execute a roll and pitch-over to follow its intended trajectory. The ASIS activated the rockets to separate the capsule before the range safety officer destroyed the Atlas, with the spacecraft coasting upward, deploying its parachutes, and landing in the Atlantic, where the Navy recovered it for use on the next mission after refurbishment. This was Project Mercury's last major flight failure. Apparently a voltage surge involving the autopilot caused the problem, so General Dynamics modified the control system to prevent a recurrence.[60]

Mercury-Atlas 4, with the second "thick-skin" launch vehicle, tested this modification and generally validated the redesign. Launched on September

Figure 39. Mercury-Atlas 3 launch vehicle being unloaded from an aircraft at Cape Canaveral on April 23, 1961. Courtesy of NASA.

13, 1961, MA-4 went into orbit with another simulated astronaut in the refurbished spacecraft, which then separated and splashed down in the Atlantic, demonstrating that the Atlas and its systems were working adequately to boost a human into orbit. Meanwhile, the Soviets had already recovered cosmonaut Gherman S. Titov from an impressive seventeen-orbit flight, showing that they were still ahead in the cold-war space race.[61]

Before it would allow an astronaut to be launched on a Mercury-Atlas, however, as with Mercury-Redstone, NASA opted to have a chimpanzee check out the system. On November 29, 1961, Mercury-Atlas 5 launched from Cape Canaveral with chimp Enos in the spacecraft. Despite some minor problems that caused the spacecraft to be returned from orbit after circling the globe only twice instead of three times, Enos and his capsule splashed down and were recovered without incident. An astronaut aboard probably could have corrected for the problems and continued on a third orbit. In any event, from February 20, 1962, to May 15, 1963, four astronauts—John Glenn, Scott Carpenter, Walter "Wally" Schirra, and Gordon Cooper—completed successful orbital missions, boosted by Mercury-Atlases 6 through 9. They experienced no harmful effects from weightlessness, and Cooper completed 22 orbits before manually firing the retro-rockets to reenter the

Figure 40. Launch of MA-6, *Friendship 7*, on February 20, 1962. Boosted by the Mercury-Atlas vehicle, a modified Atlas missile, *Friendship 7* carried astronaut John H. Glenn into orbit on the first U.S. manned orbital flight. Courtesy of NASA.

atmosphere and land off the Japanese coast. Both the Atlas launch vehicle, after its modifications, and Project Mercury had been successful. But the United States still trailed the USSR, which had orbited two cosmonauts simultaneously in mid-August 1962, with Andrian G. Nikolayev completing 64 orbits and Pavel R. Popovich completing 48 and landing within six minutes of Nikolayev.[62]

For Gemini, Atlas did not require all of the modifications needed in Project Mercury, since it was not required to carry astronauts and was already being developed as a launch vehicle. Although Agena was also being used for other programs, it did require considerable modification for Gemini to serve as a rendezvous and docking, or target, vehicle. The necessary adaptations included tracking equipment, an engine that would start five times (instead of twice for the standard Agena), improved controls, command and communications systems compatible with those on the Gemini spacecraft and the ground control network, greater ability to maneuver and stabilize in space, and a unit for the actual docking.[63] But such developments concerned its role as a spacecraft, not as an upper stage for a launch vehicle, and properly belong in a history of Gemini rather than in this book.

To use Titan II for launching the Gemini astronauts, Martin and Aerojet engineers had to do more than fix the pogo, the combustion instability, and the gas generator problem. Procedures developed for Project Mercury strongly influenced those used for Gemini, especially as many NASA and Aerospace Corporation people who had worked on Mercury also worked on Gemini. Such matters as quality control over parts and critical components as well as configuration management carried over from the earlier project. Indeed, Gemini saw an expansion of such practices and a successful application of Mercury's pilot-safety programs. Engineers responsible for the Gemini version of Titan II observed 12 of the final 13 successful flights of the Titan II research-and-development series for the missile, noted minor weaknesses such as abnormally slow buildup of thrust, and corrected them. This created enough confidence that the unmanned Gemini 1 flight launched on April 8, 1964, the day before the final Titan II test flight. As George E. Mueller, NASA's associate administrator for manned spaceflight from 1963 to 1969, stated in February 1964, the 28 launches of Titan II missiles to that date afforded "invaluable launch operations experience and actual space flight test data directly applicable to the Gemini launch vehicle which would be unobtainable otherwise." This comment speaks volumes about the success of information sharing between the Air Force and NASA, a point not always noted in space-flight literature.[64]

For man-rating the Titan II, perhaps the most significant modification was the addition of the malfunction detection system (MDS). This was comparable to ASIS on the Mercury-Atlas except that ASIS automatically initiated an abort once it sensed a serious malfunction, whereas MDS alerted the astronauts and allowed them to decide whether to continue or abort the mission. This change resulted from the experience of Project Mercury, including simulation studies, which indicated that an astronaut could effectively serve as a pilot in space, not just a passenger. But the simulation studies also mandated some exceptions to this practice. For instance, if the stage-1 engines gimballed to their extreme position at the point of maximum dynamic pressure during the launch, the studies showed that structural failure of the vehicle would ensue within a single second, too quickly for an astronaut to detect the malfunction on the instrument panel and respond appropriately. In that situation, MDS automatically switched to a secondary flight control system in 15 milliseconds. High turning rates and low hydraulic pressure also automatically triggered a switch to the backup flight control system, which was another important feature of man-rating and included a backup hydraulic system. MDS too was endowed with complete redundancy. During flight, it monitored pressures in propellant tanks and thrust chambers, stage separation, voltages in the electrical system, and turning rates. An additional prelaunch MDS made sure the vehicle did not launch if the autogenous propellant tank pressurization system was not functioning properly.[65]

Not only was there a redundant flight control system to provide an extra measure of safety not present on the missile, but there was a change to a radio-intertial guidance system for primary control of the mission during the second-stage burn only. The inertial system in the spacecraft provided attitude reference during stage-1 operations. Adapted from Titan I and flight-proven on that missile was the three-axis reference system (TARS) containing three strapped-down integrating rate gyros located in the equipment bay between the stage-2 propellant tanks. The three rate gyros from the Titan II, located in the interstage area, existed in duplicate for Gemini and measured angular accelerations during the stage-1 burn, sending data to an autopilot that also existed in duplicate for backup and sent signals to the hydraulic system controlling the gimbals on the two engines. Engineers set the pitch-over data for each Gemini mission in the TARS unit's programmer before prelaunch countdown, but they could continually update the roll program until launch as necessary for rendezvous with the Agena spacecraft.[66]

For the stage-2 guidance and control system to keep the second stage and spacecraft on course until they reached the proper injection point and orbital velocity, General Electric in Syracuse, New York, provided the radio guidance system, and the Burroughs Corporation's computer operation in Paoli, Pennsylvania, furnished the ground computers that corrected the flight path according to the guidance equations for each mission. In the process, it corrected any trajectory errors that had occurred during stage-

1 flight because of the lower precision of that stage's inertial flight control system. One reason for adopting radio-inertial guidance was that it could command larger corrections in the yaw (lateral) axis than the purely inertial system, to accommodate last-minute changes based on new information about the location of the Agena target vehicle for rendezvous and docking. Another was that the onboard portion of the radio-inertial system was 200 pounds lighter than the fully inertial system it replaced. The equipment on the ground included a tracking radar system that gathered Doppler information and transmitted it to the Burroughs computer. This then sent corrections to a receiver on stage 2 that transmitted them to the autopilot. It controlled changes in pitch and yaw via gimballing. As on the missile, a nozzle ducting gas-generator exhaust could be swiveled for roll control. Ground commands directed stage-2 engine cutoff at the appropriate time. In all, the system contained fifteen gyros to provide the desired redundancy, with the inertial guidance system for the Gemini spacecraft serving as the backup for the radio-inertial system.

Although there were problems with some components during ground testing, the flight control system functioned satisfactorily, if not perfectly, on all twelve Gemini flights. The backup systems were available but never had to be used. Small problems, such as the expected but minimal sloshing of propellants in stages 1 and 2, never interfered with carrying out the mission.[67]

Other features added to the Titan II missile for astronaut safety included redundant components of the electrical systems, and a forward "skirt" assembly to mate the spacecraft with the launch vehicle. To help compensate for all the additional weight, engineers deleted vernier and retro-rockets, which were not necessary for the Gemini mission.[68]

Presumably because of the correction of the pogo problem, the changes under the GEMSIP program, the malfunction detection systems, and the like, the engines of the twelve Gemini launch vehicles were given their own separate designations as LR87-AJ-7 for the first-stage pair and LR91-AJ-7 for the second-stage power plant, instead of the Titan II missile's LR87-AJ-5 and LR91-AJ-5. There were no major differences in thrust, specific impulse, or other parameters between the LR87-AJ-5 and LR91-AJ-5 engines, on the one hand, and LR87-AJ-7 and LR91-AJ-7 on the other. Because the spacecraft was longer than the missile's reentry vehicle, the Gemini-Titan II launch vehicle was 109 feet long to the missile's 104 feet.[69]

On April 8, 1964, following all these preparations, the acid test for Gemini-Titan came with its launch from Cape Canaveral on the *Gemini 1* mission, basically to test its performance, guidance system, and structural integrity together with that of the Gemini spacecraft. There were no astronauts on board. Both stages worked almost to perfection, putting the spacecraft with the second stage still attached into an orbit whose perigee was 173 nautical miles above Earth, less than 13 miles higher than programmed. Expected to orbit for three and a half days, it actually stayed up almost four days before plunging back into the atmosphere to burn up as planned.[70]

On January 19, 1965, *Gemini 2* launched, again without astronauts, to test reentry heat protection and the structural integrity of the spacecraft on a trajectory that carried it to an altitude of 92 nautical miles. Separated from the second stage, it hurled back through the atmosphere, falling into the South Atlantic on its parachute for recovery by the Navy's aircraft carrier *Lake Champlain*. Both heat protection and structural integrity withstood the test.[71]

From March 23, 1965, to November 15, 1966, *Gemini 3* through *12* followed, all with two astronauts on the Gemini spacecraft. Before these "manned" Gemini flights began, however, the Soviets staged another first. On March 18, 1965, cosmonauts Pavel Belyayev and Alexei Leonov launched on *Voskhod 2*. Leonov executed history's first "spacewalk"—extravehicular activity, or EVA, in NASA parlance. Although he experienced great difficulty reentering the spacecraft, and there were other problems on the mission, the two men did return to Earth successfully.[72]

The Gemini missions also had their problems as well as their triumphs. With them, the United States nevertheless finally assumed the lead in the space race with its cold-war rival. *Gemini 3* simply demonstrated human orbital flight with two astronauts in the spacecraft, but on *Gemini 4* from June 3 to 7, 1965, astronaut Edward White became the first American to walk in space, although he too had difficulty getting back into the spacecraft. The *Gemini 5* mission from August 21 to 29 was not perfect. A planned rendezvous with an evaluation pod had to be cancelled because of problems with fuel cells, but Gordon Cooper and Charles "Pete" Conrad stayed in space for eight days to evaluate the effects of near weightlessness for a long enough period to go to the Moon and back, by now a national priority under Project Apollo. They returned and passed a medical examination after surpassing a Soviet record of five days in space.[73]

A serious setback occurred on October 25, 1965, with the attempted launch of *Gemini 6* and an Atlas-Agena to inaugurate a rendezvous and docking mission. While the astronauts waited for their launch, the Atlas-Agena roared from the launching pad at Cape Canaveral. As the Agena up-

Figure 41. Atlas-Agena target vehicle lifting off from Pad 14 at Kennedy Space Center on September 12, 1966, as part of the *Gemini 11* mission. Courtesy of NASA.

per stage and target vehicle separated from the Atlas, its engine started, but then there were undetermined problems that telemetry data failed to clarify immediately. Although there had been more than 140 flights of Agena vehicles since 1959, this Agena D apparently exploded. Efforts to piece together what had happened from the skimpy data took four months. The investigators finally concluded the problem lay with NASA's specifications for starting and stopping the engine five times. In previous Agenas, with only a two-start capability, oxidizer flowed into the combustion chamber first, with a pressure switch not releasing the fuel until enough oxidizer was present for smooth combustion. But the Agena engine had too little oxidizer capacity to start the ignition sequence that way five times, so the manufacturer, Bell Aerosystems, removed the pressure switch and allowed the fuel to enter the chamber first. Concluding that, in space on the *Gemini 6* mission, too much fuel accumulated before the oxidizer reached the combustion chamber, causing an explosion, engineers modified the Agena D engines so that oxidizer entered the combustion chamber first after all. Following altitude tests at the Arnold Engineering Development Center, they requalified the engine for flight on Gemini missions.[74]

Meanwhile, from December 4 to 18, Frank Borman and James Lovell flew on *Gemini 7* for two full weeks to evaluate the effects such a lengthy exposure to near weightlessness. There were also plans for a rendezvous between their spacecraft and *Gemini 6-A*. Wally Schirra and Thomas Stafford, who had been unable to launch on *Gemini 6*, were in *Gemini 6-A* atop a Titan II when its engines started and then stopped. An electrical plug that came loose from its socket had led the malfunction detection system to shut the engines down. After deciding, at some peril, not to eject, the previously unlucky astronauts found the third time a charm and launched successfully on December 15. They then maneuvered their vessel to within a foot or so of *Gemini 7*.[75]

Gemini 8 brought a successful docking with an Agena, but this was followed by uncontrollable tumbling. On *Gemini 9* the astronauts went into orbit, but the Atlas booster delivering the Agena experienced control problems and went into the ocean with its target vehicle. Finally, on *Gemini 10* from July 18 to 21, 1966, Michael Collins and John Young were the first astronauts to complete a fully successful docking. They then used the Agena engine to rendezvous with the other Agena left behind by *Gemini 8*, whose astronauts had disengaged from it to recover from their tumbling and return to Earth. *Gemini 11* and *12* were also successful. Thus, with lots of problems, Gemini had nevertheless prepared the way for the Apollo Moon landings and achieved all essential objectives. The Gemini–Titan II launch vehicle had experienced one hiccup, on *Gemini 6-A*, but even that had shown that the malfunction detection system worked. And generally the vehicle had performed as expected.[76]

Following its successful Gemini missions, the Titan II did not serve again as a space launch vehicle until after it was taken out of service as a missile in the mid-1980s. Since the Titan II launch vehicle used components from the Titan III as well as the original Titan II missile,[77] that story will be covered in the companion volume to this book.

Figure 42. Gemini-Titan 11 lifting off at Cape Kennedy on September 12, 1966. The *Gemini 11* mission included a rendezvous with the Agena target vehicle launched by an Atlas (see figure 41). Courtesy of NASA.

Reflection and Conclusions

Development of Titan I and Titan II did not require a lot of new technology. Instead it adapted technologies developed either earlier or simultaneously for other missile or launch vehicle programs. Nevertheless, the process of adaptation for the two Titan missiles generated problems requiring engineers to use their fund of knowledge for cut-and-try solutions. These did work, and Titan II became the nation's longest-lasting liquid-propellant missile, with the greatest throw weight of any vehicle in the U.S. inventory. It also became the base on which Titan III and Titan IV were developed in building-block fashion using technologies from Polaris and Minuteman to produce huge strap-on solid-rocket motors to increase capabilities for launching satellites and other payloads. Thus the Titan program as a whole was an outstanding success for the Air Force, contributing not only an important missile but a number of different heavy space-launch vehicles.

The problems that engineers experienced with combustion instability in the Titan engines clearly were neither new nor peculiar to those engines. Following destruction of an operational Atlas missile during the summer of 1961 because of combustion instability, the Air Force had convoked a committee under the chairmanship of Clark Millikan of Caltech to investigate the problem. The committee's recommendation of increased effort to understand combustion instability led to a meeting at the Air Force Ballistic Systems Division on August 31, 1961, with attendees from the Air Force Office of Scientific Research (AFOSR), the Atlas and Titan project offices, the Aerospace Corporation, and the test group at Edwards Air Force Base that became the Rocket Propulsion Laboratory in 1963. Representing this last organization was Richard R. Weiss, a combustion technology project engineer who had worked on Jupiter engine development for Rocketdyne before becoming the Thor propulsion group engineer at Edwards.

One task of the group was to prepare for a technical conference at Princeton University on September 25–26 to review work on instability problems. Participants represented not only the Air Force but the Army, Navy, NASA, and contractors. At the conference, Weiss reported, papers presented by specialists from universities, industry, NASA, and other organizations made it clear that combustion instability was still "in the realm of the unknown" despite years of research. Industry had to resort to "crutches" and "cut-and-try" techniques to overcome combustion problems. As with the Titan and earlier engines, researchers reported some success using baffled injectors, but these were hardly final answers to the overall problem. And engineers did not fully understand why baffles worked. As Weiss himself told the conference, engineers still considered injector design an art rather than a science, requiring trial and error to produce stable combustion.

Engineers typically encountered combustion instability comparatively late in an engine's development, during testing of the complete system. By this time the design was relatively set, so only limited modifications could be made. As Weiss concluded in his own paper delivered at Princeton, "Only when the mechanisms of instability are thoroughly understood can there be high confidence of elimination."[78]

In January 1962, Weiss's organization started a Combustion Technology Program to help understand the process of combustion in rocket engines. Meanwhile, by the end of 1962 it had become apparent that combustion instability plagued not only the Navaho, Redstone, Atlas, Thor, Jupiter, Titan I, and Titan II engines but also two engines, called H-1 and F-1, that Rocketdyne was developing for the Apollo launch vehicle, the Saturn. So widespread was the problem, and so great the budget overruns and program delays it caused, that an Ad Hoc Working Group on Liquid Propellant Combustion Instability was constituted by the newly chartered Interagency Chemical Rocket Propulsion Group. Supporting the parent group was the Chemical Propulsion Information Agency (CPIA), formed by combining the Solid Propellant Information Agency and a more recent Liquid Propellant Information Agency on December 1, 1962. The CPIA operated under the auspices of the Applied Physics Laboratory at Johns Hopkins University. Maj. A. O. Bouvier Jr. from USAF headquarters headed the Ad Hoc Working Group, which included Weiss from the Edwards rocket test group, Edward W. Price from the Naval Ordnance Test Station, and representatives from ARPA, NASA's Lewis Research Center and Marshall Space Flight Center, the CPIA, the Army Missile Command, and AFOSR.[79]

By the end of 1963, the Rocket Propulsion Laboratory (AFRPL) had made some progress toward understanding combustion instability, and the Rocketdyne Division of North American Aviation was also working on the problem. At a December meeting between AFRPL and Rocketdyne at Edwards, the contractor's representatives reported that although the Atlas experience had offered some general guidelines for overcoming instabilities, the particular solutions used in Atlas did not apply to the problem on the F-1 engine. Cut-and-try methods were still necessary.

Then, in March 1964, the Ad Hoc Working Group published a report based on the knowledge and experience of its members, personal interviews, questionnaires, and inquiries to determine the status of research on combustion instability throughout the country. Recommending more research and further exchange of information, the group wrote, "Combustion instability should be recognized for what it is, a natural, unpredictable phenomenon common to most engine development programs. Reluctance to acknowledge such difficulties should not exist; realistic confrontation of the problem leads to better allocation of funds to the most effective areas of research, special development work, information activities, etc." The report did identify three types of instability. One was "chugging," normally a low-frequency phenomenon in which chamber pressure oscillation was coupled with an oscillation in the propellant feed system. A second kind was a "screaming" oscillation that followed a path similar to an acoustical wave. It could be either longitudinal or transverse. The third kind was an "entropy" wave that traveled with the gases produced by combustion and changed in frequency in relation to the gas velocity or the length of the combustion chamber.[80]

Three years after this report, Dieter K. Huzel and David H. Huang of Rocketdyne published a volume on designing liquid-propellant rocket engines that discussed combustion instability in some detail. They stated, "It has been found experimentally that the amplitude of the chamber pressure oscillations which will cause detrimental physical or operational effects varies widely for different thrust chambers and engine systems. . . . Thus it is difficult to assign a quantitative value to the amplitude at which the combustion chamber should be considered as running unstable." Besides tangential and longitudinal modes of instability, they identified another tangential mode, which operated neither along the length of a combustion chamber nor across it but in a circle around the inside of the chamber's circumference. They too identified the "screaming" or high-frequency instability, in which serious damage to the engine hardware almost invariably occurred within a fraction of a second. It could burn through an injector, causing propellant mixing of such quantity that an explosion would occur. They also identified "chugging" or low-frequency instabilities that could loosen bolts and vital connections or cause ruptures. Another type they called "intermediate-frequency" instability, which occurred only occasionally in large engines. The propellant pump was apparently the principal source of this type of instability, which could cause material fatigue. The authors said that injector modifications or damping mechanisms like baffles could tame combustion instability. Successful application of the damping was "based on criteria established empirically," since "understanding of the fundamental physical principles of the damping process" was "still limited." There had been lots of studies, but "the basic processes which trigger and sustain the various types of instability have not been isolated." Thus, "while a parameter which controls one type of instability may have been established on an engineering basis, this same design criterion may be enhancing another type of instability."[81]

Clearly, then, in the area of combustion instability—and, as Titan development showed, also in several other areas including the effects of altitude upon propulsion systems and structures and the pogo effect—engineers were making progress in understanding and fixing problems. But engineering for rockets was still partly an empirical process in which vehicles had to be built and tested, either on the ground or often in flight, before problems could be found and confronted. In some cases, engineers were able to solve the problems without fully understanding their causes or parameters. But there was no complete body of engineering science that would enable design of a problem-free missile or launch vehicle before a great deal of testing and problem solving had been done.[82]

Progress in developing successful technologies for missiles and launch vehicles occurred in part from cooperation among agencies and services. This was notably true in the mysterious area of combustion instability, but cooperation between NASA and the Air Force was responsible for the success of Project Gemini, as it had been for Project Mercury with the addition of the Army for use of the Redstone as a launch vehicle. A further source of progress was the migration of rocket engineers among firms and government agencies, bringing with them experience and knowledge. In these ways and others, missile and launch vehicle technology was moving forward.

9

Polaris and Minuteman

The Solid-Propellant Breakthrough, 1955–1970

Until Polaris A1 became operational in 1960, all intermediate-range and intercontinental missiles in the U.S. arsenal had employed liquid propellants. These had obvious advantages in terms of performance (specific impulse) but required extensive plumbing and large propellant tanks that made protecting them in silos difficult and expensive. Such factors also virtually precluded their efficient use on board ships, especially submarines. Once Minuteman I became operational in 1962, the U.S. military began to phase out liquid-propellant strategic missiles. To this day, the Air Force's Minuteman III and the Navy's solid-propellant fleet ballistic missiles continue to play a major role in the nation's strategic defenses because they are simpler and cheaper to operate than liquid-propellant missiles.[1]

Liquids, because of their generally higher performance and because they can be throttled and turned off and on with valves, remained the primary propellants for space launch vehicles. But since solid-propellant boosters, with a higher thrust-to-weight ratio, could be strapped to the sides of liquid-propellant stages for an instant addition of high thrust, solid propellants became important in launch vehicle technology. The technologies used on Polaris and Minuteman transferred to such boosters and also to upper stages of rockets used to launch satellites. Thus the solid-propellant breakthrough in missiles had important implications for launch vehicle technology as well.

Meanwhile, given the performance advantages that liquid propellants enjoyed, their head start within the defense establishment, and the disinclination of most proponents of liquids to entertain the possibility that solid propellants could satisfy the demanding requirements of the strategic mission, how did this solid-propellant breakthrough occur? The answer is complicated and technical. But fundamentally, solid propellants replaced liquids in strategic missiles because a number of heterogeneous engineers believed in and promoted them, a variety of partners in their development achieved significant technical innovations, and, while interservice rivalries encouraged the three services to develop separate missiles, interservice cooperation ironically helped them to do so. Yet despite such cooperation and the accumulating knowledge about rocket technology, missile designers could not foresee all the problems that their vehicles would develop during ground and flight testing. Thus, when problems did occur, rocket engineers still had to gather information to help identify the causes, and exercise their ingenuity to develop solutions.

By the time Polaris got under way in 1956 and Minuteman in 1958, solid-propellant rocketry had already made tremendous strides beyond the use of extruded double-base propellants in tactical missiles during World War II. But there were still enormous technical hurdles to overcome before a solid-propellant missile could hope to launch a strategic nuclear warhead far enough and with sufficient accuracy to serve effectively as a deterrent or as a retaliatory weapon in case of enemy aggression.[2] Unmet needs included (1) higher performance; (2) more stable combustion; (3) nozzle materials that would stand up to heat and to corrosive chemicals from the exhaust of longer- and hotter-burning propellants; (4) lighter combustion chambers, to reduce the weight being launched; (5) a small but still powerful warhead, also to reduce weight; (6) a way to terminate combustion as soon as the desired velocity was achieved, to improve accuracy; (7) a way to control the direction (vector) of the thrust, for steering; and (8) a small, light, and accurate inertial guidance system.[3]

Besides technical hurdles, Polaris—and later Minuteman—faced bureaucratic barriers. By the time the Navy decided to support a strategic ballistic missile, it was almost too late. Milton Rosen and Commanders Truax, Robert F. Freitag, and Grayson Merrill had all urged the sea service to use its experience with the Viking rocket in developing a ballistic missile for use from a surface vessel, a submarine, or even land. The director of the Navy's guided missile program and the chief of naval operations rejected the proposals. It was only after the appointment of Adm. Arleigh Burke as chief of naval operations in August of 1955 that proponents of a fleet ballistic missile got sufficient support at the highest level within the Navy to enter the ballistic missile competition. Burke decided to support such a missile, but his service needed to find a partner so it could develop a seagoing variant of one of the IRBMs already under consideration. Navy representatives talked with the Air Force, but that service was unwilling to make adjustments to permit a seagoing version of its IRBM, especially as the air arm's role in missile development was secure. The Army had a much weaker position in the roles-and-missions controversy that had been going on since before the Air

Force became a separate service in 1947, so Army representatives agreed to collaborate on a liquid-propellant missile (the Jupiter) that would provide a land-based alternative to the Air Force's IRBM (Thor) and also a shipboard capability. The upshot of these negotiations was the memorandum on November 8, 1955, from Secretary of Defense Charles E. Wilson to the service secretaries authorizing the two IRBMs, with the Jupiter having a dual role on land and sea.[4]

In this way, the Army's rivalry with the Air Force helped to give the Navy a role in missile development and thereby aided the solid-propellant breakthrough in rocketry. The Navy had a strong aversion to liquid oxygen, one of the propellants for Jupiter. This antipathy was partly traceable to Operations Sandy and Pushover in 1947–48 (see chapter 4). But regardless, cryogenic propellants like liquid oxygen had to be loaded immediately before launch so that little of the propellant boiled away. Even when storable liquid propellants became available with the Titan II program, they were impractical on a submarine, or indeed on a surface vessel, where pitching and rolling would create unacceptable stresses on pipes, valves, and fittings in the missile's engine and might cause propellant leaks.[5]

From Jupiter to Polaris A1

For these and other reasons, the Navy worked reluctantly with the Army on a seagoing version of Jupiter and made no secret of its intention to shift to a solid-propellant missile as soon as technological advances permitted. Such a prospect did not worry the Army, because most of its experts saw little likelihood that this would happen soon.[6]

To work with the Army on Jupiter, Admiral Burke on November 17, 1955, created a Special Projects Office in a temporary building on Constitution Avenue in Washington that belonged to the Bureau of Aeronautics. To head it he picked Captain (soon to be Rear Admiral) William F. Raborn. Burke wanted an aviator, since the missile would take over an aviation mission and involved aerodynamic problems, and he wanted a nontechnical person, who he thought would be less narrow-minded than a missile expert. He also wanted someone with the ability to organize, to work hard, to be persuasive, and to distinguish what was important from what was not. "Red" Raborn, as he was called from the color of his hair, proved a brilliant choice. He was a prototypical heterogeneous engineer with extraordinary political and motivational skills, and he was given unusual authority to develop the missile the Navy now wanted. At the same time, Burke told him that if he invoked that authority, he would destroy the program; he had to get people's willing support.[7]

Raborn won that support. He also recruited key officers and civilians for the SPO, using his special authority. One of the early selections was Merrill, now a captain, who became the first technical director and who helped organize the office around elements of the system such as propulsion, fire control, tests and instrumentation, a submarine to launch the missile, a launching system, and the warhead. Unlike Schriever's organization, and even Medaris's, Raborn's remained comparatively small, with an initial authorization of 45 officers and 45 civilians. The total grew to 160 people by July 1, 1957, then to 325 in December 1960, and about 500 a decade later.[8]

Once the Navy decided that the right vessel to launch a missile was a submarine, which could hide from enemy observation and attack beneath the trackless seas, the Navy pressed ahead with solids, because Jupiter, even the shorter and wider version the Army agreed to, was too large, too heavy, and too weak in its structure to withstand the effects of storms at sea. The effort to find the right technologies for a submarine-launched solid-propellant missile soon paid off with breakthroughs in a number of different technical areas. In early January 1956, some senior naval officers—without specific prior approval from either DoD or the secretary of the Navy—sought the assistance of Lockheed's Missile and Space Division and Aerojet General in developing a solid-propellant ballistic missile. One of the officers was Captain Levering Smith, who as a commander at the Naval Ordnance Test Station, Inyokern, California, had led the effort to develop a 50-foot solid-propellant missile named Big Stoop that had flown 20 miles in 1951. Smith joined the Special Projects Office in April 1956. It turned out that he was senior to Merrill, the technical director, but since he had expertise in solid propellants and did not insist on his seniority, he, Merrill, and Raborn agreed among themselves that Smith would be in charge of the propulsion system and would direct these two contractors' efforts, supported by studies at NOTS.

The initial missile the two contractors and the SPO conceived was the so-called Jupiter S (for solid). It had enough thrust to carry a standard atomic warhead the requisite distance, a task it would achieve by clustering six solid rockets in a first stage and adding one for the second stage. The problem was that Jupiter S would be about 44 feet long and 10 feet in diameter. An 8,500-ton vessel could carry only four of them, compared to sixteen of the later Polaris missiles. Not yet conceiving of Polaris, the Navy and its contractors nevertheless were dissatisfied with Jupiter S and continued to search for an alternative.[9]

One contribution to a better solution came from the Atlantic Research Corporation (ARC), a chemical firm founded in 1949. The Navy Bureau of Ordnance had contracted with ARC to improve the specific impulse of solid propellants. Keith Rumbel and Charles B. Henderson, ARC chemical engineers with degrees from MIT, had begun theoretical studies of this problem in 1954. They learned that other engineers, including some from Aerojet, had projected an increase in specific impulse from adding aluminum powder to existing ingredients. But these other engineers, basing their calculations on contemporary theory and doing the cumbersome math without the aid of computers, figured that once the proportion of aluminum exceeded 5 percent, performance would again decline. So they abandoned aluminum as an additive except for a role it played in damping combustion instability. Rumbel and Henderson, however, refused to be deterred by theory, and tested polyvinyl chloride containing considerably more than 5 percent aluminum. They found that with additional oxygen in the propellant and a flame temperature of at least 2310 Kelvin, a large percentage of aluminum by weight yielded a specific impulse of 230–247 lbf-sec/lbm, significantly higher than previous composite propellants such as those used in the Army's Sergeant missile.[10]

It was not ARC's polyvinyl chloride, however, that served as the binder for Polaris. Rather it was a polyurethane developed by Aerojet. This development began under a small Navy nitropolymer program funded by the Office of Naval Research about 1947 to seek high-energy binders for solid propellants. A few Aerojet chemists synthesized a number of high-energy compounds, but the process required levels of heating that went beyond what was safe for potentially explosive compounds. Then one of the chemists, Rodney Fischer, found "an obscure reference in a German patent" suggesting "that iron chelate compounds would catalyze the reaction of alcohols and isocyanates to make urethanes at essentially room temperature." This discovery started the development of polyurethane propellants in many places besides Aerojet.

Meanwhile, in 1949 Dr. Karl Klager, then working for the Office of Naval Research in Pasadena, suggested to Aerojet's parent firm, General Tire, that it begin work on foamed polyurethane, leading to two patents held by Klager with Dick Geckler and R. Parette of Aerojet. In 1950 Klager, who had earned a Ph.D. in chemistry from the University of Vienna in 1934 and had come to this country as part of Project Paperclip, began working for Aerojet. By 1954 he headed the rocket firm's solid-propellant development group. Once the Polaris program began in December 1956, Klager's group decided to reduce the percentage of solid oxidizer as one component of the propellant by including oxidizing capacity in the binder itself, using a nitromonomer as a reagent to produce the polyurethane plus some inert polynitro compounds as plasticizers, or softening agents. In April 1955 the Aerojet group found out about the work of Rumbel and Henderson. Overcoming explosions due to cracks in the grain and profiting from other developments from multiple contributors, they came up with successful propellants for both stages of Polaris A1.

These consisted of a cast, case-bonded polyurethane composition including different percentages of ammonium perchlorate and aluminum for stages 1 and 2, both of them featuring an internal-burning six-point star configuration. With four nozzles for each stage, this propellant yielded a specific impulse of almost 230 lbf-sec/lbm for stage 1 and nearly 255 lbf-sec/lbm for stage 2, the latter being higher in part because of the altitudes at which it operated.[11]

The addition of aluminum to Aerojet's binder essentially solved the problem of performance for Polaris—and, as it turned out, for Minuteman with a different binder. A separate process resolved the issue of the size and weight of the warhead, the guidance system, and the overall missile. Levering Smith had previously served at NOTS as director of the Rockets and Explosives Department, eventually becoming associate technical director of the station under its first technical director, Dr. L. T. E. Thompson. It was natural, therefore, that when he assumed his duties as head of the SPO's Propulsion Branch in April 1956, he asked Dr. Frank Bothwell, head of NOTS's Weapons Planning Group, to study how to reduce the size of Jupiter S enough so that it could be launched from a reasonable-sized submarine. Looking at available information, making projections about where technology would be by the time the missile was mature, and examining in great detail the criteria to determine necessary warhead yields from intelligence data on Soviet cities, Bothwell's group concluded that a solid-propellant missile could be reduced in weight from the roughly 160,000 pounds of Jupiter S to about 30,000 pounds.

Meanwhile Ivan Getting—an engineering scientist heading the National Academy of Sciences' Committee for Undersea Warfare, meeting at Woods Hole, Massachusetts, in the summer of 1956—suggested that Bothwell study "prosubmarine" weapons. Bothwell joined the Nobska Study, named for a nearby lighthouse on Nobska Point, where he came into contact with nuclear physicist Edward Teller of Lawrence Livermore Laboratory. When Bothwell told Teller of his group's ideas for reducing the size of a solid-propellant missile, Teller said the trend in warhead development indicated that a warhead of sufficient strength could be ready within the time frame then contemplated for the missile's completion. Bothwell went to Washington on Labor Day 1956 and briefed Thompson (by then a consultant to the SPO) and Smith, followed by Raborn the next day. The day after that, Raborn got permission from Burke to shift the focus of the SPO to developing the missile that Raborn named Polaris, after the North Star. Raborn subsequently briefed Secretary of Defense Wilson on the smaller missile's implications for the size and number of ships needed to deploy it—at an estimated saving of half a billion dollars. On December 8, 1956, Wilson issued the directive that began the Polaris program. Thus did a couple of technical estimates and the heterogeneous engineering of Raborn inaugurate a new strategic missile that, in turn, started the solid-propellant breakthrough in missilery.[12]

All of this accomplished more than the creation of a new missile project. The Bothwell group's work on size reduction led to the smaller guidance computer.[13] And together with Teller's projection, later confirmed by the Atomic Energy Commission, it also led to the smaller warhead actually used on Polaris. Even before approval of the Polaris program, the SPO had contracted with MIT's Instrumentation Laboratory to look at ship stabilization for the Jupiter missile. Later, Raborn's group contracted with that laboratory to design an inertial guidance system for Polaris. While the system the lab was designing for Thor used an analog onboard computer, the Navy opted for a digital computer, even though digital computers were then slower and larger. The lab's Eldon C. Hall, who designed the digital computer for Polaris, convinced the Navy that it would be more accurate than an analog system. Also important was the Q-guidance Battin and Laning had developed for the Air Force. With most of the Q-matrix computation done ahead of time, few calculations had to be done by an onboard computer.

Because the Instrumentation Lab had no production capability, the Navy chose General Electric to build the guidance system. The Mark 1 system featured a 2.5-inch-diameter inertial rate-integrating gyroscope (IRIG) with a pendulous integrating gyro accelerometer (PIGA) using the same type of gyro. The onboard computer was reportedly to be the first fully transistorized, digital differential analyzer and the first digital computer used for an inertial guidance system in a missile. (Polaris flew its first guided flight on January 7, 1960, two months and a day before Atlas's first guided flight with a different digital computer and guidance system.) Thanks to Q-guidance, transistors, and printed circuit boards, the digital computer and guidance system were quite small. The entire "black box" with its inertial components and electronics weighed only 225 pounds, as against the Jupiter system's

250–300 pounds and Thor's 650–700. Because it was pioneering, the system had innumerable problems with testing, wiring, gyroscopes, transistors, and core memory. But it was ready on what turned out to be an accelerated schedule, and even though it had a problematical mean time between failures, it proved adequate for the two minutes of guided flight it had to support. This guidance system illustrates the ways in which technologies and methods developed for one service's missile proved fruitful for a very different missile produced for another service.[14]

The warhead that complemented the light guidance system derived from a program undertaken at the Livermore Laboratory under its first director, Herbert York, to make a lighter payload. The major innovation came from Teller and involved use of highly enriched uranium. The Atomic Energy Commission tested a weapon design in July 1957 that essentially proved the feasibility of a warhead for Polaris. Out of it arose a weapon that yielded a half-megaton capability, well over the 300-kiloton minimum the SPO had called for. It did so in a combined warhead and reentry body weighing about 850 pounds, less than a third of the weight carried by other services' missiles. The warhead was not without problems, but like the guidance system it was relatively small yet powerful enough to do the job, and it was ready in time for early deployment of Polaris A1. To achieve its integration with the reentry vehicle, and to ensure compatibility between the support structures on the missile and the payload's shielding against the effects of reentry, a unique cooperative arrangement was required among the warhead designers, the SPO, and Lockheed, which had become designer of the reentry body as well as systems contractor for the missile.[15]

Lt. Robert Wertheim headed the SPO's reentry section, and his group set up the cooperative arrangement to pursue a reduction in payload weight. This allowed the missile itself to be as small as evolving technology allowed. While Livermore was responsible for the warhead, the Naval Ordnance Laboratory at White Oak, Maryland, developed the devices to arm and fuse the warhead. Wertheim's group intended to endow the warhead with the bare minimum of hardware it needed to survive reentry and proceed to its target. The group chose beryllium as the material for the nose cone, evidently because it offered the best integration between warhead and reentry vehicle. This necessitated a heat-sink approach rather than the ablative one used on Jupiter and proposed by Bothwell's group at NOTS. It still enabled a very light overall structure that stood up to structural, shock, vibration, and radiant-heat testing as well as flight testing of a one-third-scale model using the Lockheed X-17, redesignated FTV-3 (a further example of one service

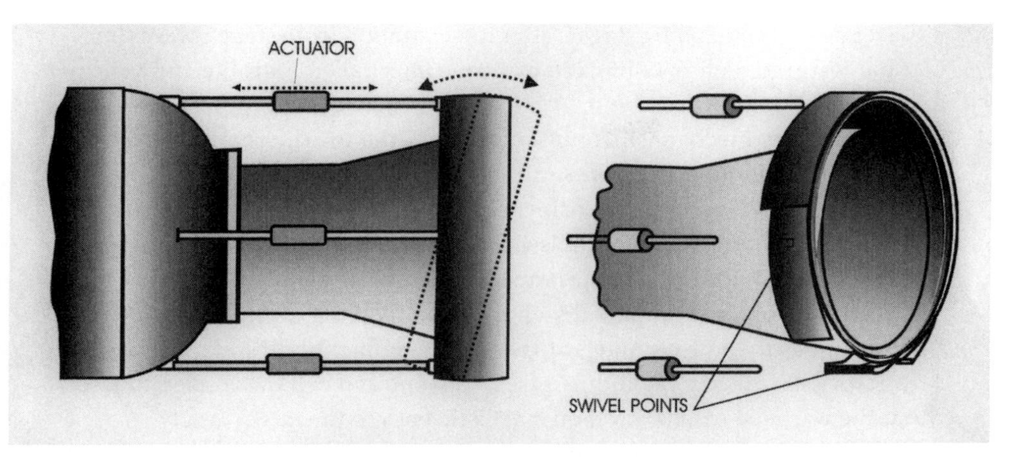

Figure 43. Diagram showing how jetavators rotated into the exhaust stream from the nozzles of the Polaris motors to deflect thrust and provide steering for the first sea-launched ballistic missile. Drawn for author by Alfred Signs based on image in *Missiles and Rockets*, February 9, 1959, 23.

benefiting from another's efforts, since Lockheed had built the X-17 for the Air Force). Finally, like some earlier reentry bodies, the one for Polaris used two small rocket motors to impart spin so it would remain stable on its flight to the target.[16]

Taken together, these innovations allowed Polaris A1 to be only 28.6 feet long and 4.5 feet in diameter, as compared with Jupiter S at 44 feet by 10 feet. The actual weight reduction was from 162,000 pounds for Jupiter S to a scant 29,000 for Polaris, slightly less than Bothwell's estimate.[17] The combustion chambers (cases) for both stages of Polaris A1 consisted of rolled and welded steel. This had been modified according to Aerojet specifications based on extensive metallurgical investigations.

Each of four nozzles for stage 1—and evidently stage 2 as well—consisted of a steel shell, a single-piece throat of molybdenum, and an exit cone liner made of "molded silica phenolic between steel and molybdenum." A zirconium oxide coating protected the steel portion. The missile's mechanical guidance-and-control system used jetavators designed by Dr. Willy Fiedler of Lockheed, a German who had worked on the V-1 program during World War II and had developed the jetavator concept while working for the U.S. Navy at Point Mugu Naval Air Missile Test Center. He had patented the idea and then adapted it for Polaris. Jetavators for stage 1 were molybdenum rings with curved inside surfaces that rotated into the exhaust stream of the four nozzles and deflected the flow to provide pitch, yaw, and roll control.

The jetavators controlling stage 2 after its separation from stage 1 were similar but featured a silica-reinforced exit cone liner called Astrolite and a shell housing of titanium.[18]

Besides this mechanical control (steering), the missile required precise thrust termination at the point in the second stage's trajectory where its direction and velocity were correct for it to hit its target. This could be achieved on liquid-propellant missiles simply by stopping the flow of propellants. For solids, the task was much more demanding. The Polaris team chose a technology that opened six ports in front of the second stage. It used pyrotechnics to blow out plugs at the proper moment in the second stage's trajectory, permitting expanding gases to escape and halt the acceleration so that the warhead would travel on a ballistic path to the target area.[19]

The SPO had to oversee and manage not only the development and deployment of the many components of the missile but also the design and building of the submarines to launch it, the ship's inertial navigation system (SINS) to determine the sub's position so it could launch the missile accurately, a related fire control system, a communications system so the Navy could command and control the launching of a missile (fortunately never necessary yet except for testing), the launching system, contractor efforts, training of submarine crews, and support facilities for the deployed fleet. All of this had to be completed in a time span that Sputnik greatly shortened. Before the launching of the Soviet satellite, the deployment schedule called for Polaris to be ready in 1963. After October 1957, the deployment date moved to 1960. Given all the elements that had to be developed concurrently, Raborn's task was more demanding from the systems-engineering standpoint than was Schriever's for any single missile, although Schriever had a much larger task with all the different missiles he had to develop, plus satellites.

To manage his diverse tasks, Raborn and a civilian manager named Gordon O. Pehrson adopted a system called Program Evaluation and Review Technique (PERT), which bore some similarity to Schriever's system but apparently had an independent genesis. In simplest terms, PERT used charts to track the multitudinous elements of the overall program. Usually the key managers met on Saturday to present the charts to Raborn. The intent was not to overwhelm Raborn with data but to give him a clear picture of how each major subsystem was progressing toward the goal. This he wanted in terms indicating whether the subsystem was progressing satisfactorily and on schedule or whether there were minor, major, or critical problems. In 1958 a PERT computer was added to the mix, but the system was only as good as the time estimates and evaluations of Navy technical division managers within the SPO and their counterparts in industry. There is no clear evidence of how well the PERT system worked or how costly it was. Technical division managers complained about it, as did industry. But everyone agreed that it was an effective tool for selling the program to outside visitors, and PERT came to be widely imitated.[20] As will be discussed in this volume's sequel,[21] both PERT and Schriever's management system later helped NASA to develop the Saturn Moon rockets for the Apollo effort, showing the importance of the two systems, even though some people in NASA also complained about them.

Flight testing of Polaris at the Air Force's Cape Canaveral, beginning in 1958 with a series designated AX, revealed a number of problems, the solutions to which illustrate the interservice cooperation necessary for the solid-propellant breakthrough. For example, flights AX-3 and AX-4 experienced loss of control when aerodynamic heating and a backflow of hot exhaust gases caused electrical wiring at the base of stage 1 to fail. The program obtained the help of "every laboratory and expert," using sled tests, wind tunnels, static firings, data from flights AX-5 through AX-8, "and a tremendous analytic effort." Among the labs contributing help and analysis were NOTS, Edwards Air Force Base with its sled tracks, and the Air Force's Arnold Engineering Development Center with its high-altitude chambers. The heating problem was solved by placing fiberglass flame shielding supplemented by silicone rubber over hydraulic elements, cabling, and the motor dome to provide a barrier to hot gases and flame.[22]

A variety of tests in a variety of places contributed to the success of the Polaris missile. Static tests showed that early carbon throat and exit cone liners for the nozzles were inadequate, so molybdenum throats and liners were employed in the production version of the missile. NOTS conducted propellant formulation studies in conjunction with Aerojet to assure the Navy that the path Aerojet was following made technical sense. By October 1959 the Navy had completed construction of a huge static test facility at NOTS called Skytop, built to test Polaris motors but capable of handling thrusts of up to a million pounds. NOTS's Michelson Laboratory had a new IBM 709 computer to evaluate the data from the static firings. Lockheed's Sunnyvale Missiles and Space Division also had a 709, on which team members used the FORTRAN language to evaluate aerodynamic data and arrive at the shape for Polaris A1.[23]

For Polaris, as for earlier missiles, combustion instability was a major concern. Levering Smith credited Edward W. Price of the Research Depart-

Figure 44. Polaris operations at Cape Canaveral in April 1957, a year before actual flight testing of the missile began. Official U.S. Air Force photo, courtesy of the 45 Space Wing History Office, Patrick AFB, Fla.

ment at NOTS with helping to understand the phenomenon. By this time Price had earned a B.S. in physics and math at UCLA, and in February 1960 he completed a major, then-classified paper on combustion instability in which he stated, "This phenomenon results from a self-amplifying oscillatory interaction between combustion of the propellant and disturbances of the gas flow in the combustion chamber." It could cause vibrations that interfered with the guidance and control system, erratic performance, even violent destruction of motor components. To date, "only marginal success" had been achieved in understanding the phenomenon. But some things were becoming known, such as that energy fluxes could amplify pressure disturbances, which had caused them in the first place.[24] As the subsequent success of Polaris showed, enough progress had been made by this time for unstable combustion not to be a major problem for this missile's motors.

On July 20, 1960, the USS *George Washington* launched the first functional Polaris A1. Both it and a second missile launched three hours later were successful over the 1,200-nautical-mile range set for Polaris A1 under the accelerated schedule, and the fleet deployed the missile on November 15. Later, on May 6, 1962, the USS *Ethan Allen* launched a Polaris A1 with an active warhead from a submerged position in the Pacific. The reentry vehicle landed "right in the pickle barrel" (on target) and detonated.[25]

Deployment of the entire Polaris system in such a short period was a remarkable achievement. It entailed compromises on range, the quality of the guidance-and-control system's electronics, and the warhead, but the missile met its revised specifications. Much of the credit goes to Raborn, who was an extraordinary salesman and inspirational leader of the Navy-industrial team that put this complicated system together. Although Burke claimed the distinction of getting Polaris started—and certainly proponents of naval rocketry had been unsuccessful in winning support for a Navy ballistic missile program until he became chief of naval operations—Burke himself credited Raborn with the success of Polaris and for keeping it within budget and on schedule. Raborn was rightly known as the father of Polaris. Among the many other people who made critical contributions to the project, the one who, as Burke said in 1963, "possibly more than any one other individual, provided the professional 'spark' that generated one after another of the technological breakthroughs without which there would be no Polaris on patrol today" was Levering Smith. Raborn agreed, calling Smith the indispensable man in the program, even more than himself.[26]

Smith had headed the propulsion branch of the Technical Division within the SPO until approval came to separate from the Jupiter project and develop Polaris, whereupon a special projects missile branch was created with Smith in charge. He became deputy director of the Technical Division under Merrill in December 1956 and technical director the next June when Merrill retired. Merrill "wondered how Levering would do under Raborn because he was almost a shy person. But he was also very down-to-earth, calm, cool, deliberate and very convincing in his own way." In February 1965 Smith became head of the SPO. Robert Wertheim, who had headed the work on the warhead for Polaris, worked with and for him for many years. In 1977 Wertheim, by now a rear admiral, succeeded Smith as head of the Strategic Systems Project Office, as the SPO had been renamed in July 1968. Of the older admiral, Wertheim said: "Levering was an absolutely remarkable person. He was quiet, soft-spoken, a very thoughtful person. He did not seek publicity or the limelight." But, "In my judgment, he was . . . if not the most accomplished engineering technical manager that the Navy has ever had, . . . one of the top two or three."[27]

For his work on Polaris, Smith was awarded the Distinguished Service Medal on January 5, 1961, but he was not promoted through the Navy's normal system. In August 1961, President John F. Kennedy bypassed the selection board and nominated Smith for the temporary rank of rear admiral. At the same time, Kennedy asked the secretary of the Navy to seek a better system for promoting officers like Smith who had special talents.[28]

Polaris A2

Long before Polaris A1 was operational, in April 1958 the Department of Defense had begun efforts to extend the missile's range from the 1,200 nautical miles of the deployed A1 to the 1,500 nautical miles originally planned. The longer-range missile, called Polaris A2, was at first slated to achieve the goal through use of higher-performance propellants and lighter cases and nozzles in both stages. But the Navy Special Projects Office decided to confine these improvements to the second stage, on which they would have greater effect since the vehicle was already at a high speed and altitude when stage 2 separated and began firing. Also, in the second stage, if the high-energy propellant detonated after ignition, it would not compromise the submarine. The SPO invited the Hercules Powder Company to offer ways to produce a second stage with higher performance than Aerojet's second stage for A1. Hercules was in a favorable position to do this because it was accustomed to working with higher-energy (but more explosive) double-base propellants and also because in 1958 it had acquired the Young Development Laboratories of Rocky Hill, New Jersey, with a capacity to produce light fiberglass cases.[29]

For Polaris A2, then, Aerojet provided the first stage and Hercules the second. Aerojet's motor was 157 inches long, 31 inches longer than for the A1. (The Navy had foresightedly designed the submarines' launch tubes with room to spare.) It contained the same propellant used in both stages of Polaris A1 with the same grain configuration. Hercules' second stage had a filament-wound case and a cast double-base grain that contained ammonium perchlorate, nitrocellulose, nitroglycerine, and aluminum, among other ingredients. The grain configuration was an internal-burning twelve-point star, and the specific impulse under firing conditions was more than

260 lbf-sec/lbm. The motor was 84 inches long and 54 inches in diameter. It featured four swiveling nozzles with exit cones made of steel, asbestos phenolic, and Teflon as well as a graphite insert.[30]

This second-stage motor benefited from the innovative addition of ammonium perchlorate to the cast double-base process used in Hercules' third stage for the Vanguard launch vehicle. Hercules' Allegany Ballistics Laboratory developed this new kind of propellant, known as composite-modified double base (CMDB), by 1958, evidently with the involvement of John Kincaid and Henry Shuey, developers of the earlier cast double-base process.[31]

Even before 1958, Atlantic Research Corporation had developed a laboratory process for preparing CMDB. In its manufactured state, nitrocellulose is fibrous and unsuitable for inclusion in the ingredients being intermixed to create a propellant. Arthur Sloan and D. Mann of ARC, however, developed a process that dissolved the nitrocellulose in nitrobenzine and then separated it out by washing it with water under high shear, a process known as elutriation. The nitrocellulose was now in compact, spherical particles no more than 20 microns in diameter. Such particles combined readily with liquid nitroplasticizers and crystalline additives in propellant mixers. The mixture could be cast into cartridge-loaded grains or case-bonded rocket cases and then converted to a solid by applying moderate heat. Sloan and Mann patented the process and assigned it to ARC. In 1955, ARC's Keith Rumbel and Charles Henderson began scaling the process up to larger grain sizes. They also began developing propellants, coming up with two CMDB formulations beginning in 1956. When ARC's pilot plant became too small to support the firm's needs, production shifted to the naval facility at Indian Head, Maryland. Because the plastisol process they had developed was simpler, safer, and cheaper than other existing processes, Henderson said, Hercules and other producers of double-base propellants eventually adopted his firm's basic method of production.

CMDB was not, however, used for upper stages of missiles and launch vehicles until quite a bit later, and then only after chemists at several different laboratories had learned to make the propellant more elastic. They did this by extending the chains and cross-linking the molecules. Hercules' John Allabashi at Allegany Ballistics Laboratory began in the early 1960s to work on chain-extenders and cross-linking, with Ronald L. Simmons at the Hercules plant in Kenvil, New Jersey, continuing the work. By the late 1960s, chemists had mostly abandoned the plastisol process in favor of dissolving nitrocellulose and polyglycol adipate together, then adding a cross-linking agent such as isocyanate. This produced a highly flexible CMDB propellant

Figure 45. Polaris A2X test missile on the launchpad at the Atlantic Missile Test Range, Cape Canaveral, Florida. The A2X was the prototype for the 1,500-nautical-mile Polaris A2 missile that became operational in 1962. NASA Historical Reference Collection, Washington, D.C. Courtesy of NASA.

of the type used on Trident submarine-launched ballistic missiles beginning in the late 1970s.[32]

The rotatable nozzles on the second stage of Polaris A2, which were hydraulically operated, were similar to those already being used on the Air Force's Minuteman I, and Hercules reportedly owed its Polaris second-stage contract to the performance of its third-stage motor for Minuteman I, once again illustrating technology transfer between services. Meanwhile, stage 1 of the A2 kept the jetavators from A1. The A2 also retained the basic shape of the A1, and its guidance and control system, the principal change being more reliable electronics.

Flight testing of the A2 missiles started in November 1960, with the first successful launch from a submerged submarine occurring on October 23, 1961, off the Florida coast in the USS *Ethan Allen*. There were 28 test flights, with 19 successes, 6 partial successes, and 3 failures. Polaris A2 became operational in June 1962. Before that happened, Raborn became a vice admiral and deputy chief of naval operations for research and development. In February 1962, Rear Adm. I. J. "Pete" Gallantin succeeded him as director of the Special Projects Office, with Rear Adm. Levering Smith remaining as technical director.[33]

Polaris A3

As a follow-on to Polaris A2, in September 1960, Secretary of Defense Thomas S. Gates Jr. approved development of a 2,500-nautical-mile version that became Polaris A3. To create a missile that would travel an additional 1,000 nautical miles when launched from the same submarine launch tubes as the A2 required not only new propellants but also a higher mass fraction. The new requirement also dictated a change from the "bottle" shape of the A1 and A2 to a shape resembling a bullet. To help increase the mass fraction, Aerojet, the first-stage manufacturer, acquired the Houze Glass Corporation at Point Marion, Pennsylvania, and moved that firm's furnaces, patents, technical data, and personnel to the Aerojet Structural Materials Division in Azusa, California, in early 1963. Aerojet could now make filament-wound cases like those used by Hercules on stage 2 of Polaris A2. The new propellant Aerojet used was a nitroplasticized polyurethane containing ammonium perchlorate and aluminum, configured with an internal-burning six-point star. This combination raised the specific impulse by somewhat less than 10 lbf-sec/lbm.[34]

Unfortunately, even so, the new propellant burned hot enough to destroy the first-stage nozzles. Aerojet had to reduce the flame temperature from a

Figure 46. Polaris launch on April 24, 1972. Notice the change from the "bottle" shape of Polaris A1 and A2 to the bullet shape of this Polaris A3. Official U.S. Air Force photo, courtesy of the 45 Space Wing History Office, Patrick AFB, Fla.

reported 6,300°F to slightly less than 6,000°F and make the nozzles more substantial, using silver-infiltrated tungsten throat inserts to withstand the high heat and chamber pressure. The additional weight of the nozzle offset the weight saving of the filament-wound case, so the mass fraction for the A3's first stage was actually slightly lower than for the A2, meaning that the same quantity of propellant in stage 1 of the A3 had to lift slightly more weight than did the lower-performance propellant for stage 1 of the A2.[35]

For stage 2 of the A3, Hercules employed a propellant containing an energetic high explosive named HMX, a smaller amount of ammonium perchlorate, nitrocellulose, nitroglycerine, and aluminum, among other ingredients. It configured this propellant into an "internal burning cylindrical center port" with an equal number of major and minor slots in the aft end of the stage, creating a cross section that resembled an ornate snowflake Christmas-tree ornament. This offered a significantly higher specific impulse than stage 2 of Polaris A2.

Moreover, the new stage 2 used a different method of achieving thrust vector control (steering). It injected Freon into the exhaust coming out the nozzles, creating a shock pattern that deflected the stream and achieved the same results as movable nozzles at a much smaller weight penalty. A further advantage of this system was its lack of sensitivity to flame temperature. The Naval Ordnance Test Station performed the early experimental work on this use of a liquid for thrust vector control. Aerojet, the Hercules-operated Allegany Ballistics Laboratory, and Lockheed did analytical work, determined the ideal locations for the injectors, selected the particular fluid to be used, and developed the injectors as well as a system for expelling excess fluid. The Polaris A3 team first successfully tested the new technology on the second stage of flight A1X-50 on September 29, 1961. This and other changes increased the mass fraction from 0.871 for Polaris A2 stage 2 to 0.935 for Polaris A3 stage 2, a major improvement that, together with the increased performance of the new propellant, contributed substantially to the greater range of the new missile.[36]

The 2,500-mile range also required a new heat shield for the warhead. A heat-sink reentry body would have had to be larger, hence heavier. So engineers shifted to an ablative design featuring a nylon-phenolic material, designed like the earlier nose cone by Lockheed. It was both lighter and cheaper than a heat-sink nose cone, but the Navy arrived at it only after contracting with more than one company to look into the use of pyrolitic graphite and after studying Air Force and Army research. There were eight flight tests in 1961–62 of reconfigured A2 test vehicles carrying the new nose cones, but only a single test provided solid data, and even then there were questions about the decision to proceed with the nylon phenolic. Polaris A3 also carried three separate warheads in a multiple reentry vehicle system developed by the Lawrence Livermore Laboratory before a nuclear test moratorium went into effect under the Partial Test Ban Treaty signed in 1963.[37]

A final major change from Polaris A2 to the A3 model was the guidance and control system. As early as 1959, MIT's Instrumentation Laboratory began work on an improved computer that would be smaller, more easily maintained, and more reliable than the one for Polarises A1 and A2. Better transistors and welded, not soldered, wires contributed to these goals. Smaller gimbals made of aluminum instead of beryllium and smaller accelerometers also reflected improved technologies, given the need for lower weight. Pulsed integrating pendulous accelerometers (PIPAs) replaced the earlier and larger pendulous integrating gyro accelerometers in two of the three axes. Since the PIPAs were held in position by electronic pulses and contained pendulous mass instead of gyros, they were simpler and cheaper, but not quite as accurate as the most sensitive PIGAs. Accordingly, the system kept a PIGA for forward acceleration, which was more critical than the others. The result of all these developments was the Mark 2 guidance system, which was about half as large and a third as heavy as the Mark 1 and accurate enough to carry the warhead the additional distance with considerably greater precision. Polaris remained, however, a weapon directed against cities rather than enemy weapons. Thus it served as a deterrent against enemy use of nuclear weapons, not primarily as a counter-force weapon, although it was more effective than Polaris A2 against enemy weapons, if only those that were large and not hardened. Seven A2 flight tests in a year beginning in November 1961 indicated the effectiveness of the Mark 2 guidance and control system for A3 use.[38]

Flight tests of the actual A3 missile began at Cape Canaveral on August 7, 1962. The first six flights all experienced comparatively minor problems, and the seventh launch brought full success. In the interim, various ground tests identified malfunctions of individual pieces of hardware on what was, by this time, "an 85% new missile." The testing also showed that the overall design was sound. On October 26, 1963, the submarine USS *Andrew Jackson* successfully launched a Polaris A3 from a submerged position off Cape Canaveral. In a total of 38 flight tests by July 2, 1964, the longest range was 2,284 nautical miles. Despite this 216-mile deficit from the specifications, the missile became operational on September 28, 1964. On Christmas Day of that year, the inaugural Polaris patrol in the Pacific Ocean made the Fleet Ballistic Missile System a global deterrent for the first time.[39]

Table 9.1 compares the three versions of Polaris.[40] Presumably because Polarises A1 through A3 were developed in overlapping time frames and used many of the same contractors, separate cost data for each of the three models are not readily available. The cost for the entire Polaris effort came to $5,058,700,000. Cost per missile was $5.535 million, a good $1.1 million less than for Thor, for example, of which many fewer missiles were produced—160 production units, compared with 1,191 for Polaris.[41] On the other hand, since Thor was a single-stage missile, it lacked interstage linkages, although it was much larger than Polaris and certainly more complicated in its plumbing. Hence comparing the two costs is misleading.

Table 9.1. Comparison of Polaris A1, A2, and A3

	A1	A2	A3
Length (feet)	28.6	31.0	32.3
Diameter (feet)	4.5	4.5	4.5
Weight at launch (pounds)	28,949	32,924	36,796
Nominal range (nautical miles)	1,200	1,500	2,500
Speed (Mach number)	7.9	10.0	10.0
Guidance system (inertial)	Mark 1	Mark 1	Mark 2
Accuracy/CEP (nautical miles)	2	2	0.5
Number of warheads	1	1	3
Nose cone	Heat shield	Heat shield	Ablative

Sources: Nicholas and Rossi, *Missile Data Book*, 3-6, 3-9 to 3-10; Fuhrman, "Polaris to Trident," 15, 17, 18; Saxton, "Introduction to PMS (A3P)," 3-12, NASM.

Note: Fuhrman is evidently giving rounded figures, so where there is disagreement I have followed those in Saxton's "Introduction," since it is a Lockheed publication and Lockheed was the system contractor. The length, diameter, and launch weight came from Saxton. Nicholas and Rossi provided range and speed. Presumably the accuracy was only approximate.

Minuteman I

While Polaris was still in development, the Air Force had officially begun work on Minuteman I. Its principal architect was Edward N. Hall, a heterogeneous engineer who helped to begin the Air Force's involvement with solid propellants as a major at Wright-Patterson Air Force Base in the early 1950s, when he was assistant chief of the Non-Rotating Engine Branch of the Power Plant Laboratory. Following a search for a less expensive JATO unit for the B-47 than was available from Aerojet and Hercules, in late 1951 the Non-Rotating Engine Branch began to study and fund solid-propellant rocketry. Hall and his associates used funds acquired in "various semi-legitimate ways over several years," establishing "a very low-profile project with several contractors in the rocket field" to develop reliable thrust-termination devices, lightweight cases, nozzles able to withstand the hot gases expelled from the combustion chamber, thrust-vector-control mechanisms with low drag, and high-performance propellants.[42]

This list looks familiar from the similar list of issues at the time Polaris got started. As Karl Klager of Aerojet has stated, Hall "deserves most of the credit for maintaining interest in large solid rocket technology [during the mid-1950s] because of the greater simplicity of solid systems over liquid systems." Hall's efforts "contributed substantially to the Polaris program," Klager added, further illustrating the extent to which interservice cooperation contributed to the solid-propellant breakthrough.[43]

Details of the program Hall and others funded through "semi-legitimate" means are unsurprisingly sketchy, but clearly some of the work was done from Wright-Patterson before the Air Force set up the Western Development Division in a former schoolhouse in Inglewood to support its ballistic missile efforts. At WDD, Hall, by then a lieutenant colonel, continued work on solids while directing propulsion development in the liquid-propellant Atlas, Titan, and Thor programs. In December 1954 he invited major manufacturers in the solid-propellant industry—Aerojet, Thiokol, Atlantic Research, Phillips Petroleum, Grand Central Rocket,[44] and Hercules Powder—to discuss prospects for solids. The apparent result was the Air Force Large Solid Rocket Feasibility Program (AFLRP). This involved a competition starting in September 1955, with specific companies looking at different technologies. It seems that the propellant for Polaris was a beneficiary of Aerojet's participation in this Air Force program.[45]

Despite this preparatory work for Minuteman, the missile could not begin formal development until the Air Force secured final DoD approval in February 1958, more than a year later than Polaris.[46] The reasons for the delay are convoluted. On the one hand, while the Navy was hardly united in support of a fleet ballistic missile, opposition was probably muted by the realization that failure to support a missile would clearly benefit the Air Force and thus harm the Navy in the interservice competition for roles and missions. Moreover, the Navy could at least unify in opposition to liquid propellants on shipboard, whereas the upper management of ballistic missiles at WDD within the Air Force was already committed to liquid-propellant missiles. Although complicated and more expensive to maintain and launch than solids, they posed fewer impediments on land than at sea. Consequently, while Schriever at WDD enjoyed authority roughly comparable to Raborn's, Hall and others at WDD who wanted him to convert to solids had a tough sell. They were aided, however, by development of Polaris because it provided what Harvey Sapolsky has dubbed "competitive pressure" for the Air Force to develop its own solid-propellant missile.[47]

Besides Hall himself, the principal proponent of what became Minuteman was Barnet R. Adelman, a chemical engineer who directed vehicle engineering for the Ramo-Wooldridge Corporation at WDD.[48] The details and precise dates of the struggle by Hall and Adelman to win WDD and Air Force acceptance of the solid-propellant missile are problematical, with agreement among available sources only on the broad outlines. One source claims that by March 1957 the two heterogeneous engineers agreed that the technology for such a missile now existed. That August, Hall turned over responsibility for Thor to others and accepted an assignment from Schriever to work on the solid ICBM.[49]

By this time, WDD had been redesignated the Air Force Ballistic Missile Division. It appears that Schriever did not initially have much faith in a solid ICBM, because he wanted Hall simply to study the idea, whereas Hall insisted on a development effort. Hall proceeded with it from a windowless office previously used for storage and with no formal staff support, although Adelman continued to work with him, as did a few others from Ramo-Wooldridge, despite opposition from many in the firm.[50]

The two heterogeneous engineers were helped significantly by Air Force civilian engineers back at the Non-Rotating Engine Branch at Wright-Patterson, who had continued efforts to develop large solid-propellant motors. This group found out how aluminum increased solid-propellant performance, perhaps from sources at Aerojet. They put this together with other information they had gathered on solid propellants while Hall was still with the branch, and Bill Fagan carried it all from Ohio to Hall at the BMD.[51]

The specifics of Minuteman technology continued to evolve, but its basic concept took advantage of the lesser difficulty of handling solids to cut the number of people required to launch a missile. Each Minuteman could be remotely launched from a control center using redundant communications cables, without a crew of missileers in attendance. The Air Force initially considered using mobile Minutemen but ultimately decided to launch them exclusively from silos. Because solids were more compact than liquids, Minuteman IA was, by one account, only 53.8 feet long to Atlas's 82.5, Titan I's 98, and Titan II's 108. Its diameter was just over half as great as the other three missiles at 5.5 feet to their 10, and its weight was only 65,000 pounds to Atlas's 267,136, Titan I's 220,000, and Titan II's 330,000. All of this made the cost of silos much lower and substantially reduced the total thrust needed to launch the missile. Consequently, the initial costs of Minuteman were a fifth and the annual maintenance costs a tenth those of Titan I. Moreover, a crew of two could launch *ten* Minutemen, whereas it took a crew of six to launch *each* Titan I—a thirtyfold advantage in favor of Minuteman.[52]

Such facts existed only as projections in 1957–58 (and not even projections for Titan II), but Hall and Adelman used similar information and other arguments to win their superiors' support for Minuteman development in a series of briefings. These occurred mainly on and shortly before February 8, 1958, when Hall was in Washington, D.C., in the company of Schriever and Terhune. Details of the trip differ among sources, but what apparently prompted it was the question whether Polaris could serve as a land-based missile. Schriever evidently still was not completely sold on an Air Force solid missile, but he did recognize the need for a viable Air Force alternative to Polaris. Terhune brought Hall in to brief Schriever on the solid-propellant missile he was developing, which did not yet bear the name Minuteman. Hall presented the cost-effectiveness and simplicity of the concept, and Schriever decided to take him to Washington to make his case.

Accounts agree that the two key briefings were of the Air Force vice chief of staff, Gen. Curtis LeMay, and of Secretary of Defense Neil McElroy. According to Hall, although LeMay had opposed Atlas and Titan, he saw that a solid-propellant missile answered his objections and he joined in the successful effort to convince McElroy of the advantages of Minuteman, a name Schriever chose from several Hall proposed.[53]

Despite approval by McElroy, some influential people at Ramo-Wooldridge's Space Technology Laboratories continued to oppose Minuteman. Hall and Adelman overcame this opposition, and on February 20, 1958, the DoD approved $50 million for research and development of the missile, an amount later increased at least partly through the efforts of Schriever, who now supported the missile.[54]

Soon after the program was approved, both Hall and Adelman left the Ballistic Missile Division. From then until turnover of the operational Minuteman I to the Strategic Air Command in October 1962, there continued to be changes in program directors and in the name of the Air Force organization responsible for the missile's development. From August 1959 to 1963, the program director was Col. (soon Brig. Gen.) Samuel C. Phillips. The Air Force Ballistic Missile Division split in two in 1961, with the Ballistic Systems Division having responsibility for Minuteman but moving to Norton Air Force Base near San Bernardino. Meanwhile, Ramo-Wooldridge had merged with a firm named Thompson Products to become Thompson Ramo Wooldridge, later shortened to TRW.[55]

Ed Hall was a controversial figure, and perhaps some of the opposition to Minuteman stemmed from this fact. Adolf Thiel, who had worked with him on Thor and continued to do so on Minuteman, called him "an obnoxious guy with an engine background" who "was not the easiest guy to work with." Truax said, "He was irascible and at times abrasive, but he knew his facts, and was capable of laying them on the table with telling effect. Gen. Schriever used to take Ed along with him to interservice discussions, partly because Ed always knew what he was talking about, but I think also to stick the knife into his fellow flag officers without having them resent him personally." At one such meeting about turbopumps for Thor and Jupiter, with Medaris and Schriever present, Thiel recalled, "The meeting got hotter and hotter. . . . And Ed Hall . . . jumped up and accused Medaris of being a damn liar, completely disregarding rank." Medaris threatened to have Hall court-martialed. "Then Schriever rose up slowly and looked at Medaris, saying, 'Bruce, if anybody is going to be court-martialed, I am going to decide.'" Then he said, "Now, Ed, calm down."[56]

Col. (later Lt. Gen.) Otto J. Glasser, who succeeded Hall as project manager for Minuteman before turning the responsibility over to Phillips, also commented about Hall's propensity for getting into trouble. But Glasser gave Hall credit for the design of Minuteman and commented that "Hall was an absolutely superb propulsion guy, despite his miserable personality." Terhune, who clearly liked Hall better than did Glasser, agreed that "Ed is not exactly a smooth character. . . . He is an expert on propulsion. He is also very good in all the other technical areas. He defined the Minuteman, he and the people that he had talked to in detail in preparing the specification." And Terhune also credited Hall with giving the briefings to Schriever, LeMay, and McElroy that got Minuteman approved. Finally, Phillips, who brought the Minuteman design to the operational phase, said, "Ed Hall deserves a lot of credit for drumming up interest in solid rockets, for keeping close track of what other people were doing, and for seeing some of the potential and the possibilities for building sufficiently large rockets to make it possible to base an intercontinental-range system on solids."[57] In short, it seems that without Hall's efforts, Minuteman might never have been developed and certainly would not have entered the inventory as soon as it did.[58]

Development of Minuteman involved multiple approaches to arrive at individual technologies, as had earlier been the practice with Atlas. BMD had contracts with Aerojet and Thiokol to work separately on all three stages of the missile. A later contract assigned Hercules Powder to work on the third stage as well. As the firms developed the relevant technologies, parallel development gave way to specific responsibilities, with Thiokol actually building the first stage, Aerojet the second, and Hercules the third. The Space Technology Laboratories division of Ramo-Wooldridge retained its role in systems engineering and technical direction, while BMD selected Boeing for missile assembly and testing—a role later expanded to include integration and checkout of operational facilities. BMD chose North American Autonetics for guidance and control and AVCO for the reentry vehicle, although GE later won the contracts to provide the reentry vehicles for Minuteman II and III.

These companies and organizations all sent representatives to frequent program-review meetings and quarterly gatherings of top corporate officials. Participants examined progress and discovered problem areas that required solution. Then the relevant organizations reached solutions to keep the program on track.[59]

One particularly severe technical challenge involved materials for the nozzle throats and exit cones. The addition of aluminum to the propellant provided a high enough specific impulse to make Minuteman feasible, as it had done for Polaris, and in burning, it produced aluminum oxide particles that damped combustion instabilities. But the hot product flow degraded the nozzle throats and other exposed structures. It seemed that it might not be possible to design a vectorable nozzle that would last the necessary 60 seconds. Solving this problem required "many months and many dollars . . . spent in a frustrating cycle of design, test, failure, redesign, retest, and failure." The Minuteman team used many different grades and exotic compounds of graphite, which seemed the most promising material, but all of them were subject to blowouts or such severe erosion that performance suffered, since the nozzle angle and integrity were essential to the proper expansion of exhaust products from the motor. One solution was a tungsten throat insert, a compromise in view of its higher weight and cost. For the exit cones, the team tried a substance called Fiberite molded at high pressure and loaded with silica and graphite cloth. It provided better resistance to erosion than graphite, but it still was subject to random failures.[60]

Another significant concern was the vectorable feature of the nozzles. Polaris had solved its steering problem with jetavators, but flight control studies for Minuteman showed that the nozzles in stage 1 needed to vector the thrust 8 degrees from the plane of the missile's fuselage, more than Minuteman engineers thought jetavators could deliver. In 1959 when Thiokol tested the motor—the largest solid-propellant power plant yet built—all four nozzles ejected after 30 milliseconds of firing, well before the full stage had ignited. There followed five successive explosions of the motors and their test stands in October 1959, each with a different failure mode. BMD halted first-stage testing in January 1960. A conference among BMD, STL, and Thiokol personnel pointed to two problem areas, internal insulation and the nozzles themselves, as masking other potential problems.

Two concurrent programs of testing were instituted. Firings with battleship-steel cases tested movable nozzles, while Thiokol used flight-weight cases to test a single fixed nozzle massive enough to sustain a full-duration firing. Thiokol solved the insulation problem by summer, but it did not resolve the nozzle issue until fall. This was close to the date set for all-up testing of the entire missile. However, Phillips had ordered Thiokol to begin manufacturing the first stage except for the nozzles, permitting these to be installed whenever they were pronounced ready.[61]

Figure 47. First launch of the Minuteman I missile from Cape Canaveral's Pad 31, February 1, 1961. Official U.S. Air Force photo, courtesy of the 45 Space Wing History Office, Patrick AFB, Fla.

A third major concern was launching from a silo. The first successful launch of the missile occurred in February 1961, on a date that General Phillips referred to as December 63rd since the original date had been in December 1960. This and the two succeeding launches took place from a surface pad and were so successful that the team advanced the first silo launch to August 1961. The missile blew up in the silo, lending credence to critics in STL who had contended from the start that firing a missile from a silo was impossible. As Phillips said, the missile "really came out of there like a Roman candle."[62]

Fortunately, team members recovered enough of the guidance system from the explosion to learn that the problem was not the silo launch itself but quality control. Solder tabs containing connections had vibrated together, causing all of the stages to ignite simultaneously. Knowing this, the team was able to prepare the fifth missile for flight with the problem solved by mid-November, when it had a successful flight from the silo. Previous testing of silo launches aided this quick recovery. In early 1958 engineers from BMD and STL had arranged for development of underground silos. They divided the effort into three phases, with the first two using only subscale models of Minuteman, while the third tested full-sized models. Most of the testing occurred at the Air Force's rocket site at Edwards Air Force Base in the remote Mojave Desert, but Boeing did some of it on subscale models in Seattle.

At the rocket site on Leuhman Ridge at Edwards, subscale silos tested heat transfer and turbulence in some fifty-six tests by November 1958. Meanwhile, Boeing modeled the pressures the rocket's exhaust gases imparted to the missile and silo. It also examined acoustic effects of the noise generated by the rocket motor in the silo on such delicate systems as guidance. Armed with such data, engineers at Edwards began one-third-scale tests in February 1959 and inaugurated construction of full-scale silos. The first full-scale tests used a mock-up made of steel plate with ballast to match the weight and shape of the actual missile, and only enough propellant in the first-stage motor to provide about three seconds of full thrust—enough to move the missile, on a tether, out of the silo and to check the effects of the thrust on the silo and missile. The engineers configured the tether so that the missile would not drop back on the silo and damage it. These tests ensured that the second silo launch at Cape Canaveral on November 17, 1961, was fully successful.[63]

The all-up testing done on Minuteman was itself a significant innovation later used in other programs including the Saturn launch vehicle for Apollo. It abandoned the usual practice for liquid-propellant missiles of gradually testing different capabilities over a series of ranges and conditions—first-stage or booster propulsion, second-stage or sustainer-engine propulsion, guidance and control, and so forth. For the first time, the all-up method performed a single test of all of the missile's functions over the full operational range. Although the practice accorded well with the general procedures of BMD, such as concurrency, its use came about in a curious way. According to Otto Glasser, he was briefing Secretary of the Air Force James H. Douglas on Minuteman, with Gen. Curtis LeMay, then vice chief of staff of the Air

Force, sitting next to Douglas. Douglas insisted that Glasser had moved the first flight of the missile to a year later than originally scheduled. Glasser's denials were of no avail, and the only way he could see to cut a year out of the development process was all-up testing. "Boy, the Ramo-Wooldridge crowd came right out of the chair on that," Glasser said. "There was no way they would stand for a test program that was being done with that sort of lack of attention to all normal, sensible standards." But they had to give in, and all-up testing "worked all the way."[64]

The Minuteman team delivered the first Minuteman I to the Strategic Air Command in October 1962, almost exactly four years after the first development contracts were signed. This was, as Glasser insisted, a year earlier than initially planned.[65] The missile in question was the A-model of Minuteman I, later to be succeeded by a B-model. The former was 53.7 feet long and consisted of three stages plus the reentry vehicle.[66]

The first stage, developed and produced by Thiokol, included a new propellant binder developed by the company's chemists between 1952 and 1954. Thiokol first tried a binder called polybutadiene-acrylic acid or PBAA, an elastomeric (rubberlike) copolymer of butadiene and acrylic acid that permitted higher concentrations of solid ingredients and greater fuel content than previous propellants. It also had a higher hydrogen content than earlier Thiokol polysulfide polymers, which had not had sufficient hydrogen to take full advantage of the aluminum for increasing the specific impulse. With PBAA, a favorable reaction of oxygen with the aluminum generated significant amounts of hydrogen in the exhaust gases, reducing the average molecular weight of the combustion products (since hydrogen is the lightest of elements) and thus adding to performance.[67]

In testing, however, PBAA proved to have a lower tear strength than polysulfide, so Thiokol added 10 percent acrylonitrile to the PBAA, creating polybutadiene-acrylic acid-acrylonitrile (PBAN). The binder and curing agent constituted only 14 percent of the propellant, with ammonium perchlorate (the oxidizer) and aluminum (the major fuel) being the two other main ingredients. The combination yielded a theoretical specific impulse of more than 260 lbf-sec/lbm, with the actual specific impulse at sea level and 70°F somewhat lower than 230.[68]

The case consisted of a double vacuum-melted steel of great strength. In the center of the case, which was insulated with a rubber liner, a Teflon-coated mandrel created the tapered six-point star that formed the internal-burning cavity. The four vectorable nozzles were the largest yet used for a solid-propellant rocket. The control unit that rotated them operated through a battery-powered hydraulic system. On the outside of the case, an Avcoat plastic insulation made by AVCO ablated to protect the exterior from aerodynamic heating. The resultant motor delivered a total impulse of 10,324,400 pound-seconds (lbf-sec).[69]

For stage 2 of Minuteman I, Aerojet used the polyurethane binder it had employed in Polaris, with ammonium perchlorate as the oxidizer and aluminum powder as the principal fuel. It used two slightly different propellant grains, with a faster-burning inner grain and a slower-burning outer one. As the propellant burned, the four-point star-shaped internal burning cavity was converted to a cylindrical one, leaving no slivers of propellant that failed to burn and contribute to the thrust of the missile. The propellant yielded a vacuum specific impulse of nearly 275 lbf-sec/lbm at temperatures ranging from 60°F to 80°F.[70]

The stage-2 case was made of the same steel used for stage 1. The filament-wound glass material used for stage 3 would not work here because, of the three stages, the second was subject to the most severe bending moments (forces) after launch and needed the stiffness the steel provided. The four second-stage nozzles pivoted 6 degrees in pairs for pitch, yaw, and roll control—about 2 degrees less than the first stage's nozzles and 2 degrees more than the third's.[71]

For stage 3, the Hercules Powder Company used the filament-wound case just mentioned and a very different propellant with a different configuration from those for stages 1 and 2.[72] The casing material was Spiralloy, about 80 percent glass filament and some 20 percent epoxy resin. The stage featured four phenolic-coated aluminum tubes for thrust termination and a grain consisting of two separate compositions. The one used for the largest part of the grain included the high-explosive HMX, which had greater energy than ammonium perchlorate but was detonable, hence less safe. HMX was combined with ammonium perchlorate, nitroglycerine, nitrocellulose, aluminum, a plasticizer, and a stabilizer. The second composition had the same basic ingredients minus the HMX and formed a segment at the front (top) of the motor, that was horizontal to the ground when the missile was on the launchpad. A hollow core ran from the back of the motor almost to the non-HMX segment. Described as a "core and slotted tube-modified end burner," it was roughly cone-shaped before tapering off to a cylinder. It yielded a specific impulse of more than 275 lbf-sec/lbm at temperatures ranging from 60°F to 80°F.[73] A hydraulic control unit supplied by NAA's Autonetics Division rotated the four nozzles for Minuteman I stage 3 in pairs up to 4 degrees in one plane to provide pitch, yaw, and roll control.[74]

AVCO produced the reentry vehicle for Minuteman IA, designated Mark 5. It carried a 1-megaton nuclear warhead. Autonetics provided the Minuteman guidance system, with a stable platform supported by two G6B4 gas-bearing, two-degree-of-freedom gyroscopes. VM4A gas-bearing accelerometers measured acceleration. Minuteman I used semiconductors in its electronics and, like Polaris, a digital computer.

For testing the Minuteman I guidance and control system, the Air Force used a new Central Inertial Guidance Test Facility (CIGTF) at the service's Missile Development Center at Holloman Air Force Base in New Mexico. The organizational establishment of the CIGTF dated from 1956, but plans for an actual laboratory awaited a formal announcement in September 1959, with the lab not expected to be fully operational until 1963. In 1959, however, a rocket sled had already tested a North American Aviation inertial guidance system built for the Navaho and similar to the one used on Minuteman I. In January 1960 the Holloman sled began testing an engineering model of the actual Minuteman inertial guidance equipment. There were twenty-four sled tests of this equipment in 1960. Even after the first flight test of Minuteman I on February 1, 1961, on which the guidance and control system was fully operational, sled testing at Holloman continued. Indeed, sled tests could gather better data on the guidance equipment than could flight tests in which the equipment was lost in the ocean, leaving only telemetry data available. The sled tests did not yield as much information as engineers would have liked, but they did "indicate that the system functioned satisfactorily."

Meanwhile, stage 3 of Minuteman I completed its preliminary flight-rating test in January 1961, with stage 1 on the same basic schedule and stage 2 a month behind. Minuteman I Wing I became operational at Malmstrom Air Force Base in Montana in October 1962.[75]

It would be unproductive to follow the evolution of Minuteman through every improvement that went into its later versions, but some discussion of the major changes is appropriate. Wings II through V of Minuteman I (each located at a different base, also called Minuteman IB) featured several measures to increase the missile's range. This was shorter than initially planned because of the accelerated schedule. The range was not a problem at Malmstrom because the base was so far north, close to the Soviet Union, but it would become a problem starting with Wing II. For these missiles, more propellant was added to the aft portion of stage 1, and the exit cones included contouring that made the nozzles more efficient. In stage 2, the material for the motor case changed from steel to titanium, which is con-

Figure 48. Air Force Minuteman I missile, which along with the Navy's Polaris inaugurated the solid-propellant revolution and contributed technology to launch vehicles. Here the missile is being positioned at a test stand on Pad 31 at Cape Canaveral, August 11, 1960. Official U.S. Air Force photo, courtesy of the 45 Space Wing History Office, Patrick AFB, Fla.

siderably lighter. It is also more expensive, but since each pound of reduced weight yielded an extra mile of range, it seemed worth the cost. The nozzles also were lighter. Overall, the weight reduction equaled nearly 300 pounds despite the extra propellant weight. The increase in propellant plus the decrease in overall weight yielded a range increase of 315 miles to a figure usually given as 6,300 nautical miles. There were no significant changes to stage 3.[76]

Minuteman II

For Minuteman II, as with IB, the major improvements occurred in Aerojet's stage 2 (and in the guidance system, which will be discussed below). In stage 1 there had been problems with the nozzles and aft closure involving cracking and ejection of graphite. These were solved through an Air Force reliability improvement program for which details are lacking. On stage 3 there had been instances of insulation burning through in the aft portion. Here too, unspecified design changes were made, inhibiting the flow of hot gases in that region. Stage 2, however, featured an entirely new rocket motor with a new propellant, a slightly greater length, a substantially larger diameter, and a single nozzle that did not swivel but used a liquid-injection thrust vector control system for directional control.[77]

The new propellant was carboxy-terminated polybutadiene (CTPB), which, interestingly, was developed not by Aerojet but by other propellant companies. Some accounts attribute it to Thiokol, which first made it in the late 1950s and developed it into a useful propellant in the early 1960s. Initially, Thiokol chemists used an imine known as MAPO plus an epoxide for curing the CTPB. It turned out that the phosphorous-nitrogen bond in the imine was susceptible to hydrolysis, causing degradation and softening of the propellant. According to Thiokol historian E. S. Sutton's account, "the postcuring problem was finally solved by the discovery that a small amount of chromium octoate (0.02%) could be used to catalyze the epoxide-carboxyl reaction and eliminate this change in properties with time."[78]

A history of Atlantic Research Corporation agrees that Thiokol produced the CTPB but attributes the solution of the curing problem to ARC, which is not incompatible with Sutton's account. According to the ARC history, "ARC used a complex chromium compound, which would accelerate the polymer/epoxy reaction, paving the way for an all epoxy cure system for CTPB polymer. Thus, an extremely stable binder system was developed which would withstand severe environmental conditions, exhibit excellent strain capacity, possess a casting life (pot life) of greater than 12 hours, and cure completely in five days at moderate temperatures."[79]

It frequently happens in the history of technology that innovations occur to different people at approximately the same time. This appears to have been the case with CTPB, which Aerojet historians attribute to Phillips Petroleum and Rocketdyne without providing details.[80] These two companies may have been the source for the information about CTPB that Aerojet used in Minuteman II stage 2. Like Thiokol, in any event, Aerojet proposed to use MAPO as a cross-linking agent. TRW historians state that their firm's laboratory investigations revealed the hydrolysis reaction. They say that "working with Aerojet's research and development staff, a formulation that eliminated MAPO was developed and implemented in the Stage II qualification design."[81] The CTPB that resulted had better fuel values than previous propellants, good mechanical properties such as the long shelf life required for silo-based missiles, and a higher solids content than previous binders. The propellant consisted primarily of ammonium perchlorate, CTPB, and aluminum. It yielded a vacuum specific impulse more than 15 lbf-sec/lbm higher than the propellant used in stage 2 of Minuteman I Wing II.[82]

Although CTPB represented a significant step forward in binder technology, its use was limited by its cost, which was higher than PBAN. Another factor was the emergence in the late 1960s of an even better polymer with both lower viscosity and lower cost, known as hydroxy-terminated polybutadiene (HTPB). This became the industry standard for newer tactical rockets. HTPB had many uses as an adhesive, sealant, and coating, but to employ it in a propellant required, among other things, the development of suitable bonding agents. These tightly linked the polymer to such solid ingredients as ammonium perchlorate and aluminum. Without such links, the propellant's solid particles could separate from the binder network when subjected to temperature cycling, ignition pressure, and other forces, producing voids in the grain, cracks, and sometimes even structural failure.

A key figure in the development of HTPB for use in a propellant was Robert C. Corley, who served as a research chemist and project manager at the Air Force Rocket Propulsion Laboratory at Edwards from 1966 to 1978 and rose through other positions to become the lab's chief scientist from 1991 to 1997. But many other people from Thiokol, the Army at Redstone Arsenal, Atlantic Research, Hercules, and the Navy were also involved. Even HTPB, once it was developed, did not replace PBAN for all uses, including the solid rocket motors for Titan III, Titan IVA, and the space shuttle, because PBAN could be produced in the 1980s for $2.50 per pound at a rate of four million pounds a year, a much higher rate than any other propellant could claim.[83]

To return to Minuteman II, however, the second major change in the stage-2 motor was the shift to a single nozzle with liquid thrust vector control, replacing movable nozzles for control in pitch and yaw. Static firings had shown that the same propellants produced appreciably less specific impulse (7 to 8 lbf-sec/lbm) when fired from four nozzles instead of from a single nozzle. With the four nozzles, liquid particles agglomerated in the approach section of the nozzles and eroded the exit cones. The solution was not only the single nozzle on Minuteman II's second stage but also the change in thrust vector control. The Navy had begun testing a Freon system for thrust vector control for the second stage of Polaris A3 in September 1961, well before the Minuteman II stage 2 program began in February 1962. Low in weight and not sensitive to propellant flame temperature, the system posed negligible constraints on the design of the nozzle. The Minuteman engineers adopted it—one more example of borrowings back and forth between the Polaris and Minuteman projects.

Despite this pioneering work by the Navy and its contractors, according to TRW historians, their firm still had to determine how much "vector capability" stage 2 of Minuteman II would require. TRW analyzed the amount of injectant that could be used before sloshing in its tank permitted the ingestion of air, and the firm also determined what the system performance requirements were. Since Aerojet was involved in the development of the system for Polaris, however, it seems likely that its participation in this process was also important. In any event, the Minuteman team, like the Navy, used Freon as the injectant, confining it in a rubber bladder inside a metal pressure vessel. Both TRW and Aerojet studied the propensity of the Freon to "migrate" through the bladder wall and become unavailable for its intended purpose. They found that only 25 of the 262 pounds of Freon would escape, leaving enough to provide the necessary control in pitch and yaw. A separate roll-control system used a solid-propellant gas generator to correct unexpected roll tendencies.

In addition to these changes, stage 2 of the missile increased in length from 159.2 inches for Minuteman I to 162.32 for Minuteman II. The diameter increased from 44.3 inches to 52.17, resulting in an overall weight increase from 11,558.9 pounds to 15,506. Of this additional 3,947.1 pounds, 3,382.2 consisted of propellant weight. Even so, the propellant mass fraction decreased slightly from 0.897 to 0.887.[84]

The guidance system for Minuteman II saw significant change. Minuteman I had featured what J. M. Wuerth of Autonetics, designer of the Minuteman guidance and control system, claimed was "the first mass production application of semi-conductors to high reliability military electronics."

Minuteman II, he wrote, was "the first program to make major commitment" to semiconductor integrated circuits and "miniaturized discrete electronic parts." The integrated circuits made the computer more powerful, affording more refined computations of trajectory and thus greater accuracy. And miniaturization gave the Minuteman II computer more than two and a half times the memory of Minuteman I in one-quarter the space. The new computer was also lighter and required less power. At the same time, the electronics were less vulnerable to nuclear effects.

A further improvement in the Minuteman II guidance and control system was the substitution of beryllium for aluminum in the stable platform. Beryllium was very expensive and had an extremely toxic dust when machined, but its greater lightness and rigidity plus its less extreme thermal expansion made it far superior to aluminum, according to Wuerth. (Curiously, the Navy had switched from beryllium to aluminum for the gimbals in the guidance and control system for Polaris A3, suggesting that Wuerth's opinion was not universally accepted.) Additionally, the gas-bearing gyros were filled with hydrogen instead of helium, reducing drag and therefore drift for a further improvement in accuracy. Among other changes, Minuteman II's system replaced the VM4A accelerometer on Minuteman I with a pendulous integrating gyro accelerometer designed by MIT's Instrumentation Lab. It too employed gas-bearing gyros. The overall result was that Minuteman II was three times as accurate as Minuteman I—a requirement resulting from a shift in U.S. strategy from massive retaliation to flexible response, necessitating that land-based missiles be accurate enough to destroy enemy missiles in their silos.

One disadvantage of the new guidance and control system was the infancy of the technology involved in integrated circuits. Many failed soon after deployment and had to be pulled out of service and replaced. But continued development ultimately solved the reliability problem and reduced the costs of system maintenance.[85]

The Strategic Air Command put the first Minuteman II squadron on operational alert in May 1966, with initial operational capability declared that December. In the next few years, the Air Force began removing the Minuteman I from service and putting the Minuteman II in its places. By May 1969 there were 500 of each type on alert in silos.[86]

Minuteman III

Minuteman III featured multiple, independently targetable reentry vehicles with a liquid fourth stage to deploy them. This last feature was not particularly relevant to launch-vehicle development except that the added weight necessitated higher booster performance. Stages 1 and 2 did not change from Minuteman II, but stage 3 became larger. Hercules did not win the contract for the larger motor. Instead, Aerojet got the award for design and development of the new third stage. Subsequently, Thiokol and the Chemical Systems Division of United Technologies Corporation won contracts to build replacement motors. Stage 3 featured a fiberglass motor case, the same basic propellant Aerojet had used in stage 2 but in slightly different proportions, an igniter, a single nozzle that was fixed in place and partially submerged into the case, a liquid-injection thrust vector control system for control in pitch and yaw, a separate roll control system, and a thrust termination system.

Aerojet had shifted its filament-wound case production from Azusa to Sacramento. It produced most of the Minuteman fiberglass combustion chambers there but ceased filament winding in 1965. Meanwhile, Young had licensed Hercules' Spiralloy technology to Black, Sivalls, and Bryson in Oklahoma City, and it became a second source for the Minuteman stage-3 motor case. This instance and the three firms involved in producing the third stage illustrate the extent to which technology transferred among the contractors and subcontractors for government missiles and rockets.

The issue of technology transfer among competing contractors and the armed services, which were also competing over funds and mission responsibility, is a complex one about which a whole chapter—even a book—could be written. To address the subject briefly, an effort to gather knowledge about rocket propulsion technology had started in 1946 when the Navy funded a Rocket Propellant Information Agency (RPIA) within Johns Hopkins University's Applied Physics Laboratory. The Army added its support in 1948, whereupon the RPIA became the Solid Propellant Information Agency (SPIA), with the Air Force adding its support in 1951 and NASA in 1959. With liquid-propellant rocket activity increasing, the Navy created the Liquid Propellant Information Agency (LPIA) in 1958. As further development of rockets and missiles ensued, it became obvious that better channels for information exchange were needed. In 1962 the DoD created an Interagency Chemical Rocket Propulsion Group, later renamed the Joint Army/Navy/NASA/Air Force (JANNAF) Interagency Propulsion Committee. And SPIA and LPIA merged on December 1, 1962, to form the Chemical Propulsion Information Agency, which also effectively promoted sharing of technology. In addition, "joint-venture" contracts—by one account, pioneered by Levering Smith of the Navy—often mandated the sharing of manufacturing technology between companies. These contracts served to eliminate the services' dependence for a given technology on sole sources that could be destroyed by fire or possible enemy targeting. The practice also ensured that there was competitive bidding on future contracts. The Air Force had a similar policy.[87]

To return to Minuteman III, Aerojet's third stage used the same propellant as stage 2, but the grain configuration consisted of an internal-burning cylindrical bore with six "fins" radiating out in the forward end, an arrangement referred to as "finocyl." This yielded a vacuum specific impulse slightly lower than for stage 2. The igniter used black-powder squibs to start some of the CTPB propellant, which in turn spread the burning to the rest of the grain.

The 50-percent-submerged nozzle had a graphite phenolic entrance section, a forged tungsten throat insert, and a carbon phenolic exit cone. Where Minuteman II's third stage was 85.25 inches long and 37.88 inches in diameter, Minuteman III measured 91.4 inches by 52 inches. The mass fraction improved from 0.864 to 0.910, and with a specific impulse greater by 9 lbf-sec/lbm, the new third stage had more than twice the total impulse of its predecessor—2,074,774 lbf-sec as compared with 1,006,000.[88]

The thrust vector control system for the new stage 3 was similar to that for stage 2 except that strontium perchlorate replaced Freon as the injectant into the thrust stream to control motion in the pitch and yaw axes. Helium gas provided the pressure to insert the strontium perchlorate instead of the solid-propellant gas generator used in the second stage. Roll control again came from a gas generator supplying gas to nozzles pointing in opposite directions. When both were operating, there was neutral torque in the roll axis. When roll torque was required, the flight control system closed a flapper on one of the nozzles, providing unbalanced thrust to stop any incipient roll.

To ensure accuracy in delivery of the warheads, Minuteman had always required precise thrust termination for stage 3 as determined by the flight control computer. On Minuteman I, the thrust termination system consisted of four thick carbon-phenolic tubes integrally wound in the sidewall of the third-stage case and sealed with snap-ring closures to form side ports. Detonation of explosive ordnance released a frangible section of the snap ring, venting the combustion chamber and causing a momentary negative thrust that allowed the payload to separate from the third stage. The system for Minuteman III involved six circular charges on the forward dome. Using high-speed film and strain gauges, the Minuteman team learned that within 20 microseconds, this arrangement cut holes that resulted in a rupture of the case within 2 additional milliseconds. Subsequent examination of the

Figure 49. Minuteman III rising on its final launch from Cape Canaveral on December 14, 1970. Official U.S. Air Force photo, courtesy of the 45 Space Wing History Office, Patrick AFB, Fla.

case showed cracks that radiated from the edge of the holes. TRW used structural analysis software called NASTRAN on its computers to define the propagation of the cracks. It then determined the dome thickness needed to eliminate failure of the fiberglass. Aerojet wound "doilies" integrally into the dome of the motor case under each of the circular charges. Testing showed that these eliminated the rupturing. Thereafter the system reliably vented the pressure in the chamber and produced momentary negative thrust, causing separation between the postboost vehicle and the third stage.[89]

Table 9.2. Comparsion of Minuteman I, II, and III

	IB	II	III
Length (feet)	55.9	57.6	59.9
Diameter (feet)	5.5	5.5	5.5
Launch weight (pounds)	69,920	72,810	78,000
Speed (Mach number)	19.7	19.7	19.7
Range (miles)	6,300	7,021	8,083
Accuracy/CEP (nautical miles)	1.1	0.26	0.21
Reentry vehicle	ablative	ablative	ablative
Unit cost	$5,783,000[a]	$7,269,000	$8,005,000

Sources: Lonnquest and Winkler, *To Defend and Deter*, 241–43; Nicholas and Rossi, *Missile Data Book*, 3-5, 3-8, which gives costs only for the A model of Minuteman I; MacKenzie, *Inventing Accuracy*, 428; ICBM System Program Office, "Minuteman Weapon System," 14; R. Anderson, "Minuteman to MX," 33.
a. Cost is for Minuteman IA.

Minuteman IIIs achieved their initial operational capability in June 1970, with the first squadron of the upgraded missiles turned over to an operational wing at Minot Air Force Base in North Dakota in January 1971. By July 1975 there were 450 Minuteman IIs and 550 Minuteman IIIs deployed at Strategic Air Command bases, every Minuteman I having evidently been retired. Table 9.2 compares the three versions of the missile.[90]

Conclusions

The deployment of the Air Force's Minuteman I in 1961 marked the completion of the solid-propellant breakthrough. Innovations and improvements to the basic technology continued, of course, to be made. But a gradual phaseout of liquid-propellant missiles followed almost inexorably from the appearance on the scene of the first Minuteman. The breakthrough in solid-rocket technology required the extensive cooperation of a great many firms, government laboratories, and universities, only some of which could be mentioned here. It occurred on many fronts ranging from materials science and metallurgy through chemistry to the physics of internal ballistics and the mathematics and physics of guidance and control. It was partially spurred by interservice rivalry for roles and missions. Less well known, however, is the contribution of interservice cooperation. Funding for advances in and the sharing of technology came from all three services, the Advanced Research Projects Agency, and NASA. Technologies such as aluminum fuel, methods of thrust vector control, and improved guidance and control transferred from one service's missiles to another. Also crucial were the roles of heterogeneous engineers like Raborn, Schriever, Draper, Hall, and Adelman.

But a great many people with more purely technical skills, such as Levering Smith, Sam Phillips, Price, Rumbel, and Henderson made vital contributions. Without all of these elements, the breakthrough would not have been possible.

Innovation was clearly a key ingredient in the progress of rocketry, whether the propellants were liquid or solid. Occasionally the available sources yield clues to how innovations occurred. In the case of Rumbel and Henderson's discovery of aluminum as a key to higher performance, the main ingredient appears to have been their simple refusal to accept existing theory and a concomitant willingness to press ahead with empirical testing, shaped by their technical training and knowledge. Other times, no doubt, existing theory was helpful in solving problems that cropped up in testing. In most cases, as with Fiedler's jetavators, we don't know exactly how innovations came about. Sometimes we don't even know the names of the individuals who came up with them or which unit of a given missile team suggested a novel approach. In large organizations or teams, there often were simply too many people involved in discussions to remember who came up with which suggestion. The Air Force's Otto Glasser likened the question of responsibility for individual breakthroughs to the conundrum "If you were to back into a buzz saw could you tell me which tooth it is that cut you?"[91]

As technology advanced and increasing numbers of people and organizations became experienced in the development and testing of rockets, confidence and competence increased. Without both, the Polaris program would not have launched two missiles from the *George Washington* on July 20, 1960, risking failure with a second launch when success on the first would have sufficed. And without both confidence and competence, when BMD was coerced into all-up testing for the Minuteman, it would have been unable to meet the challenge. Missile designers still could not foresee all of the problems that would occur with their "birds," though, so empirical solutions continued to be frequent and necessary when systems failed to perform as expected. Elaborate testing also continued to be necessary, with new facilities such as the inertial guidance sled-testing laboratory at Holloman Air Force Base and the Skytop static test stand at NOTS coming on the scene during the course of the Polaris and Minutemen programs.

The significant shift from liquid to solid rockets in the missile programs was not the only result of the solid-propellant breakthrough. It also had important effects on launch vehicles. The propellants for the large solid-rocket boosters on Titan III, Titan IVA, and the space shuttles were derived from the one used on the Air Force's Minuteman stage 1. Without Atlantic Research's discovery of aluminum as a fuel and Thiokol's development of PBAN as a binder, it is not clear that the huge Titan and Shuttle boosters would have been possible. Another possible spinoff from the Polaris program, or maybe just a parallel development with shared technology, was the Altair II motor, produced by the Hercules Powder Company under the sponsorship of the Bureau of Naval Weapons and NASA and used as a fourth stage for the Scout launch vehicle. Hercules produced Altair II almost simultaneously with the Polaris A3 stage 2. Although Altair II was a smaller motor with a different grain configuration, it used a fiberglass/resin case and the same basic propellant ingredients as the larger Polaris A3 stage.[92] Many other solid-propellant formulations used aluminum, HMX, and other ingredients of the Polaris and Minuteman motors as well. Although some or all of them might have been developed even if there had been no urgent national need for solid-propellant missiles, it seems highly unlikely that their development would have occurred as quickly as it did.

Conclusion

This book has followed the history of rocket technology, especially missile technology, from its early beginnings in the research and writings of Robert Goddard and Hermann Oberth through the development of the Minuteman missile. The sequel, *U.S. Space Launch Vehicle Technology: Viking to Space Shuttle*, will pick up the evolution of the technology, especially as it applies directly to space launch vehicles, from the Viking rocket and the Vanguard launch vehicle through the space shuttle. Already in early missile development, many of the themes of *Viking to Space Shuttle* had begun to emerge.

One such theme is the diversity of contributors to rocket technology in America. The German V-2 and the team of German rocket developers under Wernher von Braun undoubtedly made major contributions to American rocket development. The Redstone missile that they helped to develop became, with modifications, the first stage of America's first launch vehicle, the Juno I. But the engine for Redstone included many contributions from Americans, including work done for the Air Force's Navaho missile by North American Aviation and its later Rocketdyne Division. And the upper stages for Juno I came from the Jet Propulsion Laboratory, which also contributed many other innovations such as the composite solid propellants and storable, self-igniting liquid propellants that greatly influenced the development of U.S. missiles and rockets.

James H. Wyld, associated with the American Rocket Society in the 1930s, developed the United States' first regeneratively cooled combustion chamber, to prevent the burn-throughs that plagued early rocket engines. His work led to the founding of the first U.S. rocket company, Reaction Motors, in December 1941. Among its innovations was "spaghetti" construction for combustion chambers, invented by Edward A. Neu Jr. Other important contributions came from the Naval Engineering Experiment Station at Annapolis, including the first American use of self-igniting propellants, later explored further by JPL and its offshoot Aerojet Engineering, the second U.S. rocket development firm.

In solid-propulsion technology, in addition to JPL's advances, a separate rocket effort by the California Institute of Technology in Eaton Canyon in southern California made important discoveries. Other technological developments came from the Naval Ordnance Test Station (NOTS) in the sparsely populated desert region around Inyokern, California, the Allegany Ballistics Laboratory in western Maryland, Atlantic Research Corporation, Aerojet, and the Thiokol Chemical Corporation. These included improved composite propellants, castable double-base propellants, and solid propellant grains with internal burning cavities. Such cavities protected the case from overheating and enabled sophisticated control of the thrust the rocket would yield over time. Together with a large increase in the size of solid-propellant rockets, these innovations furthered the development of missiles and launch vehicles in important ways.

The Air Force, beginning with its MX-774B project in 1946–48, made major contributions to missile and later launch-vehicle development, with Atlas as the first ICBM, Thor becoming the basis of the later Delta launch vehicle, Titans I and II serving as significant early missiles, and Atlas and Titans III and IV becoming critical launch vehicles. And the Navy significantly expanded solid-propellant technology through Polaris, which together with the Air Force's Minuteman contributed in important ways to various stages of launch vehicles.

Another theme that emerges in this book is the empirical nature of rocket development. Often referred to as rocket science, the process of developing complex missiles and early launch vehicles such as the Juno I and the Mercury-Redstone entailed a great many improvisational solutions to problems, and often downright trial and error. Although the technical literature and data that informed rocket development grew quickly, the processes used in designing and developing rockets and missiles hardly constituted a mature science capable of making accurate predictions. Early rockets were notoriously prone to failure, and even the kinds and causes of failure were often unpredictable. So what emerged in the process of rocket development was not a science (at least by some definitions) but rather an engineering culture marked by the ability of its practitioners to create innovative solutions to problems as they occurred in ground or flight testing. Often a solution arose despite uncertainty about the cause of failure, as engineers from a variety of disciplines and organizations were able to gather available information and use it to fix the problem through engineering art and intuition as well as more purely scientific procedures.

We know much less about such innovation than would be desirable. Many discoveries or fixes for problems resulted from group efforts that are not well documented, with no record of who made critical suggestions and how they were implemented. Specifications for individual missiles became quite precise, but the process behind that precision remains something of a mystery. What we do know are the results. By way of example, we have seen that for the Air Force's Atlas missile, important innovations included: (1) the steel-balloon tanks of Convair's Charlie Bossart; (2) numerous new engine technologies at Rocketdyne; (3) digital computing at and inertial guidance at several firms and organizations; (4) nose cone development at ABMA, AVCO, Cornell University, and General Electric; and (5) testing apparatuses ranging from shock tubes to rocket sleds. Rocketdyne, especially, showed that a willingness to take risks could pay off in engine improvements.

But some of those improvements caused problems in flight testing, which remained the final adjudicator of whether systems engineering on a complicated rocket had been successful. In both ground and flight tests, engineers found out that there were limitations in their ability to use theory and available information. Combustion instability, turbopump failures, and propellant sloshing, among other problems, revealed the unpredictabilities that remained inherent in rocket engineering. Cut-and-try or related empirical methods continued to be necessary to overcome unexpected failures.

Another element in the mix was cooperation and a sharing of information across competing services and industries. Although the Thor-Jupiter controversy highlighted intense competition between the Army and the Air Force, the two rivals also shared a great deal of information. The same was true between the Navy and the Air Force in the development of Polaris and Minuteman. These two solid-propellant missiles altered the course of missile development and made crucial contributions to future launch vehicles, as solid-propellant upper stages continued to evolve and strap-on solid-propellant motors added their thrust to that of liquid-propellant stages. Technologies such as aluminum fuel, methods of thrust vector control, and improved guidance and control transferred from one service's missiles to another. Also crucial were the roles of heterogeneous engineers, who engineered the social as well as the physical world, such as Bernard Schriever, William F. "Red" Raborn, Charles Stark Draper, Edward Hall, and Barnet Adelman. But a great many people with more purely technical and managerial skills, including Levering Smith, Samuel Phillips, Edward Price, Keith Rumbel, and Charles B. Henderson, made vital contributions as well.

Meanwhile, even as technology evolved, some problems remained. One was combustion instability, which affected both liquid- and solid-propellant power plants. Following the destruction during the summer of 1961 of an operational Atlas missile due to combustion instability, the Air Force investigated the problem. Specialists from universities, industry, the Navy, NASA, and other organizations reported that combustion instability was still "in the realm of the unknown" despite years of research. Industry had to resort to "crutches" and "cut-and-try" techniques to overcome combustion problems. As with the Titan and earlier engines, researchers had some success using baffled injectors, but these were hardly fully developed solutions capable of across-the-board application. Engineers did not completely understand how baffles worked, and they still considered injector design to be an art, necessitating trial-and-error methods to come up with stable combustion. Some progress occurred in ensuing years, but engineering for rockets was still partly an empirical process in which vehicles had to be built and tested, on the ground and in flight, before problems could be found and solved.

A further theme that emerged in missile development was technology transfer. The rapid development of the Jupiter missile, for instance, owed much to preceding technologies developed on the V-2, Navaho, and Redstone. Thor borrowed from Navaho and Atlas. The process would continue with technology transferring from missiles to launch vehicles, although occasionally the flow of information went in the opposite direction. Thus, as *Viking to Space Shuttle* discusses, Viking and Vanguard pioneered the use of gimbals for directional control, with missiles appropriating the technology.

A final factor in missile and launch-vehicle development was management. As missiles became larger and more complex, it became increasingly necessary to develop management systems to ensure that their electrical, propulsion, structural, and guidance/control systems worked together effectively; that the multitude of firms and engineers engaged in producing the components of the missiles delivered their products on schedule and complied with specifications; that effective testing occurred at the appropriate times; that teams of engineers and technicians resolved problems as they occurred; and that costs were contained as much as possible. Threats from the Soviets during the cold war ensured that vast amounts of funding were available—itself a crucial factor in the development of missile and launch-vehicle technology—but there were still limitations that necessitated cost control. Each organization adopted its own solution to these systems-engineering issues, but the two most important management systems were those of the Western Development Division for the Air Force and the Program Evaluation and Review Technique (PERT) of the Polaris team. Different in detail, they were similar in their goals and methods. As will be seen in *Viking to Space Shuttle*, both went on to influence NASA's development of the giant Saturn launch vehicles.

Which of these themes and factors were most important? Any answer to that question must be prefaced by the recognition that history is not a laboratory science.[1] We cannot repeat historical events numerous times, on each occasion deleting one factor to determine its importance relative to the others. That said, it seems likely that the most important single ingredient in the successful development of missile and rocket technology was funding. Without the enormous expenditures lavished on missile and launch-vehicle development, it seems unlikely that progress on the technologies involved would have been nearly as rapid as it was. Perhaps tied for second in importance were information sharing (technology transfer) and innovation. Both were dependent upon funding, and they worked together to further technological progress.

But without management systems like those of the Schriever and Raborn teams, none of these factors alone would have been effective. All of them interacted in immeasurable ways to promote progress. On balance, probably management systems were third in importance. Integral to both funding and management systems was the role of heterogeneous engineers. It was they who generated funding from Congress and the DoD and they who encouraged participating engineers, technicians, and others to work together to achieve important technological breakthroughs.

Scarcely less important than management and heterogeneous engineering were the huge numbers of contributing agencies—the armed services, a variety of industrial concerns, universities, and the like—and the engineering culture that took available theory and data from previous projects and combined them to resolve unexpected problems as they occurred. Both of these factors contributed to innovation, and it was the sheer number of contractors and subcontractors, starting with Atlas missile development, as well as the complexity of the missiles themselves and the relatively undeveloped state of the technology in the early years, that necessitated the management systems Schriever and Raborn pioneered.

What, finally, can be said about the relative importance of the many organizations that contributed to missile development? Obviously, the Air Force, the Army, and the Navy were all central to missile development. It is virtually impossible to say that one was more important than another. For example, despite the Army's early start with the WAC Corporal and the Corporal E, the engine for the Redstone evolved from a power plant for the Air Force's Navaho project. In the process of developing the Air Force's Atlas and Thor plus the Army's Jupiter, the two services each made important contributions to one another's development efforts, as also happened between the Air Force and the Navy in the development of the Polaris and

Minuteman missiles. To this day, both the Navy and the Air Force supply important missiles for the strategic defenses of the United States. Both also added to launch-vehicle technology, although the Air Force undoubtedly has contributed more to launch vehicles over the years than either the Army or the Navy.

On the Army side of the ledger, the von Braun group transferred to the NASA Marshall Space Flight Center and contributed very significantly to the Saturn launch vehicles, as *Viking to Space Shuttle* relates. Some of the former Army personnel later contributed directly and through the engineering culture they established at Marshall to the development of the space shuttle main engine. What is critical here is not the issue of which service was most important but the fact that all three contributed in various ways to a national effort. Interservice rivalries stimulated efforts to succeed in individual projects like Thor, Jupiter, Polaris, and Minuteman, and without those rivalries, progress might have been slower. But sharing of information was even more crucial.

Among the contractors, Rocketdyne was almost certainly the most important rocket-development entity, followed by Aerojet, Martin, Thiokol, and, in the early years, Reaction Motors. But General Electric made early contributions too, and Atlantic Research was the company that discovered the important addition of aluminum to solid-propellant grains. JPL, a contractor by turns to the Army and NASA, also advanced technology in significant ways, as did Lockheed. Again, the point is not to rank individual contributors (of which this list is only partial) but to emphasize their sheer number and the fact that these and many other firms contributed in a huge variety of ways to the development of missiles and launch vehicles. *Viking to Space Shuttle* will elaborate on all of these points.

Appendix

Chronology

Before 1100	Gunpowder invented in China.
Mid-1100s	Probable first use of military rockets by Chinese.
Mid-1200s	Probable first use of military rockets by Arabs and Europeans.
1780–99	Indian troops used rockets against British in Mysore Wars.
1805	William Congreve adapted Indian rocket technology for European use.
1805–15	British troops used Congreve rockets against Danes, French, and United States.
1820s	Congreve-type rockets adapted for use in lifesaving and whaling.
1840	William Hale invented first spin-stabilized rocket.
February 9, 1909	Experiment by Robert Goddard on exhaust velocity of a black-powder rocket.
January 1920	Smithsonian Institution published Goddard's "A Method of Reaching Extreme Altitudes."
1921	Goddard switched to liquid propellants.
1923	Hermann Oberth's *Die Rakete zu den Planetenräumen* (The rocket into interplanetary space) published.
March 26, 1926	Goddard launched first liquid-propellant rocket known to fly.
April 4, 1930	Creation of American Interplanetary Society (after 1934, American Rocket Society).
1936–39	Small group of researchers around Frank Malina at Caltech began rocket research that led to later creation of Jet Propulsion Laboratory.
December 10, 1938	James H. Wyld tested first American engine to apply regenerative cooling to entire combustion chamber.
June 1940	Former Goddard associate Clarence N. Hickman submitted rocket proposals to Frank B. Jewett, president

	of Bell Labs and a division chairman of recently created National Defense Research Committee, leading to establishment of Section H (for Hickman) of NDRC's Division of Armor and Ordnance. Section H engaged in rocket research and development during World War II.
December 16, 1941	Reaction Motors, America's first rocket company, founded.
1942	John W. Parsons of Malina group at Guggenheim Aeronautical Laboratory, California Institute of Technology (GALCIT) developed first castable composite solid propellant.
March 1942	Aerojet Engineering Corporation (later Aerojet General Corporation), America's second rocket firm, founded.
October 3, 1942	German A-4 (V-2) missile successfully flight- tested.
1942–45	Caltech, urged by its Charles Lauritsen, operated a rocket research effort in Eaton Canyon, Calif.
November 1943	Naval Ordnance Test Station (NOTS) established near Inyokern, Calif.
1944	GALCIT began calling itself Jet Propulsion Laboratory (JPL).
November 15, 1944	Army Ordnance contracted with General Electric (GE) for Hermes project.
October 11, 1945	JPL's WAC Corporal reached an altitude of 230,000–240,000 feet.
Fall 1945	At North American Aviation (NAA), William Bollay created what became Aerophysics Laboratory, forerunner of Rocketdyne Division.
1946	Caltech's Edward W. Price developed internal-burning, star-shaped perforation for White Whizzer rocket, allowing higher propellant load and protecting case from heat of burning.
March 29, 1946	Army Air Forces contracted with NAA for MX-770 missile, which evolved into Navaho.
December 3, 1946	Navy funded Rocket Propellant Information Agency, a forerunner of Chemical Propulsion Information Agency.
1946–48	Air Force's MX-774B furthered ballistic missile technology.

mid–late 1940s	Charles Bartley at JPL adopted Thiokol's polysulfide as a solid-propellant binder.
1947	Bartley and associates switched from potassium perchlorate to ammonium perchlorate as oxidizer.
May 22, 1947	First launch of Army's Corporal E missile.
September 18, 1947	U.S. Air Force became separate military service, severing ties with U.S. Army, in which it had been Army Air Forces.
1948	Thiokol entered rocket business in Elkton, Md. Army began supporting Rocket Propellant Information Agency, renamed Solid Propellant Information Agency.
1949	Thiokol designed T-41 motor for Hughes Aircraft's Falcon missile.
February 24, 1949	Two-stage Bumper WAC climbed a reported 244 miles in first successful American ignition of a second stage at altitude.
November 1949	Construction began on Air Force's rocket testing site at Edwards Air Force Base, Calif.
1950	Reaction Motors' Edward A. Neu Jr. applied for patent on "spaghetti" construction of combustion chambers.
July 10, 1950	Development of Army's Redstone missile began.
1951	Aerojet under Navy contract began studying hydrazine as rocket propellant.
January 23, 1951	Air Force contracted with Convair for MX-1593 project, which soon became Atlas.
December 1951	Successful static test by Thiokol and GE of RV-A-10 missile's 31-inch solid-propellant motor.
Late 1951	Air Force's Non-Rotating Engine Branch began to study and fund solid-propellant rocketry.
20 August 20, 1953	First flight test of Army's Redstone missile.
1953–57	At Air Force insistence, NAA conducted Rocket Engine Advancement Program, which led to development of RP-1 (kerosene) as rocket fuel.
1954	Army's Corporal missile declared operational.
February 1, 1954	Teapot Committee, commissioned by Air Force, issued important report stressing urgency of intercontinental ballistic missiles (ICBMs).

July 1, 1954	Western Development Division (WDD) established in Inglewood, Calif., to manage the development of ICBMs.
August 2, 1954	Air Force Brig. Gen. Bernard A. Schriever assumed command of WDD.
1954–55	Keith Rumbel and Charles B. Henderson of Atlantic Research Corporation discovered that aluminum as main fuel in solid propellant could significantly increase specific impulse.
1954–62	Army's Sergeant solid-propellant missile developed.
1955	Rocketdyne Division of NAA established.
January 6, 1955	Air Force contracted with Convair Division for development and production of Atlas airframe, plus integration, assembly, and testing of its other subsystems.
January 14, 1955	WDD and collocated Special Aircraft Project Office (SAPO) of Air Materiel Command contracted with Aerojet General for development of alternate liquid-oxygen–hydrocarbon propulsion system for Atlas missile (later used in Titan I).
April 28, 1955	Air Force Secretary Harold E. Talbott preliminarily approved development of a second ICBM, which became Titan I.
May 2, 1955	Air Force gave formal approval to the development of an ICBM with two full stages, later called Titan I.
ca. September 1955	Air Force began Large Solid Rocket Feasibility Program.
September 14, 1955	Air Materiel Command made Glenn L. Martin Aircraft Company overall contractor for two-stage missile that became Titan I.
November 8, 1955	Defense Secretary Charles E. Wilson formally directed Air Force to proceed with intermediate range ballistic missile (IRBM) number 1, which became Thor, and told Army to develop an alternate IRBM that became Jupiter, with Army and Navy collaborating on a sea-launched version of that missile.
November 17, 1955	Navy's Special Projects Office established to work with Army on adapting Jupiter for sea use (but ultimately, as it turned out, to oversee Polaris).
December 23, 1955	WDD and SAPO contracted with Douglas Aircraft for airframe and assembly of Thor IRBM.

February 1, 1956	Army Ballistic Missile Agency activated.
November 26, 1956	Defense Secretary Wilson limited Army jurisdiction to missiles up to a 200-mile range for tactical battlefield use, giving operational control over all IRBMs to Air Force.
December 8, 1956	Secretary Wilson directed Navy to begin Polaris program, adding it to Jupiter and Thor as IRBMs under development.
January 25, 1957	First (unsuccessful) flight test of Air Force's Thor IRBM.
March 1957	Air Force Col. Edward N. Hall, assisted by Barnet R. Adelman, began work on solid-propellant ICBM that became Minuteman.
March 1, 1957	First (unsuccessful) flight test of Army's Jupiter IRBM.
June 1, 1957	Redesignation of Air Force's WDD as Ballistic Missile Division (AFBMD).
June 11, 1957	First (unsuccessful) flight test of Air Force's Atlas A missile.
July 1, 1957–December 31, 1958	International Geophysical Year.
October 4, 1957	Soviet launch of Sputnik 1.
December 6, 1957	First (failed) launch of Vanguard with a satellite.
ca. 1958	Hercules' Allegany Ballistics Laboratory developed composite-modified double base (solid) propellant.
1958	Navy established Liquid Propellant Information Agency.
January 31, 1958	Army's Juno I launched Explorer I, America's first satellite.
Febraury 7, 1958	Department of Defense (DoD) activated Advanced Research Projects Agency.
February 20, 1958	DoD approved what in June became Minuteman solid-propellant missile project.
April 23, 1958	First (unsuccessful) launch of a Thor-Able vehicle.
June 18, 1958	Army deployed Redstone Missile.
October 1, 1958	National Aeronautics and Space Administration (NASA) created.
October 9, 1958	Air Force contracted with Boeing Airplane Company for assembly and testing of Minuteman missile.
October 11, 1958	A Thor-Able vehicle launched first successful space probe, NASA's Pioneer I.

December 18, 1958	Successful launch into orbit of entire Atlas missile with a communications repeater in Project Score.
December 20, 1958	First (unsuccessful) flight test of a Titan missile.
January 19, 1959	General Schriever approved in-silo launch for future Titan II.
February 6, 1959	Successful launch of Titan I with dummy second stage.
April 28, 1959	NASA contracted with Douglas to develop a Thor-Vanguard launch vehicle named Delta.
May 1959	Initial operational capability for Jupiter IRBM.
June 1959	First Thor squadron on operational alert in Great Britain.
September 1, 1959	Air Force's Atlas D, first U.S. ICBM, operational.
November 1959	DoD authorized Air Force to develop Titan II.
January 7, 1960	First guided flight of Polaris A1, marking first use of a digital computer for an inertial guidance system on a missile.
July 1, 1960	First launch of four-stage Scout solid-propellant launch vehicle with all stages live (from Wallops Island).
July 29, 1960	First launch of functional Polaris missile.
	AFBMD contracted with Martin for production of Titan II.
	First (unsucessful) flight test of Mercury-Atlas 1 launch vehicle.
September 21, 1960	First launch of Air Force Blue Scout (from Cape Canaveral).
February 1, 1961	First launch of Minuteman missile and first test with all stages functioning on initial flight (all-up testing).
February 21, 1961	Successful launch of Mercury-Atlas 2.
April 1, 1961	Air Force reorganization, creating Air Force Systems Command with subordinate Ballistic Missile Division split into Ballistic Systems Division and Space Systems Division.
May 5, 1961	Astronaut Alan Shepard launched on Mercury-Redstone 3 into successful suborbital flight.
September 29, 1961	Successful test of liquid-injection thrust-vector-control system for Polaris A3.
November 17, 1961	First successful launch of Minuteman missile from underground silo.

February 20, 1962	Astronaut John Glenn launched into orbit on Mercury-Atlas 6.
March 16, 1962	First launch of Titan II.
April 18, 1962	First Titan I squadron deployed.
June 1962	Polaris A2 operational.
October 1962	First Minuteman I delivered to Strategic Air Command.
December 1, 1962	Creation of Chemical Propulsion Information Agency from Solid Propellant Information Agency and Liquid Propellant Information Agency.
June 13, 1963	First Titan II squadron operational.
October 26, 1963	First launch of Polaris A3 from a submarine.
April 8, 1964	Successful test flight of *Gemini 1* on a Titan II.
September 28, 1964	Polaris A3 operational.
March 23, 1965	Successful launch of manned *Gemini 3* on a Titan II.
December 8, 1965	DoD approval for development of Minuteman III.
July 18–21, 1966	*Gemini 10* mission, with astronauts Michael Collins and John Young successfully docking with Agena spacecraft launched by an Atlas launch vehicle.
December 1966	Initial operational capability for Minuteman II.
May 26, 1967	Air Force Systems Command consolidated Ballistic Systems Division and Space Systems Division into Space and Missile Systems Organization (SAMSO).
February 12, 1969	Final Minuteman I removed from silo, ending deployment of that missile.
June 1970	Initial operational capability for Minuteman III.

Sources: Dates from narrative, except for entries before 1900, which are based on Van Riper, *Rockets and Missiles*, xiii–xiv; initial deployment dates of Titan I and Titan II, taken from Stumpf, *Titan II*, 31, 134; scattered entries from "BMO Chronology."

Notes

Notes are in shortened format, with full information provided in the list of sources. Archival and private sources are identified by an abbreviation at the end of the citation denoting the repository, such as NHRC for the NASA Historical Reference Collection or NASM for the Smithsonian National Air and Space Museum; these abbreviations are given at the head of the Archival and Private Sources section of the list of sources. Oral history interviews, denoted OHI, are also listed in this section. All other cited works, without such a repository abbreviation, are listed under Published Sources.

Preface

1. The two books are being published simultaneously by the University Press of Florida. A third book from another press, *The Development of Propulsion Technology for U.S. Space-Launch Vehicles, 1920-1991* focuses more narrowly on propulsion and is a separate undertaking.

Introduction

1. Tatarewicz, "Telescope Servicing Mission" (quotation, 365); Bilstein, *Testing Aircraft, Exploring Space*, 156–57; Crouch, *Aiming for the Stars*, 278–79; Launius, *NASA*, 126.

2. See Layton, "Mirror-Image Twins," 562–63, 565, 575–76, 578, 580; Layton, "Technology as Knowledge," 40; Layton, "Presidential Address," 602, 605; Vincenti, *What Engineers Know*, 4, 6–7, 161; Ferguson, *Mind's Eye*, xi, 1, 3, 9, 12, 194. Of course, there are many ways in which science and engineering overlap, as emphasized in Latour, *Science in Action*, 107, 130–31, 174, and by Layton himself as quoted in Bijker, Hughes, and Pinch, *Social Construction of Technological Systems*, 20.

3. Simmones, résumé, n.d., and e-mail messages, July 15, 2002 (quotations), mid-2002; e-mails, Simmons to author, July 15, 2002. Actually, black powder has probably been known since before 1100.

4. Even rocket engineers often refer to liquid- or solid-fuel rockets. But the liquids or solids in question include not just fuel but an oxidizer. This is what distinguishes rockets from jet engines; jets use oxygen from the atmosphere to combine with their fuel and permit combustion, while rockets carry their own oxidizer. Hence, the proper terminology is liquid- or solid-*propellant* rockets.

5. Incidentally, the technical designation of liquid-propellant combustion systems is engines, while their solid-propellant counterparts are usually called motors. A rocket is a type of engine or motor that burns fuel and oxidizer to produce thrust; also, a vehicle propelled by a rocket engine or motor.

6. This section is based on far too many sources to cite here. One source that covers much the same material in language comprehensible to nonexperts is NASA Education Division, *Rockets*, 12–18. See glossary entries, "control" and "guidance."

7. M. Neufeld, "Wernher von Braun, the SS, and Concentration Camp Labor."

8. Stuhlinger and Neufeld, "Concentration Camp Labor" (quotation, 126). For a defense of von Braun, see Stuhlinger's comments in this article and also Stuhlinger and Ordway, *Crusader*, esp. 42–53.

Chapter 1. The Beginnings: Goddard and Oberth

1. Bainbridge, *Spaceflight Revolution*, 15; Winter, *Rockets into Space*, 6–13; McDougall, *Heavens and the Earth*, 20. The Goddard portion of this chapter borrows extensively from Hunley, "Enigma." The journal in which this article appeared, *Technology and Culture*, published at the time by the University of Chicago Press, is now published by the Johns Hopkins University Press, which has kindly allowed me to republish material from the article.

2. Barth, *Oberth*, 18–25, on Krasser and his prediction; for Kennedy's more famous words, couched rather as a goal than as a prediction, *Public Papers*, 404.

3. For the title, von Braun and Ordway, *Space Travel*, 43; Winter, *Rockets into Space*, 34, on the other hand, raises the question whether he deserves even the title Father of U.S. Rocketry, while granting that his "determination and vision" inspired "countless researchers" in the decades following his death.

4. Lehman, *Goddard*, 10–19, 21, 57–58, 79–80; on Worcester, see also Rice, *Worcester*, 173, 189, 443, 455; Baedeker, *United States*, 60–61; Rosenzweig, *Eight Hours*, 12.

5. Lehman, *Goddard*, 16, 19; "Material for an Autobiography of R. H. Goddard," *Goddard Papers*, 6. Speculation that Goddard already suffered from tuberculosis is mine; for his lasting affliction with the disease, see Lehman, 61, 71, 242–43, 252.

6. "Who I Am and Why I Came to the Institute: Theme in English at the Worcester Polytechnic Institute," November 22, 1904, *Goddard Papers*, 67; "Material," ibid., 10; Goddard to Edward F. Bigelow, June 19, 1901, ibid., 54; Lehman, *Goddard*, 28.

7. "Material," *Goddard Papers*, 9–10 (also quotations); Goddard to H. G. Wells, April 20, 1932, ibid., 821, where Goddard says he read *War of the Worlds* in 1898; Lehman, *Goddard*, 22, 28.

8. See, e.g., diary entry for 1929, *Goddard Papers*, 708. Other such entries are listed in the index in volume 3.

9. "Material," *Goddard Papers*, 10.

10. Diary and Graduation Oration, *Goddard Papers*, 63; "Citation for R. H. Goddard . . . June 2, 1945," ibid., 1603–4; "Who I Am," ibid., 67; extract from Worcester Polytechnic Institute (WPI) yearbook for 1908, 159, in Goddard Papers, box 4, CU. A call to the WPI archivist confirmed that Goddard's B.S. was in general science. Other material in box 4, including his "Record of Scholarship" for the term ending June 1, 1908, and experiment reports in box 14 suggest that his concentration in science was already in physics.

11. Cf. H.D.S., "Granville Stanley Hall," 128; Koelsch, *Clark University*, 24–25.

12. Koelsch, *Clark University*, 24–31, 37–41, 48–51, 62, 66–67, 70, 77, 133, 137–38, Koelsch, "Michelson Era," 135, 139–40; H.D.S., "Granville Stanley Hall"; Goddard to Albert C. Erickson, September 25, 1934, *Goddard Papers*, 890; Wills, "Webster, Arthur Gordon." On the experimental rather than theoretical or mathematical nature of

American physics in this period, see Holton, "Quantum Physics Research," esp. 181–82, 194.

13. Winter, *Rockets into Space*, 14–18; "Material," *Goddard Papers*, 11–14. See note 68 for documentation of his 214 patents.

14. "Material," *Goddard Papers*, 18; "A Method of Reaching Extreme Altitudes," ibid., 337, including notes thereto.

15. Affidavit, *Goddard Papers*, 1030; Koelsch, *Clark University*, 144; Rynin, *Interplanetary Flight and Communication*, 99, which uses information supplied by Goddard.

16. "Material," *Goddard Papers*, 13, 22 (quotation, 22); cf. appendix 4, "U.S. Patents Issued to R. H. Goddard, 1914–1956," ibid., 1651; Winter, *Rockets into Space*, 16, 27.

17. Winter, *Rockets into Space*, 27–28; Goddard to President, Smithsonian Institution, September 27, 1916, *Goddard Papers*, 172–74. On de Laval, see Baker, *Rocket*, 18. It should have been clear from Newton's laws that reaction against air would not be a factor.

18. Edmund C. Sanford to Goddard, March 1, 1916, *Goddard Papers*, 166, indicating the new salary; Lehman, *Goddard*, 77, which mentions only the previous salary of $1,000.

19. Goddard to President, Smithsonian Institution, *Goddard Papers*, 170–75 (quotations, 170–71); Durant, "Goddard and the Smithsonian."

20. See outline of article "The Navigation of Interplanetary Space," September 10–October 11, 1913, *Goddard Papers*, 117–23; cf. "Material," ibid., 24–25.

21. *Goddard Papers*, 413–30.

22. "Material," *Goddard Papers*, 24–25.

23. Goddard to C. G. Abbot, March 9, 1940, *Goddard Papers*, 1302. See also other entries in *Papers* index under "Interplanetary navigation," 1688.

24. C. D. Walcott to Goddard, October 11, 1916, and January 5, 1917; Goddard to Walcott, October 19, 1916, *Goddard Papers*, 176, 190, 177.

25. S. M. Friedman, "Inflation Calculator."

26. Lehman, *Goddard*, esp. 91–97, 303–6, 352; reports and letters too numerous to list, *Goddard Papers*, 199–316 and passim; Hickman, "History of Rockets," 367.

27. Goddard to Abbot, April 7, 1919, *Goddard Papers*, 320–21.

28. "A Method of Reaching Extreme Altitudes," *Goddard Papers*, 337–38. I owe the insight about engineering method to John Anderson—an emeritus professor of engineering at the University of Maryland, a part-time curator at NASM, and author of numerous books and articles about aerodynamics—in a discussion on October 10, 2001.

29. Ibid., 338–43, 347–73. Although on page 343 Goddard mentioned only two stages, on 382–83 he discussed the use of rockets in bundles employed as multiple stages, and on 403–404n16 he also discussed multiple stages.

30. Ibid., 343–95; for comment about the Moon, 393.

31. Ibid., 404–405n19 for comment on hydrogen and oxygen, 396 for quotation.

32. Ibid., 397.

33. Winter, *Rockets into Space*, 18; Winter, *Prelude*, 14.

34. Lehman, *Goddard*, esp. 104, 108–12; *Goddard Papers*, esp. 406–13; Winter, *Rockets into Space*, 29. The *New York Times* editorial of January 18, 1920, is conveniently reprinted in Clarke, *Space Age*, 66–67. In this connection, see also Carter, *Politics, Religion, and Rockets*, esp. 182–83.

35. Sources differ over the month he made this switch. Winter, *Rockets into Space*, 29, says January; *Goddard Papers*, 474, puts the date at July 11; Durant, "Goddard and the Smithsonian," 61, says September. As Frank Winter pointed out on reviewing these lines, a diary entry for January 29 in the *Goddard Papers* states cryptically, "Got oxygen," seeming to support the January date Winter favors.

36. Appropriately to this important achievement, the hill bore an Indian name meaning "a turning point or place"; it is also intriguing that the other two locations where Goddard tested his high-altitude rockets were Hell Pond on Camp Devens, Massachusetts, and Eden Valley near Roswell, New Mexico.

37. Diary entries, photo captions, and report, in *Goddard Papers*, 580–82, 587–90, 769; Lehman, *Goddard*, 130, for the information on "Aunt" Effie Ward, as Goddard called her.

38. Abbot to Goddard, May 8; Goddard to Abbot, May 11, 1926, *Goddard Papers*, 591.

39. Goddard, *Rocket Development*, xix, 107. Cf. *Goddard Papers*, 786, 911, 917, 983, 1053, 1185, 1200, 1207, 1663–66.

40. Von Kármán, *Wind and Beyond*, 242.

41. Malina, "GALCIT Rocket Research Project," 117. Something of this attitude that Malina describes comes through in a letter, Goddard to L. T. E. Thompson, September 3, 1940, *Goddard Papers*, 1352–53, where he states, "I certainly have no objection to the Cal Tech working on rockets," but indicates pretty clearly his (unjustifiably) low opinion of GALCIT's prospects for success. He also admits his unwillingness to share information with Malina.

42. Malina, "ORDCIT Project," 358.

43. These comments are based largely upon widely scattered evidence in the *Goddard Papers* and, to a lesser extent, Lehman's biography, but see also Durant's introduction thereto, xi, and his interesting discussion of Goddard's relationship with Abbot in "Goddard and the Smithsonian," 57–69. For further qualifications on Goddard's unwillingness to work with others, see Hunley, "Enigma," 342–43.

44. See, e.g., Winter, "Harry Bull," 293–94, for Goddard's influence on Bull.

45. See Winter, *Prelude*, 14; McDougall, *Heavens and the Earth*, 77, for similar judgments by other scholars. More recently, G. Sutton, *Liquid Propellant Rocket Engines*, 268–69, has reached similar conclusions.

46. "Liquid-Propellant Rocket Development," *Goddard Papers*, 968–84. For Guggenheim and Lindbergh urging him to publish the paper, see Goddard to Abbot, September 28, 1935, ibid., 937. A copy of the paper in its original formatting is in "1919 and 1936 Rocket Essays," NHRC, but for readers with access only to the *Papers*, I have here cited that reproduction.

47. Cf. Goddard, *Rocket Development*, xix, 112–15, and passim; *Goddard Papers*, 1076–79, 1110; Winter, *Rockets into Space*, 33; Durant, "Goddard and the Smithson-

ian," 63; Durant, "Roswell Years," 320–39; G. Sutton, *Liquid Propellant Rocket Engines,* 258–61.

48. *Goddard Papers*, 1429, 1461, 1464–65, 1570–71, 1592n, 1606–7; Lehman, *Goddard*, 350–51; Hallion, "American Rocket Aircraft," 289; Stuart, "New Rocket Engine Is in 'Final Stages'"; Clary, *Rocket Man*, 196; Matthews, *Saga of Bell X-2*, 24–26.

49. Vincenti, "Air Propeller Tests," 714–15, 746n98, 749n104. See also in this connection Hansen, *Engineer in Charge*, 78–79, 124; Vincenti, *What Engineers Know*, 251–52.

50. *Goddard Papers*, 1193.

51. Perhaps misled by Goddard's own statement (see note 50), so knowledgeable a scholar as Hacker in "Robert H. Goddard," 239, has parroted it in referring to "his methodical, one-step-at-a-time approach." Similarly, Jimmy Doolittle, after a visit to Goddard's Mescalero Ranch in New Mexico, wrote in some rough notes that he sent to Goddard for correction, "Dr. Goddard, rightly, feels that [the many problems he was still working to solve] must be attacked one at a time and each one worked on until a solution, not necessarily the best solution, is obtained" ("Major J. H. Doolittle's Notes on Visit to Mescalero Ranch," October 13, 1938, *Goddard Papers*, 1210).

52. Goddard, *Rocket Development*, 21–31 (quotation, 31); Durant, "Roswell Years," 320–21.

53. Durant, "Roswell Years," 321; Goddard, *Rocket Development*, 28–30 (first two quotations, 29); Goddard's notes from "Test of September 29, 1931," 102–14, esp. 113, box 18, Goddard Papers, CU; "Report on Rocket Work at Roswell, New Mexico," December 15, 1931, *Goddard Papers*, 815; E. Goddard, "Excerpts," 527, entry for September 29, 1931, CU. Goddard's comment about "relatively low" altitudes was a generic one from the *Goddard Papers*, 815, about the series of tests in the fall in which the rocket rose "about 2000 feet."

54. See, e.g., Goddard, *Rocket Development*, 17–18, 20, 31, 44, 90–91, 101, 104, 107, 108, 125.

55. *Goddard Papers*, 627, 631, 635, 1164, 1176, 1413, 1434; Goddard, *Rocket Development*, 64, 65, 177.

56. Lehman, *Goddard*, 260–61.

57. Durand to Rear Admiral John H. Towers, Chief, Bureau of Aeronautics, July 29, 1941, "Goddard/NACA," NHRC.

58. Cf. similar points in Lehman, *Goddard*, 2–3, 205, 231, 245, 251. On his eagerness for space travel, see also the popular but accurate article by Rhodes, "Ordeal of Robert Hutchings Goddard."

59. Cf. Lehman, *Goddard*, 80, and the partial diary entry in *Goddard Papers*, 176. The interpretation that follows makes use of material in Lehman, but it is not presented in his book in quite the same way as I envision it, although in his discussion of Goddard's "'trust-in-the-Lord' engineering" he refers to the physicist's combining of "not-quite-foolproof" components with one another "in the hope that the combination would somehow work infallibly" (286).

60. Lehman, *Goddard*, 245, 286–87, for example.

61. *Goddard Papers*, 575.

62. E. Goddard, "Excerpts," 147, 148, 724, 868, and entries for February 2, 1906, and October 28, 1944, CU. The unattributed quoteation apparently is from Joseph Wood Krutch.

63. "Material," *Goddard Papers*, 18.

64. Lehman, *Goddard*, 71, 198, 241–43, 252 (quotation, 252); *Goddard Papers*, 843, 1075–76.

65. Rhodes, "Ordeal of Robert Hutchings Goddard," 30; Lehman, *Goddard*, 390–99; *Goddard Papers*, 1607–9.

66. E. Goddard, "Excerpts," entry for August 10, 1945, 9:05–9:10, CU.

67. However, Winter, *Rockets into Space*, 24–25, argues that "by virtue of his thoroughness and the fact that his ideas were openly published and sparked a widespread movement, Oberth alone deserves the title 'Father of the Space Age.'"

68. *Goddard Papers*, 1651. For details of the settlement, see "Goddard Patent Infringement," NHRC. While the patents provided for the support of his widow, it is important for understanding Goddard's role in American rocketry to recognize that patents are no substitute for successful rocket designs. It was a huge step to take the ideas in his patents and develop them into space rockets, something Goddard never succeeded in doing himself.

69. Goddard, *Rocket Development*.

70. G. Sutton, *Liquid Propellant Rocket Engines*, 269.

71. On Goddard's activities at this time and his engagement announcement, *Goddard Papers*, 477–85; on his future bride, Lehman, *Goddard*, 106. The marriage did not take place until June 21, 1924 (Lehman, 138).

72. Barth, *Oberth: Briefwechsel*, 1: 7. The letter also appears in *Goddard Papers*, 485.

73. *Goddard Papers*, 485n.

74. Barth, *Oberth*, 15, 25, 367n1.

75. See esp. Wandycz, *Price of Freedom*, passim; Barth, *Oberth*, 266–68, 367n6.

76. Dragomir, *Ethnical Minorities in Transylvania*, 50; Teutsch, *Siebenbürger Sachsen*, 242, 245; Jekeli, *Entwicklung des siebenbürgisch-sachsischen höheren Schulwesens*, 114–25; Gerard, *Land Beyond the Forest*, 11, 55, 59–63 (quotations from this work).

77. For the facts of his upbringing but not the interpretation about his self-confidence, Barth, *Oberth*, 15–23.

78. Oberth, "Autobiography," 113; for half the population being German, Barth, *Oberth*, 24.

79. Crouch, "To Fly to the World in the Moon," 17; McCurdy, *Space and the American Imagination*, 13–14; Oberth, "Autobiography," 114 (first quotation); Oberth, "Contributions," 129 (second quotation).

80. Oberth, "Autobiography," 114–16; Oberth, "Contributions," 130 (quotation); Barth, *Oberth*, 24, 50–51; Oberth, OHI, 20.

81. Barth, *Oberth*, 26 (first quotation), 52–53; Oberth, "Autobiography," 117 (second quotation).

82. Gerard, *Land Beyond the Forest*, 65–66; Barth, *Oberth*, 57.

83. Barth, *Oberth*, 43, 59–62; Oberth, "Contributions," 132; Oberth's contribution to Brügel, *Männer der Rakete*, 42–43.

84. Oberth, "Autobiography," 117–18; Barth, *Oberth*, 61–64; Oberth, OHI, 14.

85. Elder, "Oberth," 285–86; Gartmann, *Men Behind the Space Rockets*, 48 (quotation); Barth, *Oberth*, 64.

86. Brügel, *Männer der Rakete*, 43; Barth, *Oberth*, 67–72 (quotations). Information on Wolf is from "Maximilian F. J. C. Wolf"; on Lenard from A. Hermann, "Lenard, Philipp"; on Prandtl, esp. Busemann, "Ludwig Prandtl, 1875–1953," 193–205. According to his OHI, 26, Oberth did not attend Prandtl's lectures but read a lot that the aerodynamicist wrote and got advice from him, although Oberth characteristically believed that his own ideas were better than Prandtl's.

87. Lehman, *Goddard*, 56.

88. Oberth, "Autobiography," 118; Barth, *Oberth*, 69–72; Brügel, *Männer der Rakete*, 43. See Manuila, *Aspects démographiques*, 74, for the size of Mediaş.

89. Lehman, *Goddard*, 133; cf. *Goddard Papers*, 522, 736.

90. Oberth, *Rakete zu den Planetenräumen*, passim, esp. 90–92. NASA had the book translated as TT F-9227 in 1965, but because of copyright restrictions it did not publish the translation, restricting it to internal use by NASA personnel.

91. M. Neufeld, "Weimar Culture," 731–32; Barth, *Oberth*, 94–98; Essers, *Max Valier: Ein Vorkämpfer der Weltraumfahrt*, 82–117, 166, 181–82, 308–9 (in the translation, *Max Valier—A Pioneer of Space Travel*, 56–83, 126–27, 138, 254–55). For Oberth's collaboration with Valier, see Oberth, *Ways to Spaceflight*, 378–79.

92. Bergaust, *Wernher von Braun*, 34–36; Stuhlinger and Ordway, *Aufbruch*, 46; Barth, *Oberth*, 194. See also von Braun's comment in a letter to Oberth in 1948 that since his earliest youth he had regarded Oberth as the father of modern rocket technology (Barth, *Oberth: Briefwechsel*, 1: 139). The other side of the picture is that Dornberger, and apparently von Braun, kept Oberth away from Peenemünde until 1941 because, as Dornberger said, he was difficult and technically naïve despite his theoretical accomplishments. See M. Neufeld, "The Excluded," 214–18.

93. Barth, *Oberth: Briefwechsel*, 2: 100, but see also the comment in note 92; Elder, "Oberth," 12–93; Winter, *Prelude*, 35–54; M. Neufeld, "Guided Missile," esp. 54–58; M. Neufeld, "German Rocket Engineers," 4; M. Neufeld, *Rocket and the Reich*, passim. Dannenberg directed the Jupiter project at the Army Ballistic Missile Agency in Huntsville and later became deputy director of the Saturn Systems Office, according to a biographical sheet in his file, NHRC. See also Dannenberg, "Hermann Oberth." There is an inevitable tendency for such retrospective comments on an individual to be laudatory and to overlook nuances, but there is no reason to doubt that what Dannenberg remembered about Oberth's influence was essentially true, even if exaggerated.

94. Barth, *Oberth*, 84.

95. Bainbridge, *Spaceflight Revolution*, 141–42; Rynin, *Interplanetary Flight and Communication*, 263–72 for a translation of Lorenz's article, also 273–87 for Lorenz's comments before the aviation society; on Lorenz himself, Mayr, "Lorenz, Hans"; Oberth, "Ist die Weltraumfahrt möglich"; Oberth, "Raketenflug und Raumschiffahrt"; Oberth, "Contributions," 137; Barth, *Oberth*, 101; von Kármán, *Wind and Beyond*, 243.

96. Oberth, "Contributions," 131, 133. In writing of people, Oberth normally followed the German practice of giving only last names. According to the editor of "Contributions," von Dallwitz-Wegner's comments appeared in a journal called *Autotechnik* in 1929, but standard scientific reference works do not identify him or the other figures mentioned but incompletely identified in the narrative above.

97. M. Neufeld, "Weimar Culture," 751; von Kármán, *Wind and Beyond*, 243; on the naming of JPL and the stigma attached to the word "rocket" in the United States, Koppes, *JPL*, 20. Buck Rogers was a comic-strip character who traveled about space in swashbuckling fashion, appearing in daily and Sunday papers after 1928 and on radio from 1932; see Horrigan, "Popular Culture," 54.

98. Thomas, *Men of Space*, 5: 59, on unidentified Caltech professor, based on recollections by Albert R. Hibbs; von Kármán, *Wind and Beyond*, 243, for Bush's remark, date unspecified.

99. Oberth, *Ways to Spaceflight*; Winter, *Prelude*, 39; M. Neufeld, "Weimar Culture," 738–41; Barth, *Oberth*, 125–40, 145–53.

100. Thomas, *Men of Space*, 1: esp. 3–5; Elder, "Oberth," 292; Sloop, *Liquid Hydrogen as a Propulsion Fuel*, 191–94; Mallan, *Men, Rockets, and Space Rats*, 292–93, 320.

101. Winter, *Rockets into Space*, 25; von Braun and Ordway, *Space Travel*, 57; Winter, *Prelude*, 51–52; M. Neufeld, "Guided Missile," 56–57; M. Neufeld, "Weimar Culture," 752. Von Braun also wrote that Oberth's *Rockets into Outer Space* was "the scientific foundation upon which the technical development of astronautics has since been built" ("Prophet of Space Travel").

102. Barth, *Oberth*, 173–92; Oberth, OHI, 38–60; Elder, "Oberth," esp. 300–301; Gartmann, *Men Behind the Space Rockets*, 69–71.

103. Barth, *Oberth*, 211–14, 220–27, 232–33, 239; Gartmann, *Men Behind the Space Rockets*, 72; Oberth, OHI, 57–58.

104. *Goddard Papers*, v.

105. I owe this insight to an anonymous reviewer of a different manuscript for the NASA History Office.

Chapter 2. Peenemünde and the A-4 (V-2), 1932–1945

1. All of these developments are described in much greater detail in a variety of sources, especially M. Neufeld, *Rocket and the Reich*, 5–9; M. Neufeld, "Weimar Culture"; Winter, "Birth of the VfR"; Winter, *Prelude*, 35–39; Crouch, *Aiming for the Stars*, 42–57.

2. Winter, "Birth of the VfR," 248, and M. Neufeld, *Rocket and the Reich*, 10, provide slightly different details about Winkler's efforts, but they agree on the essentials.

3. Winter, *Prelude*, 249–50; M. Neufeld, *Rocket and the Reich*, 5–6, 11–12.

4. Winter, "Birth of the VfR," 249–50; Hölsken, *V-Waffen*, 15, 226n5.

5. Hölsken, *V-Waffen*, 16; Thomas, *Men of Space*, 2: 48. Michael Neufeld tells me the actual date was early 1932, not 1930.

6. Thomas, *Men of Space*, 2: 45–46; Hölsken, *V-Waffen*, 226n6; M. Neufeld, *Rocket and the Reich*, 9; Dornberger bio, February 18, 1963, in "Dornberger, Walter R. (Gen.)," NASM. Various sources give differing dates for Dornberger's degree. M. Neufeld, *Rocket and the Reich*, 55, states that it was an honorary degree arranged by Becker.

7. M. Neufeld, *Rocket and the Reich*, 6, 12–14; Winter, *Prelude*, 39–41.

8. Sänger-Bredt and Engel, "Regeneratively Cooled Liquid Rocket Engines," esp. 220, 223–27, 244; Winter, *Prelude*, 40–48; M. Neufeld, *Rocket and the Reich*, 15–16.

9. The East Elbian landed elite, to which Magnus von Braun belonged, exercised real power before the collapse of the German Empire after World War I, and this continued to a significant degree during the Weimar Republic. Margaret Lavinia Anderson makes a good case in her *Practicing Democracy* against the idea that German history followed a special path (*Sonderweg*) in which the aristocracy maintained control of Germany and the country failed to modernize, helping lead to Hitler's rise to power in 1933. She nevertheless approvingly quotes Chancellor Bethmann Hollweg about the gentry's "effortless dictatorship" (194) over the peasants on their estates, saying that as late as 1912 the nobles faced few challenges. She also says, "Although forty years of manhood suffrage had not shaken the aristocracy's hold on the levers of bureaucratic, diplomatic, military, or local power," Weimar democracy after the war, with real parliamentary responsibility for the administration, "forced it [the nobility] to play the parliamentary game" to an even greater degree (198). Nevertheless, as Shelley Baranowski has shown in *The Sanctity of Rural Life*, the East Elbian landed elite retained substantial power throughout the Weimar period.

10. The above two paragraphs are based on Hunley, "Braun, Wernher von," and "Wernher von Braun, 1912–1977," which in turn are based in part on Ruland, *Leben für die Raumfahrt*, and Bergaust, *Wernher von Braun*, although neither is a scholarly treatment of his life. See also his father's autobiography, *Weg durch vier Zeitepochen*, 291–95, in which Magnus Freiherr von Braun talks about the family background. While Wernher was growing up, his father was director and then general director of a bank. He later served as minister of nutrition and agriculture in the last two governments of the Weimar Republic, before Hitler became chancellor in 1933 and replaced the parliamentary system with a dictatorship.

11. Ruland, *Leben für die Raumfahrt*, 61; Bergaust, *Wernher von Braun*, 37–40.

12. Von Braun, "Reminiscences" 128–31 (quotations, 129, 130). On page 131 von Braun says he started work for the army on November 1, 1932, but a document Michael Neufeld found in the Wernher von Braun Papers, USSARC, places the date at November 27. An earlier, typescript version of "Reminiscences" titled "Behind the Scenes of Rocket Development in Germany" is at USSARC, where I saw it in 1995 in a folder marked "Manuscript: Behind the Scenes . . ." in the von Braun collection (the Center's archives have since been reorganized). The wording in the typescript and published versions differ somewhat, and sometimes the original wording is valuable, as are the penned-in corrections. A copy of "Behind the Scenes" has been donated to the NHRC.

13. Dornberger, *V-2*, 27.

14. Von Braun, "Reminiscences," 131.

15. Von Braun, "Reminiscences," 131. Von Braun, "Behind the Scenes," 11, USSARC, calls the test "successful," although Dornberger says the engine was "calculated to develop a thrust of 650 pounds" (*V-2*, 25), which it failed to meet, according to von Braun's recollections.

16. Dornberger, *V-2*, 26.

17. Letter, Heinrich Grünow to von Braun, April 3, 1956, found by Michael Neufeld in the von Braun papers, USSARC.

18. Von Braun, "Reminiscences," 131 (all quotations but first and last); von Braun, "Behind the Scenes," 11, USSARC (first quotation); Dornberger, *V-2*, 28 (his quotation).

19. Von Braun, "Flüssigkeitsrakete," 30. Copies are in "von Braun, Dissertation," NHRC, and in "von Braun, Wernher," NASM.

20. M. Neufeld, *Rocket and the Reich*, 34, supplies this information but not the name of the Vereinigte Aluminiumwerke. Presumably, however, he is talking about that firm, identified by von Braun in his dissertation.

21. Von Braun, "Reminiscences," 131 (also quotation); von Braun, "Behind the Scenes," 11, USSARC; Dornberger, *V-2*, 32–33; M. Neufeld, *Rocket and the Reich*, 35; Schulze, "Technical Data," USSARC. Von Braun used the term "brute force" to apply to the A-2, but it accurately describes the operation of the A-1 control mechanism as well.

22. Von Braun, "Reminiscences," 131; von Braun, "Behind the Scenes," 12, USSARC.

23. Dornberger, *V-2*, 33, 36.

24. Cover of von Braun, "Flussigkeitsrakete" seen in "von Braun, Dissertation," NHRC; Rudolph, OHI, tape 1, side 1; Franklin, *American in Exile*, 13–27.

25. Von Braun, "Reminiscences," 131; Dornberger, *V-2*, 27–28; M. Neufeld, *Rocket and the Reich*, 33. I have accepted Neufeld's date for his employment rather than Dornberger's, which is too early.

26. Dornberger, *V-2*, 29–30; M. Neufeld, *Rocket and the Reich*, 31; Franklin, *American in Exile*, 38–41; Klee and Merk, *Birth of the Missile*, 14, which shows a picture of the engine but says that it provided 280 pounds of thrust for 50 seconds, which doesn't square with what Rudolph told Franklin.

27. M. Neufeld, *Rocket and the Reich*, 36–38; von Braun, "Reminiscences," 131; Ordway and Sharpe, *Rocket Team*, 23. Von Braun gives the altitude as a mile and a half, but Neufeld, citing a report von Braun wrote on January 28, 1935, seen in the Bundesarchiv/Militärarchiv Freiburg, gives the figure as about 1,700 meters, which is only some 300 feet above a mile.

28. Although quite old now, a still useful overview of these developments and others is provided by Bracher, *German Dictatorship*, esp. 191–214, 289. Bracher's case-history approach is no longer fashionable. Nor is his interpretation of Nazism, but his lengthy narrative is still valuable for the details it provides. For criticism of Bracher and other older histories, see Maier, *Unmasterable Past*, esp. 100–120. For another recent view of German scholarship, see Evans, *Rereading German History*, esp. 8–9, 12–17, 122, 237.

29. On this, see esp. Corum, *Luftwaffe*, 75–76, 85, 115, 125–27, 155, 157–58, 161–62. Göring became Reichskommissar for aviation in the first week of Hitler's chancellorship. He became aviation minister on May 1, 1933, with the army's Air Defense Office transferred from the minister of defense to the air minister on May 15, 1933.

30. Murray, *Luftwaffe*, 4.

31. Although the work for jet-assisted takeoff, especially in the propulsion area, was relevant to A-4 development, the scope of this book does not permit extended discussion. See M. Neufeld, *Rocket and the Reich*, esp. 43–51, 57–63, and the sources he cites for further information.

32. Copy of Col. Dr. Dornberger's comments before the Commission for Long-Distance Shooting (*Fernschiessen*), Berlin, March 9, 1943, Wa Prüf 11, PGM microfilm roll 24, FE 342, NASM; von Braun, "Reminiscences," 133–34; Corum, *Luftwaffe*, 165, 174 (on von Richthofen only); M. Neufeld, *Rocket and the Reich*, 43–54. Von Richthofen, incidentally, later rose to the rank of field marshal.

33. Von Braun, "Development of German Rocketry," 20, USSARC. The "farm" was an estate that his father had purchased on July 1, 1930, as the elder von Braun discusses in *Weg durch vier Zeitepochen*, 291–300.

34. Dornberger, *V-2*, 40–41; von Braun, "Reminiscences," 134; M. Neufeld, *Rocket and the Reich*, 54–55.

35. Dornberger, *V-2*, 50–53; M. Neufeld, *Rocket and the Reich*, frontispiece map, 56; Huzel, *Peenemünde to Canaveral*, 239.

36. See "Vortragsnotiz für den Herrn Oberbefehlshaber des Heeres," Wa A, September 18, 1941, on PGM microfilm roll 24, FE 342, NASM, which does not mention the A-4 by designation but does state that test stands for the "Einsatzgeräte als Versuchsgeräte" (operational apparatus as test apparatus) were ready on September 9, 1941.

37. Although development of the Wasserfall extended Peenemünde's rocket research and development into new areas, the scope of this book does not permit extended coverage of it. See M. Neufeld, *Rocket and the Reich*, esp. 230–37, 252–54. De Maeseneer, *Peenemünde*, also contains a good deal of scattered information about Wasserfall as well as the V-1, which is not covered in the present book since it had little demonstrable influence on U.S. launch vehicle development. In De Maeseneer see, e.g., 120–21, 145, 150, 155, 193, 196, 246, 289, and his index on Wasserfall. Unfortunately, he does not cite his sources except in a selected bibliography at the end, so it is impossible to know where he got his information on particular points.

38. Dornberger, *V-2*, 58, 126, 131–33; Huzel, *Peenemünde to Canaveral*, 239–41. "Vortragsnotiz Betr. Stand der Entwicklung Vorhaben Peenemünde, Wa A, Wa Prüf 11, Oberst Dornberger," April 28, 1941, on PGM microfilm roll 24, FE 342, NASM, states that the completion of Test Stand 7 would probably occur at the beginning of August, although evidently it occurred by June 13 when the first attempted launch of an A-4 took place. On Mittelwerk (Central Works) see, among other sources, Béon, *Planet Dora*. For a different perspective on Mittelwerk, see also Stuhlinger and Ordway, *Crusader*, 42–53.

39. Von Braun, "Behind the Scenes," 13–14, USSARC; MacKenzie, *Inventing Accuracy*, 16–19, 31–40, 45–46.

40. Dornberger, *V-2*, 34; F. Mueller, OHI, tape 1, side 1. Mueller received what apparently was an honorary doctorate of science in 1958 from Rollins College. See his obituary in the *Huntsville Times*, May 18, 2001. He died on May 15, 2001, at the age of 93.

41. F. Mueller, OHI, tape 1, sides 1 and 2; cf. F. Mueller, "Inertial Guidance," 181;

MacKenzie, *Inventing Accuracy*, 52–53; M. Neufeld, *Rocket and the Reich*, 66–67, all of which have illustrations of the system with the stable platform but none of which go into the detail provided by the OHI of Mueller. For the idea of using jet vanes in the first place, see Dornberger, *V-2*, 35. See glossary entries "control" and "guidance."

42. Dornberger, *V-2*, 34, 49, 53; von Braun, "Reminiscences," 132, 137; Kurzweg, "Aerodynamic Development," 50–53; "Memoirs of Dr. Rudolf Hermann" (unpublished, based on Hermann, OHI), 15–16, and other biographical materials in "Rudolf Hermann," NASM; Gorn, *Universal Man*, 27–34; Busemann, "Ludwig Prandtl," 193–205; Hanle, *Bringing Aerodynamics to America*, 22–27, 66–67; "Comparative History of Research and Development Policies," 136–37, NA; M. Neufeld, *Rocket and the Reich*, 86; Rotta, *Aerodynamische Versuchanstalt*, passim.

43. Von Braun, "Behind the Scenes," 13–14, USSARC; cf. "Reminiscences," 132, where the description of the valves is slightly different.

44. M. Neufeld, *Rocket and the Reich*, 64; Schulze, "Technical Data," 5, USSARC.

45. M. Neufeld, *Rocket and the Reich*, 68–70; Dornberger, *V-2*, 54–56; for the size of the Oie, Kennedy, *Vengeance Weapon 2*, 10. Curiously, von Braun, "Reminiscences," 132–33, had the launches in the summer and discussed only three A-3s.

46. F. Mueller, OHI, tape 1, side 2; Kurzweg, "Aerodynamic Development," 55; MacKenzie, *Inventing Accuracy*, 53; Dornberger, *V-2*, 55–56; von Braun, "Reminiscences," 133.

47. Dornberger, *V-2*, 56–58; von Braun, "Reminiscences," 132; Schulze, "Technical Data," 13–21, USSARC.

48. Hermann, "Memoirs" (see note 42), 3–16, and résumé, in "Hermann, Rudolf," NASM; von Braun, "Reminiscences," 137; Dornberger, *V-2*, 53–54; M. Neufeld, *Rocket and the Reich*, 86–87; Wegener, *Peenemünde Wind Tunnels*, 27.

49. Wegener, *Peenemünde Wind Tunnels*, 26–27 (quotation, 26); M. Neufeld, *Rocket and the Reich*, 89; Schulze, "Technical Data," 5, 13, USSARC.

50. Kurzweg, "Aerodynamic Development," 53, 55–57, including photo of the hand-carved model; Dornberger, *V-2*, 57.

51. This was because shock waves that formed as the speed approached that of sound caused choking in the tunnel and distorted the data on either side of the speed of sound.

52. Dornberger, *V-2*, 56–59; R. Hermann, "Wind Tunnel Installations," 48; M. Neufeld, *Rocket and the Reich*, 44, 90; [Rudolf Hermann], "Denkschrift über die Windkanäle der Heeres-Versuchsstelle Peenemünde," June 1, 1939, 5, PGM microfilm roll 19, Archiv 66/11, NASM. Incidentally, in "Denkschrift," 42, Hermann reveals that as of June 1, 1939, the aerodynamic division included 34 white-collar and 20 blue-collar employees, with 9 of the former having degrees from institutions of higher learning (*Hochschule*).

53. Dornberger, *V-2*, 56–58. He mentions only molybdenum as a material for the A-3 vanes, but it was alloyed with tungsten.

54. Von Braun, "Reminiscences," 136; Dornberger, comments (see note 32). Following the developments in a variety of documents available from Peenemünde is further complicated by the fact that many have become illegible with the passage of time.

55. F. Mueller, "Inertial Guidance," 181, for all quotations except those for the torquer and the tilt program; F. Mueller, OHI, tape 2, side 2, for those two quotations. He does not explain how the tilt program worked.

56. Schulze, "Technical Data," 13–14, USSARC; Dornberger, *V-2*, 62; "Zusammenfassender Bericht über die Starts mit ungesteurten und gesteurten Rückstossgeschossen Typ A 5 II vom 24.10.[38] bis 13.12.39 auf der Greifswalder Oie," n.d., PGM microfilm roll 27, Archiv 78/2, NASM.

57. Karner, "Steuerung der V2," 47–52; M. Neufeld, *Rocket and the Reich*, 96–99, 107; F. Mueller, OHI, tape 2, side 2. Cf. F. Mueller, "Inertial Guidance," 181; Haeussermann, "Guidance and Control," 226–28. Karner, 53, says that the flight was in 1939; Neufeld, 107, puts the flight with the Siemens system in April 1940. "Arbeitsbericht der Abteilung für Bordausrüstung, Steuerung und Messtechnik [BSM] für den Monat April 1940," on PGM microfilm roll 31, Archiv 96/4, NASM, provides the precise date.

58. M. Neufeld, *Rocket and the Reich*, 100; for the dates of the two flights, Arbeitsberichte of BSM (see note 57) for April 1940 and August 1940, PGM microfilm roll 31, Archiv 96/4, NASM.

59. Biographical sketch of Steinhoff in Stuhlinger et al., *Astronautical Engineering and Science*, 318; Bowman, "Steinhoff Dreams of Flying"; M. Neufeld, *Rocket and the Reich*, 101; von Braun, "Reminiscences," 138.

60. F. Mueller, OHI, tape 1, side 2; Dornberger, *V-2*, 15, quotation.

61. Hoelzer, OHI, tape 1, side 1; Hoelzer bio, April 1964, biographical files, MSFC/HO.

62. In the yaw axis only (with the front end of the rocket rotating left or right), not the pitch axis (with the front end rotating either above or below the desired path), which is why this was not a guide beam, although that term is sometimes loosely used to describe it.

63. Hoelzer, OHI, tape 1, side 2; Tomayko, "Hoelzer's Analog Computer," 230; Haeussermann, "Guidance and Control," 226; M. Neufeld, *Rocket and the Reich*, 103, 105; De Maeseneer, *Peenemünde*, 75, 258; Dornberger, *V-2*, 231; Arbeitsberichte of BSM (see note 57) for March–July 1941, PGM microfilm roll 31, Archiv 96/11–96/15, and "Vortragsnotiz" (see note 38), NASM.

64. Von Braun, "Reminiscences," 136; Schulze, "Technical Data," 14, USSARC; Dornberger, comments (see note 32); Walther Riedel, "Lecture on the History of Rocketry and the Development of the A4 (V2)," January 3–8, 1946, PGM microfilm roll 33, FE 814, NASM.

65. Ernst Steinhoff, 1940 BSM report, January 10, 1941, PGM microfilm, FE 769, 3, NASM.

66. Von Braun, "Reminiscences," 129–30; for other accounts of the Day of Wisdom, cf. M. Neufeld, *Rocket and the Reich*, 83–84; Ordway and Sharpe, *Rocket Team*, 35.

67. Cf. M. Neufeld, *Rocket and the Reich*, 55, 74, 206, 243, 248, 255; Klee and Merk, *Birth of the Missile*, 109. This section on personnel is very similar to what I wrote in Hunley, "Antecedents," 6.

68. Dornberger, *V-2*, 50 (quotations); *Lebenslauf* (curriculum vitae) of Thiel at the time of his doctoral degree, photocopied onto a letter to Frank Winter at NASM from I. Ritter at the Staatsbibliothek Preussischer Kulturbesitz, August 16, 1976, along with

the title of his dissertation, seen in "Thiel, Walter," NASM; Martin Schilling to K. L. Heimburg, August 27, 1976, same folder.

69. Schilling to Heimburg (Schilling's quotation); Dornberger, *V-2*, 50, 53 (other quotations); von Braun, "Reminiscences," 137; Ordway and Sharpe, *Rocket Team*, 121; M. Neufeld, *Rocket and the Reich*, 84, for Schilling's becoming Thiel's "replacement." Not everyone shared Schilling's highly favorable view of Steinhoff.

70. See Dornberger, *V-2*, 149, 151.

71. [Walter] Thiel, Memorandum, Weapons Test Organization, Kummersdorf—Target Range, "On the Practical Possibilities of Further Development of the Liquid Rockets and a Survey of the Tasks to be Assigned to Research," March 13, 1937, translated by D. K. Huzel, seen in "Thiel, Walter," NASM. In a cover letter to "Those Listed," October 11, 1960, Huzel, then working for Rocketdyne, said that the letter seemed "to show that some of our present day problems (and ideas) are not new."

72. Schilling, "V-2 Rocket Engine," 284.

73. Dornberger, *V-2*, 50–52.

74. Ibid., 52; Reisig, "Peenemünder 'Aggregaten,'" 45–46; M. Neufeld, *Rocket and the Reich*, 78–79; Schilling, "V-2 Rocket Engine," 286.

75. Oberth, *Ways to Spaceflight*, 41.

76. Reisig, "Peenemünder 'Aggregaten,'" 47; M. Neufeld, *Rocket and the Reich*, 80; Dornberger, *V-2*, 52; Schilling, "V-2 Rocket Engine," 286. Interestingly, Arthur Rudolph said (OHI, tape 1, side 1) that he had discovered film cooling well before it was used on the A-4, but that he had no part in introducing the concept into the 25-ton combustion chamber's design.

77. Reisig, "Peenemünder 'Aggregaten,'" 47; M. Neufeld, *Rocket and the Reich*, 80–81.

78. Sources differ about the number of revolutions per minute and the amounts of (75 percent) alcohol and liquid oxygen delivered. Reisig, "Peenemünder 'Aggregaten,'" 47, gives figures of 3,800 rpm, 58 kg/sec of alcohol and 72 kg/sec of oxygen. U.S. Army Ordnance Research and Development [R and D] Translation Center, "German Guided Missiles," 12, USSARC, gives the figures of 3,000 rpm, 56 kg/sec for alcohol, and 69 kg/sec of oxygen. The British Ministry of Supply's "Report on Operation 'Backfire,'" 2: 18, RATL, reports 4,000 revolutions (no indication of time frame), 56 kg/sec of alcohol, and 70 kg/sec of oxygen.

79. Ehricke, "Peenemünde Rocket Center," 60; Schilling, "V-2 Rocket Engine," 289–90; "Niederschrift über die Besprechung des Arbeitsstabes Wa A/VP am 3.3.1942," PGM microfilm roll 59, FE 692f, NASM; letter, HAP to Firma Klein, Schanzlin & Becker, January 9, 1943, PGM microfilm roll 36, FE 746c, NASM; and, from PGM microfilm roll 40, FE 737, NASM, three documents: "Niederschrift der Besprechung am 8.8.41 betreffend Aufteilung der Oddesse-Turbopumpen auf Aggregate A4"; "Entwurf, von Braun to Wa Prüf 11, Betr. Serienanlauf Turbo-Pumpe A4," March 31, 1941; letter, von Braun to Firma Klein-Schanzlin-Oddesse re "Fertigung von 600 Stück Turbopumpen," May 27, 1941. The Oddesse Pumpen und Motoren Fabrik GmbH was part of Klein, Schanzlin and Becker from 1939 to 1945—hence apparently the "Klein-Schanzlin-Oddesse."

80. Schilling, "V-2 Rocket Engine," 289–90; Reisig, "Peenemünder 'Aggregaten,'" 47.

81. Ehricke, "Peenemünde Rocket Center," 60.

82. M. Neufeld, *Rocket and the Reich*, 84; Wewerka [last name only], "Untersuchung einer Kreiselpumpe-Odesse Typ 3/18 brennstoffseitig," July 1941, Archiv 33/8, and Wewerka, "Untersuchung einer Turbopumpe, Typ TP 1 g," February 1942, Archiv 33/9, both summarized in "Accession List," 2: 598, RATL.

83. I am indebted to John Anderson for discussions in the fall of 2001 that increased my understanding of this technology.

84. Schilling, "V-2 Rocket Engine," 284; Reisig, "Peenemünder 'Aggregaten,'" 46; Wewerka, "Der günstigste Erweiterungswinkel bei Lavaldüsen," February 5, 1940, Archiv 33/2, and Wewerka, "Günstige Bemessung einer Lavaldüse bei zeitlich veränderlichem Aussendruck," April 8, 1940, Archiv 33/3, both summarized in "Accession List," 2: 617–18, RATL.

85. Schilling, "V-2 Rocket Engine," 284–85.

86. "Peenemunde East Through the Eyes of 500 Detained at Garmisch," 184–85, NHRC. This is a generally unreliable source, but in this case the information seems to be largely plausible.

87. Dannenberg, "From Vahrenwald via the Moon," 123.

88. Summerfield, "Questions," 3–4, USSARC; British Ministry of Supply, "Report on Operation 'Backfire,'" 17–34, RATL; U.S. Army Ordnance R and D Translation Center, "German Guided Missiles," 12, USSARC.

89. Dornberger, *V-2*, xvii, for the exhaust velocity (which is given in Summerfield, "Questions," Sketch 1, USSARC, as 2,070 m/sec); Schilling, "V-2 Rocket Engine," 285, for the specific impulse. Specific impulse is defined (JPL, *Mariner-Mars 1964*, 42n) as the ratio of thrust a rocket engine or motor produces to the amount of propellant needed to produce that thrust per unit of time.

90. Summerfield, "Questions," 1, USSARC.

91. Von Braun, "Reminiscences," 137; Dannenberg, "From Vahrenwald via the Moon," 122.

92. Warren, *Rocket Propellants*, 104–5.

93. Summerfield, "Questions," 1–2, sketch 1, USSARC; Schilling, "V-2 Rocket Engine," 287–88.

94. R. Hermann, "Wind Tunnel Installations," in "Hermann, Rudolf," NASM (cited instead of Launius, *Rocketry and Astronautics*, because of an error in the date of a staff count in Launius, 41); "Tätigkeitsbericht der Aerodynamischen Hauptabteilung für das Jahr 1940," January 10, 1941, PGM microfilm roll 20, Archiv 66/39, 3, NASM, for the dates accepted in the narrative; Wegener, *Peenemünde Wind Tunnels*, 25, for Hermann's research, which Hermann does not mention in his own account.

95. Archival version of Hermann, "Wind Tunnel Installations" (see note 94); Wegener, *Peenemünde Wind Tunnels*, 27, 63–67; Lehnert, Kurzweg, and Hermann, "Bericht über Dreikomponenten-Messungen am Aggregat A4-Modell im Uberschall-Windkanal der HVP," October 10, 1940, Archiv 66/33, and [Siegfried] Erdmann, "Vorläufiger Bericht. Druckverteilungsmessungen am A4 V 1P bei Uber- und Unterschallgeschwin-

digkeit. Erst Fortsetzung," April 18, 1942, Archiv 66/74, both summarized in "Accession List," 2: 841, 842, RATL.

96. Archival version of Hermann, "Wind-Tunnel Installations" (see note 94); Wegener, *Peenemünde Wind Tunnels*, 65–69.

97. Eber and Hermann, "Ueber die Berechnung der Hauttemperatur des Aggregates IV während seines Fluges," March 30, 1938, Archiv 66/4, summarized in "Accession List," 2: 846, RATL; G. R. Eber, "Hauttemperaturen des A-4 während seines Fluges," Archiv 66/4[?], second report cited in Kurzweg, "Aerodynamic Development," 69, with discussion of its findings, 66 (also quotation). That these two reports actually carried the same Archiv number seems doubtful, but a later report by Eber and Hermann on the same subject in February 1939 was numbered 66/5, so maybe the same number was mistakenly assigned to two related reports.

98. See Dornberger, *V-2*, 222; Kurzweg, "Aerodynamic Development," 65.

99. Cf., e.g., U.S. Army Ordnance R and D Translation Center, "German Guided Missiles," 13, USSARC; [Richard] Lehnert, "Bericht über Fortsetzung der Dreikomponentenmessungen am Aggregat A 4 . . . im Uberschallwindkanal der HVP," January 23, 1942, Archiv 66/65, summarized in "Accession List," 2: 837, RATL; Hugh L. Dryden, notes on interview with Dr. Hermann at Kochel, June 15, 1945, in "Rockets . . . V-2," NHRC; Kurzweg, "Aerodynamic Development," 61–62.

100. H. Friedman, "A-4 Control," 40, NHRC.

101. H. Friedman, "A-4 Control," 51, NHRC.

102. Kurzweg, "Aerodynamic Development," 53 (last quotation); Wegener, *Peenemünde Wind Tunnels*, 168 (quotation on Prandtl and other institutes). For other work by Tollmien on nozzles, see Archiv 54/1 and 54/2, and for work done at Göttingen, see Archiv 41/9, 41/12, 41/13, 41/14, and others, summarized in "Accession List," 1: 531–32, 555, 560, and 2: 615, RATL.

103. Cf. Hermann, archival "Supersonic Wind Tunnel Installations" (see note 94), where he does not claim to have done this but lists it as a requirement.

104. Dornberger, *V-2*, 214, 218–23; M. Neufeld, *Rocket and the Reich*, 204, 220–22; De Maeseneer, *Peenemünde*, 304–5.

105. Hoelzer, OHI, tape 1, sides 1–2; Tomayko, "Hoelzer's Analog Computer," 228, 230–32; H. Friedman, "A-4 Control," 129–35, NHRC; M. Neufeld, *Rocket and the Reich*, 106.

106. H. Friedman, "A-4 Control," 108–15 (quotations, 108), NHRC; cf. F. Mueller, "Inertial Guidance," 181–83 (see note 41). Friedman, 115–20, emphasizes that there were variants of this system.

107. Material since note 106 from F. Mueller, "Inertial Guidance," 183–85 (quotations, 184); H. Friedman, "A-4 Control," 100–105 (quotation, 105), NHRC; F. Mueller, OHI, tape 2, side 2, and tape 3, sides 1 and 2; MacKenzie, *Inventing Accuracy*, 59; Haeussermann, "Guidance and Control," 228. The description here is a composite of the sources above, none of which provides all of the elements discussed. Mueller does not use the words "pendulum" or "pendulous" in his description in "Inertial Guidance," saying only that the gyro was "supported in an unbalanced position" (185). But he provides the same diagram as Haeussermann, who calls it a pendulous integrating gyro

accelerometer (PIGA). MacKenzie, 147, notes that PIGAs were later used on Polaris by the U.S. Navy.

108. H. Friedman, "A-4 Control," 57–62, 73, 207, NHRC.

109. Ibid., 177; Otto Müller, "The Control System of the V-2," in Benecke and Quick, *German Guided Missiles*, 88–89.

110. Hoelzer, OHI, tape 1, side 1; Haeussermann, "Guidance and Control," 228. Haeussermann mentions other developments, including a cutoff device developed by Theodor Buchhold and Carl Wagner of the Technical Institute of Darmstadt, which may have been one of the developments that never found application during World War II, although apparently it did fly on the A-4 at least in tests. On this device, there is much more information in H. Friedman, "A-4 Control," 76–88, NHRC, but he too fails to indicate that it actually flew on the A-4. However, according to Haeussermann, such developments "provided the background for future developments." Kennedy, *Vengeance Weapon 2*, 71, suggests the device did fly on the rocket, and Farrior, "Inertial Guidance," 403, indicates the same.

111. Haeussermann, "Guidance and Control," 225 (biographical sketch), 229–31; Hoelzer, "Guidance and Control Symposium," 301–3; Tomayko, "Hoelzer's Analog Computer," 233; Haeussermann bio, January 1963, microfiche 1068, biographical files, MSFC/HO. For the various operations carried on in Steinhoff's division, see the snapshots by month in the Arbeitsberichte of the BSM department on PGM microfilm roll 31, Archiv 96/4, NASM, which also make frequent mention of the contributions of the technical institutes.

112. Hoelzer, "Guidance and Control Symposium," 303–15; Tomayko, "Hoelzer's Analog Computer," 232–36 (first two quotations, 232, 234); Hoelzer, OHI, tape 1, side 2. For dates of launches and comments from a different perspective, see Schulze, "Technical Data," 22–23, USSARC.

113. Schulze, "Technical Data," 23–24, USSARC, for example, lists a partial success on October 21, 1942; failure of the missile to tilt over into its trajectory on November 9, 1942; tumbling of the missile on November 28, 1942; an explosion four seconds into takeoff on December 12, 1942; and an explosion during ignition on January 7, 1943. H. Friedman, "A-4 Control," 3, NHRC, says none of the launches from number 6 to number 19 was successful but there was better success beginning with launch 20. As discussed above in the narrative, however, there were major problems even later.

114. Such statistics vary with the source. I have accepted those of Dornberger, *V-2*, xvii–xviii, for convenience. These tally pretty well with those in "A-4 Missile—Illustrated Synopsis," USSARC. The "effective range," as I have called it, comes from H. Friedman, "A-4 Control," 10, NHRC, which he defined as the range over which it was "advisable to engage targets."

115. MacKenzie, *Inventing Accuracy*, 299.

116. H. Friedman, "A-4 Control," 15, NHRC.

117. Hölsken, *V-Waffen*, 163, 200–201; M. Neufeld, *Rocket and the Reich*, 264, 273.

118. MacKenzie, *Inventing Accuracy*, 28. On heterogeneous engineering, see also Law, "Technology and Heterogeneous Engineering," among other sources.

119. M. Neufeld, *Rocket and the Reich*, 5–21, 119–24.

120. For the meetings attended and battles fought by the two managers, see ibid., 118, 127, 129, 139, 141–43, 144, 161–64, 167–74, 191–95, 201–3, 223, 247, 257, and passim. On polycracy as the predominant characteristic of National Socialism, see Peukert, *Inside Nazi Germany*, 30, 43–44, 81, 112–13 (quotation, 81); he also talks about the "permanent confused petty warfare among rival power groups" (43). Kershaw, *Nazi Dictatorship*, 65–66, characterizes polycracy as "leadership chaos . . . a multidimensional power-structure, in which Hitler's own authority was only one element (if a very important one)."

121. Farrior, "Inertial Guidance," noted in 1962: "It is a credit to the early German investigators that in spite of the almost insurmountable difficulties, they fully recognized the potential of inertial guidance. In synthesizing various schemes to solve the inertial guidance problem, they came up with almost all of the basic devices which are in use today" (402).

122. Quotation about "toys" from Tomayko, "Hoelzer's Analog Computer," 234; Dornberger's from *V-2*, 139–40.

123. Huzel, *Peenemünde to Canaveral*, 126 (quotation); see also the numerous reports of meetings scattered throughout the documents that survive from Peenemünde and Kummersdorf, e.g., "Niederschrift 401/37," August 6–September 9, 1937, on PGM microfilm roll 6, FE 74b, NASM, and "Niederschrift über die Besprechung am 23 Apr. 1942," in folder 3, Peenemünde Document Collection, NASM. In this connection, see also Hunley, "Antecedents," 13–14.

124. Dornberger, "Epilogue," in Stuhlinger et al., eds., Peenemunde to Outer Space, 852.

125. E.g., Wegener, *Peenemünde Wind Tunnels*, 48; Reisig, *Raketenforschung in Deutschland*, 72–77.

126. Stuhlinger et al., *Peenemünde to Outer Space*, vii–viii (first two quotations); Stuhlinger and Ordway, *Crusader*, 37 (third quotation).

127. See, e.g., Wegener, *Peenemünde Wind Tunnels*, 80–87; Hoelzer, OHI, tape 1, side 2.

128. It is revealing in this regard that when George Sutton was performing the "analysis and design" for the Redstone engine in 1947–51, he said that "my fellow designers and I had not even heard of Goddard or any of his know-how or his unique contributions to the state of the art of LPREs [liquid-propellant rocket engines] . . . Instead Rocketdyne received a lot of help and data from the Germans and their V-2 LPRE information, which was very useful." (*Liquid Propellant Rocket Engines*, 269).

Chapter 3. JPL: From JATO to the Corporal, 1936–1957

1. What became JPL started life as the more or less informal Guggenheim Aeronautical Laboratory, California Institute of Technology (GALCIT) rocket (or jet propulsion) research project. The organization involved in the research began to be called JPL in 1943, with the name formalized in 1944. The word "jet" in JPL's name was not very descriptive of what the organization did, but the word "rocket" was then in bad repute in what were considered "serious" scientific circles, so "jet" had to serve as a synonym. See, e.g., Malina, "The Rocket Pioneers," 32.

2. Newell, "Guided Missile Kinematics," 1, NA.

3. See, e.g., Holton, "Quantum Physics Research," 181–82, 194n3; M. Walker, *Quest for Nuclear Power*, esp. 74–75; Cassidy, *Uncertainty*, esp. 262–63; Weart, "Physics Business in America," 298–99, 309.

4. Seely, "Research, Engineering, and Science."

5. See, e.g., Hansen, *Engineer in Charge*, xxxiv; Hanle, *Bringing Aerodynamics to America*,, xi, 14.

6. On Klein and Göttingen, see Hanle, *Bringing Aerodynamics to America*, 22–27, 66–67; on Aachen, see Gorn, *Universal Man*, 25–28, 34–35, 38; and "Comparative History of Research and Development Policies," 136–37, NA; on reports done for Peenemünde by University of Göttingen, see Peenemünde Archives Reports 41/9, 41/12, 41/13 to 41/15, available at RATL; on Hermann's connections to other aerodynamicists, see his OHI, esp. 12–16.

7. Seeley, "Research, Engineering, and Science," 363; Hanle, *Bringing Aerodynamics to America*, 80–157; Goodstein, *Millikan's School*, esp. 74–75, 163–64; Gorn, *Universal Man*, 49–92. On the rise of Caltech "as a major center of American science," see also Geiger, *To Advance Knowledge*, 183–91. On von Kármán's students, see also Thomas, "von Kármán's Caltech Students."

8. Thomas, "von Kármán's Caltech Students," 13.

9. Geiger, *To Advance Knowledge*, 188.

10. On Sänger, see Sänger-Bredt, "The Silver Bird Story." Like von Braun, Sänger was inspired to work on rockets by science fiction and the early writings of Hermann Oberth. Sänger did not work at Peenemünde, however.

11. Malina, OHI by Hall, 1.

12. Malina, "Jet Propulsion Laboratory," 160–61. See also Malina, "Biographical Information," NHRC.

13. Winter and James, "Aerojet," 30, (supplemental) n1. I am grateful to John Bluth of the JPL Archives for calling my attention to this citation and to the FBI's file on Parsons, page 8 of which gives the dates and locations shown in the narrative, which differ slightly from those in Winter and James's note. According to the FBI file, on Parsons's own testimony, he took only correspondence and extension courses from the University of Southern California and also the University of California. Presumably the latter were also in chemistry, although the file does not say.

14. Malina, OHI by Wilson, 17.

15. Malina, "Jet Propulsion Laboratory," 161.

16. Cf. comments by Malina in "Jet Propulsion Research Project," 158, and by Lloyd Berkner, quoted in R. C. Hall, "Earth Satellites," 269n9: "During the war our early studies in the Navy showed the orbiting Earth satellite was within the range of our technology. But in 1943 it was clear that space technology, aside from short-range rockets, would not be a factor in the war, so the matter was laid aside."

17. Goodstein, *Millikan's School*, 244–60; Burchard, *Rockets, Guns and Targets*, 50–62, 83–209; Price, Horine, and Snyder, "Eaton Canyon"; Snyder, "Caltech's *Other* Rocket Project," 2–13. I am grateful to Ray Miller for loaning me the Price-Horine-Snyder paper and to Cargill Hall for calling the Snyder source to my attention.

18. See Malina, "GALCIT Rocket Research Project," 121–22. The other three early participants were Apollo M. O. Smith, Hsue-shen Tsien (Chien Hsueh-sen, as he later was known), and Weld Arnold. Their mishaps earned the group the sobriquet Suicide Club or Suicide Squad.

19. See von Braun's typescript, "Development of German Rocketry," 9–10, USSARC, where he noted that he and his fellow rocket developers at the Raketenflugplatz before 1932 did have a thrust balance but that Col. Karl Becker, then chief of Ballistics and Ammunition in Army Ordnance, complained about their almost complete lack of "scientific and accurate data on such matters as propellant consumption, specific impulse of your motor, combustion pressure, and the like." Compare that comment with figure 3 in Malina, "GALCIT Rocket Research Project," 120, showing the kind of test setup he and his colleagues were using in November 1936 and his later test stand from 1939 in figure 4, page 122.

20. Malina, "GALCIT Rocket Research Project," 114; see also other references to theory on pages 115 and 117 and elsewhere in this article and in his "Jet Propulsion Research Project," 160–61.

21. Malina, OHI by Wilson, 4, 16.

22. Malina, "GALCIT Rocket Research Project," 120–21; included were a paper by Bollay from 1935 entitled "Performance of the Rocket Plane" and one by Malina from 1937, "Analysis of the Rocket Motor."

23. J. Parsons, "Practicality," JPL. Archivist John Bluth, in a 1995 conversation, wondered if Parsons wrote this paper by himself, but even if he had help, the paper shows clearly the approach taken by the Malina group.

24. See, e.g., Malina, "Rocketry in California," tear sheets of which were found in an unnumbered folder in the Frank J. Malina Collection (originally at Caltech, which organized it before it moved to the Library of Congress, Manuscript Division), box 12, LC/MD, which cites Sänger, the Italian A. Bartocci, and Goddard; Malina, "Report on Jet Propulsion," LC/MD, which cites two National Advisory Committee for Aeronautics reports as well as Goddard and Sänger and discusses their implications; and the highly theoretical article by Malina, "Characteristics of the Rocket Motor." The Caltech archives retain a microfiche of the Malina Collection, so it is available both there and at the Library of Congress, which has the original paper documents.

25. For the complex history of these contracts and the overlapping relationships with the various Army and Navy organizations, see Koppes, *JPL*, esp. 18–19; Malina, "ORDCIT Project," 343–47; U.S. Army Missile Command, "Sergeant Weapon System," 3–4, RATL; Miles, "ORDCIT Project," 2–18, 87–91, JPL.

26. Goodstein, *Millikan's School*, 245.

27. "Facilities and Equipment of the Air Corps Jet Propulsion Research Project," LC/MD, 1; Malina et al., "Jet Propulsion Laboratory, GALCIT," 1, LC/MD; [Stanton], "Research and Development," 4, LC/MD.

28. Haley, *Rocketry and Space Exploration*, 100. See also the sources cited in note 29.

29. Gorn, *Universal Man*, 87; von Kármán, *Wind and Beyond*, 245–46; von Kármán and Malina, "Ideal Solid Propellant Rocket Motor," 94–106; Malina, "Jet Propulsion Re-

search Project," 169–76, 183–87; Carroll, "Sergeant Missile Powerplant," 123–26. For an exhaustive technical discussion of the castable asphalt propellant known as GALCIT 53 and related technologies used in the JATO units, see Parsons and Mills, "Asphalt Base Solid Propellant," JPL.

30. See, e.g., Liquid Propellant Information Agency, *Liquid Propellant Safety Manual,* chap. 12, pp. 1–2.

31. Slater, "Research and Development," 41; Malina, "Jet Propulsion Research Project," 160–61, 167 (quotation, 160). On Tsien, biographical resume in "Tsien, H. S.," NHRC; von Kármán, *Wind and Beyond,* 308. See also the biography of Tsien by Chang, *Thread of the Silkworm.*

32. Malina, Parsons, and Forman, "Final Report for 1939–40," 4–10, 22–24, JPL.

33. Most of this paragraph is based on Summerfield, OHI, 3–8, but cf. Malina, "Jet Propulsion Research Project," 161–62. For information on Wyld, see Ordway and Winter, "Pioneering Commercial Rocketry," 542–44; Winter and Ordway, "Pioneering Commercial Rocketry," 155–58. Wyld and Reactions Motors will be covered in chapter 4.

34. Malina, "Jet Propulsion Research Project," 168, 179–80; Slater, "Research and Development," 42–45; Summerfield, OHI, 8. See also Truax, "Liquid Propellant Rocket Development," 60–61. See chapters 4 and 7 for discussions of Truax and his role in the development of rocketry in the United States.

35. Winter and James, "Highlights of 50 Years of Aerojet," 2–3; Malina, "Jet Propulsion Research Project," 194–95; Malina, "Review of Developments," esp. 2, LC/MD; Haley, letter, NA, which describes the facilities at Aerojet.

36. See esp. M. Neufeld, *Rocket and the Reich,* 202–9, 223–30; Bornemann, *Geheimprojekt Mittelbau,* 30–45, 61–75, 105–7, 124–35.

37. Summaries for May 1–June 30, 1942, and July 1–July 31, 1943 [misprint for 1942], in "JPL/Monthly Summaries, 1942–1944," LC/MD. For the cooling, see the summaries for October and December 1942.

38. "Facilities and Equipment," 2–3, LC/MD; "Description of the Experiment Station," 3, LC/MD.

39. See Malina et al., "Jet Propulsion Laboratory, GALCIT," 2, LC/MD; "Conference Minutes, ORDCIT Project," LC/MD.

40. Summerfield, OHI, 25. For the roles of von Kármán and Malina, see Malina, "Jet Propulsion Laboratory," 164, 217–18.

41. Biographical sheet in "Stewart, Homer Joe," NHRC; Stewart, OHI, 5–9, 73–74, 82–85. For a similar discussion of the role and activities of Research Analysis, followed by a discussion of the other sections, see Miles, "ORDCIT Project," 52–56, JPL.

42. Malina, "Jet Propulsion Laboratory," 218; Goodstein, *Millikan's School,* 252–55.

43. "Facilities and Equipment of the ACJP," 3, LC/MD. On Pauling's other work, see Noyes, *Chemistry,* 133–35. Pauling's group developed a method for chromatic analysis of propellants later employed by Aerojet's rival, Hercules Powder Company.

44. Malina et al., "Jet Propulsion Laboratory, GALCIT," 3, 11–12. For a report of the work Sperry did on control mechanisms for the ORDCIT Project, see Miles, "ORDCIT Project," 132–49, JPL, and below in this chapter. On Puckett, see Thomas, "von Kármán's Caltech Students," 13.

45. "Conference Minutes, ORDCIT Project," August 15, 1944, LC/MD. This source does not list Klemperer's full name, but presumably it was Wolfgang Benjamin Klemperer, listed as a research engineer at Douglas in *Who's Who in World Aviation*, 173.

46. Malina, "ORDCIT Project," 353–54; "Private A," NA; Dunn and Mills, "Status and Future Program," 21, JPL; Goldberg, "Firing Tests of 'Private A,'" 16–25, JPL. On the development of the 4.5-inch booster rockets, see Burchard, *Rockets, Guns and Targets*, 54–62, and chapter 5 of this history.

47. Bragg, "Corporal," 1: 108, AMC; Seifert, "Ordnance Research," 18, JPL. As Seifert reports, JPL also tested the Private F, a winged version of Private A, at Fort Bliss, Texas, in April of 1945 to check the feasibility of stabilizing a supersonic vehicle solely by gravity and wing dihedral. This test demonstrated that roll stability was too much affected by minor structural asymmetries to be achieved without a control system.

48. Malina, "ORDCIT Project," 356–70; Slater, "Research and Development," 50–54; Malina, "Development and Flight Performance," 6–8, JPL; Sandberg and Berry, "Design and Fabrication," esp. 3, JPL; Goldberg, "Field Preparations," 41, JPL; Bradshaw and Mills, "WAC Corporal Booster Rocket," JPL; Seifert, "Ordnance Research," 18–20, JPL; Bragg, "Corporal," 1: 50, AMC (for dimensions; cf. slightly different ones Bragg provides on 1: 58); Koppes, *JPL*, 24. On the development of the Tiny Tim, see Burchard, *Rockets, Guns and Targets*, 156–64. On Aerobee, see also Winter and James, "Highlights of 50 Years of Aerojet," 15, 19. The Tiny Tim was modified to increase its 30,000 pounds of thrust for 1 second to 48,000 pounds for 0.65 second. For further details of the modifications for the Tiny Tim, see Bradshaw and Mills, "WAC Corporal Booster Rocket," JPL. The weight of the WAC Corporal given in the narrative apparently did not include the payload. Depending on the payload, the loaded weight varied from 683 to 704 pounds. Specific impulse is from Fahrney, "Pilotless Aircraft and Guided Missiles," 912, NHC.

49. Bragg, "Corporal," 1: 61–63, AMC; U.S. Army Ordnance Corps, "Corporal," 55–56, JPL.

50. Bragg, "Corporal," 1: 67–75, 113, AMC (quotations, 113).

51. U.S. Army Ordnance Corps, "Corporal," 2, JPL; Seifert, "Ordnance Research," 20, JPL.

52. Bragg, "Corporal," 1: 44, 108–9, 112, and 2: 28-2, AMC; Seifert, "Ordnance Research," 4, JPL. As Seifert points out, the project did develop a successful Corporal-sized pump but never used it.

53. U.S. Army Ordnance Corps, "Corporal," 57, JPL; Bragg, "Corporal," 1: 116–17, AMC.

54. U.S. Army Ordnance Corps, "Corporal," 2–4, JPL; Bragg, "Corporal," 1: 115–17 and 2: 19-1, AMC; Koppes, *JPL*, 39; Pickering, "Countdown to Space Exploration," 393, for "rabbit killer."

55. U.S. Army Ordnance Corps, "Corporal," 53, 57–59 (quotation, 58), JPL; Bragg, "Corporal," 1: 117, AMC; Denison and Hamlin, " Axial-Cooled Rocket Motor," 2, 4–11, 18, 21, JPL.

56. Richard C. Miles, "History of the ORDCIT Project at the Sperry Gyroscope Company to 31 January 1946," in Bragg, "Corporal," 2: 17-4 to 17-6, AMC, with portions of a letter, Millikan to Sperry, August 8, 1944, quoted on 17-4.

57. U.S. Army Ordnance Corps, "Corporal," 89 (also quotations), JPL; Bragg, "Corporal," 2: 19-1 to 19-3, AMC, for descriptions of the launches. Bragg, 1: xv, reveals that the axially cooled engine was basically the one that the program retained all the way to the tactical missile. There were no control problems on launch 2, but with the propulsion system malfunctioning, the flight did not really test the control system.

58. Koppes, *JPL*, 41–45; Bragg, "Corporal," 1: xv, 131, AMC (quotation, 131); "Major General Holger N. Toftoy." Dunn had succeeded Malina as acting director of JPL on May 20, 1946. He became director on January 1, 1947, succeeding von Kármán. Pickering had worked on telemetry and telecommunications for JPL, becoming the chief of a section on remote control in 1945 and of guidance and control in 1950. He was the manager in charge of developing the guidance and telemetry systems for Corporal, becoming the director of JPL on September 1, 1954. Notes provided by John Bluth of the JPL Archives on August 22, 1994. Cf. JPL, organization charts, July 3, 1946; April 1, 1948; January 1, 1950; November 1, 1954, JPL.

59. U.S. Army Ordnance Corps, "Corporal," 89, JPL.

60. Bragg, "Corporal," 1: 131, AMC.

61. This account is pieced together from elements in three rather confusing treatments: Seifert, "Ordnance Research," esp. 25, JPL; Bragg, "Corporal," 1: esp. 131–32, 137–38, AMC; U.S. Army Ordnance Corps, "Corporal," 16, 41–42, 89–90, JPL, with minor details from 91–100; see also Fahrney, "Pilotless Aircraft and Guided Missiles," 914, NHC. The stabilizers and rudders (aerodynamic control surfaces) were made of magnesium rib castings with a single aluminum sheet attached to them by rivets and flush screws. The inboard portions of the rudders were further covered by stainless steel for protection against the heat of the exhaust gases from the engine. The jet vanes consisted of a graphite body with molybdenum on the leading edges.

62. U.S. Army Ordnance Corps, "Corporal," 53–54, 90–91, JPL; Seifert, "Ordnance Research," 25–27, JPL.

63. U.S. Army Ordnance Corps, "Corporal," 95–98, JPL.

64. Seifert, "Ordnance Research," 10, JPL; Bragg, "Corporal," 1: 137–140 and 2: 19-4, 19-5, AMC.

65. Bragg, "Corporal," 2: 19-5, AMC.

66. Ibid., 19-6; U.S. Army Ordnance Corps, "Corporal," 7, JPL; Johnson, "Craft or System?" 19.

67. Stewart, OHI, 36–37; Stewart, "Stability Estimates," 14, JPL; Bragg, "Corporal," 2: 19-6, AMC; Seifert, "Ordnance Research," 25, JPL.

68. Seifert, "Ordnance Research," 10, JPL; U.S. Army Ordnance Corps, "Corporal," 20–26 (quotation, 21), JPL; Puckett, "12-Inch Wind Tunnel," iii, 1, 2, 11, JPL. Seifert puts the completion of the 12-inch tunnel in 1947, but Puckett has it in April 1949.

69. U.S. Army Ordnance Corps, "Corporal," 8, 59, JPL; Koppes, *JPL*, 51–53; Bragg, "Corporal," 2: 19-7, AMC; Seifert, "Ordnance Research," 30, JPL.

70. U.S. Army Ordnance Corps, "Corporal," 10–11, 16, 105–10, 137–38, JPL.

71. Bragg, "Corporal," 1: 165–67, 170, AMC; Nicholas and Rossi, *Missile Data Book*, 3-2.

72. U.S. Army Ordnance Corps, "Corporal," 16, JPL.

73. On the injector plate for the V-2 and other rockets, see chapter 2, but the sources used here are Reisig, "Peenemünder 'Aggregaten,'" 74; Stuhlinger, OHI, 31–33; Stuhlinger to author, 1995, UP.

74. U.S. Army Ordnance Corps, "Corporal," 371–73 (also quotations), JPL; Koppes, *JPL*, 52–53; Johnson, "Craft or System?" 19–20.

75. According to two sources, Bell Laboratories invented the transistor. Walter H. Brattain and John Bardeen of Bell Labs published a paper in 1948 that introduced the technology. In 1956 they, together with William Shockley, also of Bell Labs, received the Nobel Prize in Physics for the invention. However, it was some time before transistors became entirely reliable. See Day and McNeil, *Biographical Dictionary of the History of Technology*, 41, 94, 637, for the inventors and Ceruzzi, *Modern Computing*, 64, for the initial lack of reliability.

76. Bragg, "Corporal," 1: 138–39 (quotations, 138), AMC; Seifert, "Ordnance Research," 25, 28, JPL.

77. Koppes, *JPL*, 60; Pickering's description from his "Countdown to Space Exploration," 401.

78. Koppes, *JPL*, 21; JPL Executive Board Minutes, Meeting of December 28, 1944, seen in "Jet Propulsion Laboratory Conference Minutes," NASM; JPL, organization chart, January 9, 1945, JPL.

79. Wunderman, OHI, 6–7, copy generously provided by Ben Zibit; Malina, OHI by Hall, 18–19; Koppes, *JPL*, 30–31; Chang, *Thread of the Silkworm*, 82, 108, 142; Malina, "Reflections of an Artist-Engineer," 19–29, with a biographical sketch on 19. Both Chang and Wunderman stressed the friction between Millikan and Malina, and passages in Millikan, "Diaries," Caltech suggest a degree of tension but also on April 7, 1945, mention a dinner with Malina in Juarez, Mexico. For October 25, 1945, Millikan reported Malina's excited resignation and a lot of tension, saying that he phoned to apologize. For October 30, he listed a conference with Malina, saying he later discovered Malina and Liljan had separated and that this probably had caused much of Malina's stress. The statement that Malina was anxious to work for peace should not be interpreted as suggesting that rocket engineers who did not share his liberal outlook were not also anxious to promote peace.

80. Koppes, *JPL*, 31–32, 47, 64–65; Malina, OHI by Wilson, 34–35; Seifert, OHI, 2, 13–15; JPL organizational charts, September 1, 1950, through November 15, 1955, JPL. Incidentally, both Dunn and Pickering had at one time worked for Seifert. Another JPL engineer, James D. Burke, commented about Dunn that he "always seemed to have eaten a bad mushroom just before you talked to him" (Burke, OHI).

81. These generalizations are based on hints in many of the sources cited above, especially Johnson, "Craft or System?" See also Pickering, OHI, 24.

82. Lattu and Dowling, "John W. Parsons," 6, copy generously furnished by John Bluth with the knowledge of the authors; FBI, file on Parsons, FBI; Rasmussen, "Life as Satanist." Rasmussen reports a joke in the aerospace community that JPL stands for Jack Parsons's Laboratory or Jack Parsons Lives. On Parsons's inventiveness, see also a letter of April 4, 1952, to Ward E. Jewell in the von Kármán collection, box 22.25, microfiche, NASM, in which von Kármán referred to Parsons's inventiveness and his excellent knowledge and experience, especially with explosives.

83. Summerfield, OHI, esp. 26–30; JPL, organization charts, July 3, 1946, through September 1, 1948, JPL; C. W. Chillison, President, American Rocket Society, to Theodor von Karman [sic], September 18, 1952, in "Correspondence with the American Rocket Society," NASM; Martin Summerfield to William E. Zisch, General Manager, Aerojet Engineering Corporation, December 10, 1952, in "Correspondence with Martin Summerfield," NASM.

84. Chang, *Thread of the Silkworm*, 120–50, 199, 207, 211–25.

85. News release, Caltech Office of Public Relations, March 11, 1949, in "Correspondence with Clark Millikan," NASM.

86. Koppes, *JPL*, esp. 77, 95, 105. JPL did continue to do some propellant research in connection with its work on spacecraft.

87. See chapter 8 of this book and chapter 6 of the sequel, *Viking to Space Shuttle* for development of this technology.

88. Malina, "Jet Propulsion Research Project," 195–96; Klager and Dekker, "Early Solid Composite Rockets," 3, UP. Thanks to Dr. Klager for generously sending me this and many others of his papers, both published and unpublished. Copies of the printed lectures entitled *Jet Propulsion*, edited by H.-s. Tsien, exist in various places including the Malina Collection, folder 7.4, LC/MD; in the text of the lectures, the preface by von Kármán explains the background of the course, as does Malina's article.

Chapter 4. From Pompton Lakes to White Sands: Other Liquid-Propellant Rocket Developments, 1930–1954

1. Winter, *Prelude*, 73, 82; McCurdy, *Space and the American Imagination*, 24.

2. Pendray, "Early Rocket Development," 141–48 (quotations, 141, 148). Winter, *Prelude*, 74, is the source for Goddard's becoming a member.

3. Winter, "Harry Bull," 291–94, 298–302; Wyld, "Rocket Motor," 461; Wyld to David Fallon, Wilcolator Corp., June 22, 1938, and undated Wyld resumé from same period, both in "Wyld, James Hart," NASM.

4. Winter, *Prelude*, 84; Wyld, "Rocket Motor," 461; Pendray, "Early Rocket Development," 151.

5. Pendray, "Early Rocket Development," 149–51; Wyld, "Rocket Motor," 461; for Malina and Wyld, Summerfield, OHI.

6. Ordway and Winter, "Pioneering Commercial Rocketry," 36 (1983): 542–43. This article and its sequel on projects are also printed in Launius, *Rocketry and Astronautics* (15th–16th Symposia), 75–127.

7. Winter, *Prelude*, 84; Truax, "Annapolis Rocket Motor Development"; "Truax, Robert C.," NHC.

8. Truax, "Liquid Propellant Rocket Development," 58–61 ("Pioneer Rocket Project" is an earlier, slightly different version of this article); biographies of Stiff by Aerojet, 1963 and 1969, in "Stiff, Ray C., Jr.," NASM.

9. Ordway and Winter, "Reaction Motors," 79–82, 104; Ordway and Winter, "Pioneering Commercial Rocketry," 38(1985): 156.

10. Winter and Ordway, "Reaction Motors," 105–8; Truax, "Liquid Propellant Rocket Development," 61.

11. Winter and Ordway, "Reaction Motors," 113–14; Truax, "Liquid Propellant Rocket Development," 65–66; M. White, "Interpretive History," 22–25, 30–32, UP; Stiff biography, 1969, in "Stiff, Ray C., Jr.," NASM (also quotation). White's history of Point Mugu was generously provided by Capt. Grayson Merrill, USN, Ret., who conceived the Lark. One version of the missile used an exhaust-driven turbopump whose design was initiated by Truax. The Lark became, as Truax wrote, "the first U.S. anti-aircraft missile to hit a drone target" (66), but it never went into operational use.

12. Truax, "Liquid Propellant Rocket Development," 66–67; M. White, "Interpretive History," 40–42, 45, UP; Menken, "Pacific Missile Range, 30 June 1959," 1–4, NA.

13. Ordway and Winter, "Reaction Motors," 87–100.

14. This paragraph is based in part on Ordway, "Reaction Motors Division," 137, but also on research for preceding portions of this book as well as Hunley, "Evolution," 22–38.

15. Winter and Ordway, "Reaction Motors," 108–27; Winter, "Reaction Motors Division." There is an enormous literature on the X-1, D-558-2, and X-15. Winter, "Reaction Motors Division," 183–91, has a good treatment of the X-15 engine. On the power plant for the X-1 and the D-558-2, see Winter, "Black Betsy," pts. 1 and 2; on the MX-774, see J. Neufeld, *Ballistic Missiles*, 44–48.

16. Winter and Ordway, "Reaction Motors," 120; communication to author on April 9, 2002, by Frank Winter correcting the date for Neu's patent from the one in the article to a filing date of April 5, 1950, with the patent granted June 22, 1965.

17. Ordway, "Reaction Motors Division," 149 (quoting his interviews with Parker in 1983 or possibly a letter with attachments dated June 22, 1986), 164.

18. Among other sources on these developments, see M. Neufeld, *Rocket and the Reich*, 256–67; Stuhlinger and Ordway, *Crusader*, 60–61; Bower, *Paperclip Conspiracy*, 114.

19. U.S. Army Ordnance Corps/GE, "Hermes Guided Missile Research," ii, NHRC; Toftoy, "Hermes II," 1, NASM; Lasby, *Project Paperclip*, 39. Although chapter 2 of this book usually referred to V-2s under their research-and-development name of A-4, here they are called V-2s since that name was more frequently used in the United States after their employment as weapons in World War II.

20. *GE Challenge*, 2–3; Bower, *Paperclip Conspiracy*, 107–9; Lasby, *Project Paperclip*, 38–39; "Major General Holger N. Toftoy"; Porter, OHI, 26–30.

21. Stuhlinger and Ordway, *Crusader*, 67; Toftoy, "Hermes II," 1, NASM; Bower, *Paperclip Conspiracy*, 129. These and other sources differ on the numbers of Germans who initially went to Fort Bliss. Toftoy says the group with von Braun consisted of 4 people and that in November another 80 arrived. Stuhlinger and Ordway say 6 people accompanied von Braun and 118 followed between November and February. Bower agrees with the figure of 118 but has them arriving from October to December. Some of them did not go to Fort Bliss immediately but stayed at Aberdeen Proving Ground in Maryland. Huzel, *Peenemünde to Canaveral*, 222, put that number at about 6, "who had been detached to the Aberdeen Proving Grounds in Maryland, to classify, sift, and catalogue the Peenemünde secret files." He said they joined the rest of the group in summer 1946 and that the group of Germans reached a total of about 130, a figure that Stuhlinger and Ordway put at 127.

22. Toftoy, "Hermes II," 1–2 (quotations), NASM; Stuhlinger and Ordway, *Crusader*, 71. A ramjet is a simple kind of jet engine in which the air for combustion is compressed in a tube by the forward motion of the vehicle in the atmosphere. See glossary.

23. Rosen, *Viking Rocket Story*, 189.

24. Huzel, *Peenemünde to Canaveral*, 223; Stuhlinger and Ordway, *Crusader*, 71; L. D. White, "Final Report," 44, NHRC; Powell and Scala, "White Sands," 83, 87; London, "Brennschluss," 340.

25. See note 24.

26. London, "Brennschluss," 337; Powell and Scala, "White Sands," 86; Starr, "Launch of Bumper 8," 2. Where London's information differs from Powell and Scala's, I follow London, who seems more trustworthy, although both are useful. The same launch facilities were used for the WAC Corporal and the V-2 despite the difference in size. I am grateful to Fred Ordway for a copy of the Starr paper.

27. L. D. White, "Final Report," 9, 16, 31–35, 121–23, 126–28, NHRC (quotation, 123).

28. Ibid., 6; London, "Brennschluss," 337, 347–48; Powell and Scala, "White Sands," 88.

29. L. D. White, "Final Report," 144, NHRC; London, "Brennschluss," 348.

30. L. D. White, "Final Report," 9, NHRC; London, "Brennschluss," 345.

31. L. D. White, "Final Report," 9, 117, NHRC; London, "Brennschluss," 348; Powell and Scala, "White Sands," 88–90; U.S. Army Ordnance Corps/GE, "Hermes Guided Missile Research," 1–2, NHRC. London gives the launch total from White Sands as 74 instead of 73, but he includes a round that resulted in only a 6-inch rise above the launchpad. I do not count that one. There were two other launches in Florida and one from a carrier deck, as will be discussed.

32. London, "Brennschluss," 341.

33. London, "Brennschluss," 339–40; DeVorkin, *Science with a Vengeance*, esp. 59–67, 167–68; "Major General Holger N. Toftoy."

34. London, "Brennschluss," 356–57; DeVorkin, *Science with a Vengeance*, 81–82, 113–14; U.S. Army Ordnance Corps/GE, "Hermes Guided Missile Research," 2, NHRC.

35. London, "Brennschluss," 357; L. D. White, "Final Report," 37, NHRC; Powell and Scala, "White Sands," 88; DeVorkin, *Science with a Vengeance*, 140–41.

36. See esp. DeVorkin, *Science with a Vengeance*, passim, with generalizations on 3–5, 341–46. The quotation is from Newell, *Beyond the Atmosphere*, 37. Newell participated in the V-2 research, so he may be biased in its favor. He worked as a mathematician and theoretical physicist at NRL from 1944 to 1958 and became NASA's associate administrator for space science and applications, retiring in 1973. See also London, "Brennschluss," 341. On the birth of NASA, see Schoettle, "Establishment of NASA"; Glennan, *Birth of NASA*.

37. London, "Brennschluss," 349; "Upper Air Rocket Summary, V-2 No. 19."

38. Bragg, "Corporal," 1: 76–77, 89, AMC; Malina, "Origins and First Decade of the Jet Propulsion Laboratory," 65; comment by William H. Pickering in "Bumper 8," 46, NHRC.

39. L. D. White, "Final Report," 24, NHRC; Bragg, "Corporal," 1: 88–89, AMC.

40. General Electric, "Bumper Vehicle," 6–9, JPL; Bragg, "Corporal," 1: 89–92, 98–101, AMC.

41. General Electric, "Bumper Vehicle," 6–9, JPL; Bragg, "Corporal," 1: 92, 102, 105, AMC.

42. London, "Brennschluss," 353; Bragg, "Corporal," 1: 105, AMC.

43. Bragg, "Corporal," 1: 105–6, AMC; London, "Brennschluss," 353–54; U.S. Army Ordnance Corps/GE, "Hermes Guided Missile Research," 4, NHRC; General Electric, "Bumper Vehicle," 35, JPL; Starr, "Launch of Bumper 8," 4.

44. Benson and Faherty, *Moonport*, chaps. 1–3; Swenson, Grimwood, and Alexander, *This New Ocean*, 19; London, "Brennschluss," 355; Bragg, "Corporal," 1: 106, 2: 14-1, AMC; Starr, "Launch of Bumper 8," 7.

45. Bragg, "Corporal," 2: 14-1, 14-2, 14-5, 14-7, 14-8, AMC; Starr, "Launch of Bumper 8," 7, 9.

46. London, "Brennschluss," 355; L. D. White, "Final Report," 24, 26, 41, 43, NHRC; Spinardi, *Polaris to Trident*, 20, 202–3n8; U.S. Army Ordnance Corps/GE, "Hermes Guided Missile Research," 2, NHRC; "GE Reveals Hermes Missile Milestones," 30–31.

47. London, "Brennschluss," 350–51; U.S. Army Ordnance Corps/GE, "Hermes Guided Missile Research," 5–12, NHRC; Ordway and Sharpe, *Rocket Team*, 355–56; [U.S. Army Ordnance Corps/GE], "Hermes Guided Missile Systems," 3–5, 12, 21–24, 29, 61, 158, NASM (this evidently being the classified version, declassified in 1968, of their "Hermes Guided Missile Research," NHRC); Georgia Institute of Technology, "Missile Catalog," 108–9, NHRC; Bullard, *Redstone*, 12, 15. I am grateful to the Army Missile Command History Office for providing me with copies of this last source and other monographs published by that office.

48. [U.S. Army Ordnance Corps/GE], "Hermes Guided Missile Systems," 3, 7, 29–30, 57–68, 79–80, 89–90, 97–102, 121–23, NASM, which see for details; Powell and Scala, "White Sands," 91; Bullard, *Redstone*, 10.

49. *GE Challenge*, 2, 4, 6–7 (quotations, 4, 7); Green and Lomask, *Vanguard*, 43, 177; Ezell, *NASA Historical Data Book* 2: 45, 86; Nicholas and Rossi, *Missile Data Book*, 3-2, 3-3; Fuhrman, "Polaris to Trident," 8 (with thanks to Karl Klager for a copy); Spinardi, *Polaris to Trident*, 23; Herring, *Way Station to Space*, 109; Farrior, "Inertial Guidance," 408–9; G. Sutton, *Liquid Propellant Rocket Engines*, 330–31. Incidentally, *GE Challenge*, 6, put the maximum total employment on Hermes at more than 1,300, and it should be noted that the Vega was cancelled in favor of Agena B before it was ever launched.

50. Levine, *Missile and Space Race*, 13; London, "Brennschluss," 365.

51. Braun, "Legacy of Hermes" (quotations, 135–37, 141).

52. Summerfield, "Questions," USSARC; Summerfield, OHI; Koppes, *JPL*, 40.

53. Costs of V-2 firings from Bullard, *Redstone*, 11. See especially chapters 7 through 9 for the development of the thesis suggested here and the sources for it.

54. *GE Challenge*, 4.

55. For these articles, see "V-2 (A-4) Missile," NASM.

56. London, "Brennschluss," 366.

57. Quoted in McCurdy, *Space and the American Imagination*, 29. McCurdy, 35, says that American perceptions began to change in 1951, but attributes the changes to publicity efforts by German-born space and rocket enthusiast Willy Ley and others, not the launching of the V-2s and the accompanying publicity.

Chapter 5. From Eaton Canyon to the Sergeant Missile: Solid-Propellant Rocket Developments, 1940–1962

1. This excludes earlier developments such as Congreve rockets. For accounts of the early history of rockets before Goddard, see von Braun and Ordway, *Space Travel*, 22–36; Winter, *First Golden Age*; Van Riper, *Rockets and Missiles*.

2. As implied in the narrative double-base propellants were those usually based on two ingredients: nitrocellulose as a matrix and nitroglycerine (or another nitrate ester) as a plasticizer. As the narrative will explain, other ingredients come to be added to these two bases to form composite modified double-base propellants. See, e.g., A. Oberth, *Principles*, 1–17, 2–4, 2–5.

3. This account simplifies a complex series of developments. See Price, Horine, and Snyder, "Eaton Canyon," 2, 4, 8; G. B. Kistiakowsky and Ralph Connor, "Molded Solid Propellants," in Noyes, *Chemistry*, 96–97; Christman, *Sailors, Scientists, and Rockets*, 13–14, 35, 100–101, 104–6, 113; Burchard, *Rockets, Guns and Targets*, 14–22, 43–82; e-mail comments on this chapter by Ray Miller, May 15, 2002. On the bazooka, see also Green, Thompson, and Roots, *Planning Munitions for War*, 353–60.

4. Based principally on Price, Horine, and Snyder, "Eaton Canyon," 1–10, 18, 28, but see also Christman, *Sailors, Scientists, and Rockets*, 167–204, 212; "From the Desert to the Sea"; Green, Thompson, and Roots, *Planning Munitions for War*, 354; Snyder, "Caltech's *Other* Rocket Project," 3–13; Goodstein, *Millikan's School*, 251–59. Biographical information on Price is from his fax to author, September 22, 1997.

5. See Huggett, Bartley, and Mills, *Solid Propellant Rockets*, 125; Noyes, *Chemistry*, 103, 111–13; chapter 3 of this book.

6. Carroll, "Sergeant Missile Powerplant," 130–31; Koppes, *JPL*, 13, 36; Bartley, OHI, 10–28, 34–35, 43–44. Thanks to John Bluth for furnishing the last. A polymer is an elastic compound consisting of repeated, linked, simple molecules. See glossary.

7. E. Sutton, "Polymers to Propellants," 1–3; see also Sutton's *History of Thiokol*, 3–8. I am indebted to Bob Geisler for providing a copy of the latter source and also for organizing the session at the Joint Propulsion Conference where Sutton presented his paper, a copy of which Ernie kindly provided.

8. E. Sutton, "Polymers to Propellants," 4–5; Bartley, OHI, 28; Carroll, "Sergeant Missile Powerplant," 133. Sutton's paper said the polymer Bartley acquired was LP-3, but Bartley and Carroll stated it was LP-2, which Sutton later told me by e-mail was correct.

9. The above two paragraphs are based on Carroll, "Sergeant Missile Powerplant," 132–38. Carroll has researched this matter thoroughly and covered it in great detail. See also Bartley, OHI, 43–46; Shafer, OHI, 17, 20–21, 30–33. On ABL, see Moore, "Solid Rocket Development at Allegany," 1–5. The flight tests were done at the Naval Ordnance Test Station, so it is curious that Bartley found out about the star configu-

ration from an English development (rather than from Price's White Whizzer), but both he and Carroll claim that was the source. Moreover, Joseph W. Wiggins wrote in "Hermes: Milestone," 34, that both internal burning and case bonding "are generally credited to Dr. Harold Poole, an English propellant chemist. They were introduced to this country through technical exchanges with England during World War II." And Shafer, in his OHI, 20–21, 30–32, specifically mentions discussions with the British about it and Poole's development of it around 1935. (He also remembered the ABL application as the Deacon rocket, however.) As Carroll's research revealed, the report Shafer used was W. H. Avery and J. Beek Jr., "Propellant Charge Design of Solid Fuel Rockets," Allegany Ballistics Laboratory, Final Report, series B, number 4 (OSRD ref. No. 5890), June 1946, using the data in appendix B.

10. See Wiggins, "The Hermes," 345; Summerfield et al., "Applicability," 23, JPL.

11. Summerfield et al., "Applicability," 11, 23, JPL.

12. Thackwell and Shafer, "Applicability," 2, JPL. On page 5, the two researchers concluded that solid-propellant rockets with JPL118 as the propellant and a ten-point star configuration could equal the performance of the V-2 liquid-propellant German missile with only half its size and that solid rockets of greater length could double the range of the V-2. On the facilities for solid-propellant research at JPL in 1951, see also Meeks, Altman, and Shafer, "Summary," JPL.

13. Dunn and Mills, "Status and Future Program," 8, JPL; Wiech and Strauss, *Fundamentals of Rocket Propulsion*, 78; Meeks, Altman, and Shafer, "Summary," table V, JPL. None of these sources gives the precise conditions of pressure and nozzle expansion ratio for these specific impulses, but presumably the figures are at least roughly comparable with the specific impulse of the JPL118, which was given as 206 in Meeks, Altman, and Shafer. At this period there was as yet no standard for measuring and comparing specific impulses, so measurements varied.

14. On ammonium perchlorate, see Klager and Dekker, "Early Solid Composite Rockets," 13–24, UP; for the fact that Aeroplex (not so named in the source) "cured into a very hard and brittle propellant grain which was not susceptible to case bonding and needed trap supports," H. W. Ritchey (of Thiokol), "Technical Memoir," 6, kindly supplied by Ernie Sutton.

15. E. Sutton, "Polymers to Propellants," 6–7; Dooling, "Thiokol."

16. E. Sutton, "Polymers to Propellants," 9, 23; Sutton, *History of Thiokol*, 58; Wiggins, "The Hermes," 345; Shafer, OHI, 18, where he points out that JPL had loaded the first cases for the Falcon motors with the Thunderbird propellant, after which production shifted to Thiokol.

17. Edward N. Hall, "Air Force Missile Experience," 22–23.

18. Wiggins, "The Hermes," 344.

19. Memorandum to Dr. L. F. Welenetz, n.d., generously sent by Ernie Sutton along with two other early Thiokol documents from his collection. This one bears the notation "must be a copy of a trip report of Glen Nelson's written in late January or early February 1949." On JPL's contributions, see also Guzzo, "Progress Report" (from Sutton's collection, UP), which states, "Early in 1948, when Thiokol initially attempted to produce polysulfide-perchlorate propellants on a pilot-plant scale, the work of the Jet Propulsion Laboratory furnished a valuable fund of basic information" (3).

20. Sutton, *History of Thiokol*, 28, which includes the quotation from physical chemist Harold W. Ritchey, described on page 20 as "the most influential person in Thiokol's rocket history"; Guzzo, "Progress Report," 3–8, UP; Wiggins, "The Hermes," 422.

21. Carroll, "Sergeant Missile Powerplant," 140–41; Hunley, "Evolution," 25, 34n22; Denison and Thackwell, "Progress of Sergeant Program," JPL. On the Big Richard, see Price, Horine, and Snyder, "Eaton Canyon," 6–7.

22. Sutton, *History of Thiokol*, 29; Carroll, notes, JPL.

23. Wiggins, "The Hermes," 347–48, 355; Carroll, notes, JPL.

24. Wiggins, "The Hermes," 357–62. The remainder of the weight was principally taken up by curing agents. LP-33 was a polymer that had better physical performance at low temperatures than LP-3, the binder that succeeded LP-2. However, LP-3 was less inclined to "creep" (deform) at high temperatures.

25. Sutton, "Polymers to Propellants," 11.

26. Wiggins, "The Hermes," 363, 388–90.

27. Ibid., 421.

28. Ibid., 389, 396, 400–401. Wiggins does not mention the Titans and space shuttle. Quotation from Sutton, "Polymers to Propellants," 10. See also Wiggins, "Hermes: Milestone," 39.

29. [Cagle], "Sergeant," 23–25, 31, 43–47, 97, AMC.

30. Ibid., 64–65, 193–94, 199–200, 202–3; see also Koppes, *JPL*, 63–66, 71–77, which uses internal JPL sources different from those cited by Cagle.

31. [Cagle], "Sergeant," 278–79, AMC; Hunley, "Solid-Propellant Rocketry," 2. Among the problems with Sergeant development was combustion instability in the earlier motors. This was solved in the JPL500 motor by about 1958–59; see [Cagle], 104–7. Nichols, Parks, and Burke, "Solid-Propellant Development," 11, JPL, also discusses Sergeant's combustion instability but not the solution.

32. Koppes, *JPL*, 79–93; JPL, "Explorer I," 22, 83; Bullard, *Redstone*, 140–46; Grimwood and Strowd, "Jupiter," 81, AMC. Chapter 6 will continue the story of Redstone and Jupiter.

33. Koppes, *JPL*, 75–79, 95; [Cagle], "Sergeant," 60–65, AMC; Johnson, "Craft or System?" 20–21. Cagle, the Army historian for the project, states that upon JPL's effective elimination from the program as of July 1, 1960, "Sperry was woefully unprepared to carry on the program and the consequences were severe" (64). Koppes, however, evenhandedly discusses the problems on all three sides.

Chapter 6. Redstone, Jupiter C, and Juno I, 1946–1961

1. Bullard, *Redstone*, 12–14, 17–18; U.S. Army Ordnance Missile Command, *History of Redstone Arsenal*.

2. U.S. Army Ordnance Missile Command, *History of Redstone Arsenal*; K. Hughes, "Two 'Arsenals of Democracy,'" 4.

3. Bullard, *Redstone*, 19; Bilstein, *Stages to Saturn*, 390–91; Ordway and Sharpe, *Rocket Team*, 363–66; Dunar and Waring, *Power to Explore*, 3, 14; K. Hughes, "Two 'Arsenals of Democracy.'" Watercress was not, to be sure, the major source of income in the Huntsville area. The primary crops were cotton, hay, corn, fruits, and peanuts. The

local farmers also raised livestock, and many people worked in cotton mills, although strikes in 1934 had led to a decline of that industry around Huntsville.

4. Bullard, *Redstone*, 22–25; "Redstone," entry for 10 July 1950. Stuhlinger and Ordway, *Crusader*, 101, stresses the urgency of the project, given the war and the Soviet testing of an atomic bomb on August 29, 1949, but also mentions its low funding. Levine, *Missile and Space Race*, 27, says that there was little impetus behind Redstone until December 1950 when the Department of Defense began to emphasize it.

5. Bullard, *Redstone*, 27–35; von Braun, "Redstone, Jupiter, and Juno," 109. Von Braun does not mention the supplemental radio guidance, which later was dropped.

6. Bullard, *Redstone*, 35–41; Levine, *Missile and Space Race*, 27; J. Neufeld, *Ballistic Missiles*, 37–38, 66. "K. T." Keller was called the "missiles czar." Appointed by George C. Marshall, secretary of defense and former Army chief of staff, Keller was regarded with some suspicion by the Air Force because of his association with Army officers.

7. Bullard, *Redstone*, 53–54, 95.

8. Ibid., 55–59; Winter, "Rocketdyne," 9.

9. "Rockwell International," in Derdak, *International Directory of Company Histories*, 1: 78–79; Bromberg, *Space Industry*, 19–20; Gray, *Angle of Attack*, 25–26, 78–79 (quotation, 25); Heppenheimer, "Navaho Program," 6.

10. Heppenheimer, "Navaho Program," 6; Bromberg, *Space Industry*, 20; notes from telephone interview, February 13, 2002, by author with Robert Kraemer, a former NAA engineer.

11. Winter, "Rocketdyne, 2; Heppenheimer, "Navaho Program," 6; Myers, "Navaho"; Lonnquest and Winkler, *To Defend and Deter*, 3–4, 24–25. See also the popular but useful Gibson, *Navaho Missile Project*.

12. Winter, "Rocketdyne," 2–4; Ezell and Mitchell, "Engine One," 54; Winter, "East Parking Lot," 2. Recollections of the (unidentified) employee are from Karr, "Hoffman," 26.

13. Winter, "East Parking Lot," 9–10; G. Sutton, *Liquid Propellant Rocket Engines*, 405, 412. On page 13, Winter also reveals that by April 1948 NAA had employed a former JPL chemist; he must have brought knowledge of JPL's technology with him.

14. Ezell and Mitchell, "Engine One," 54–55; Heppenheimer, "Navaho Program," 8.

15. Winter, "Rocketdyne," 6–8; Heppenheimer, "Navaho Program," 7; Reed, "Rocketdyne Razing Historic Test Stand."

16. Ezell and Mitchell, "Engine One," 55–63; Winter, "Rocketdyne," 9.

17. Winter, "Rocketdyne," 8; Heppenheimer, "Navaho Program," 13, 15. I have taken the number 2,400 from Winter, who cites a larger number of sources for his information on this score. Heppenheimer gives the number of 1,600 employees. Perhaps both were valid at different times in 1951. On Hoffman's background, see his biographical sketch, business or professional record, and a sheet listing his responsibilities as chief engineer in the Lycoming Division in "Hoffman, Samuel Kurtz," NASM.

18. Bullard, *Redstone*, 59–60; Ezell and Mitchell, "Engine One," 60.

19. "Redstone;" Bullard, *Redstone*, 60; Stadhalter, "Redstone: Built-In Producibility," NASM. As explained in Chrysler, "This Is Redstone," V-6, RATL, the pneumatic system used electricity to operate valves and pressurized tanks.

20. Stadhalter, "Redstone: Built-In Producibility," 5–16, NASM; Chrysler, "This Is Redstone," II-10, RATL.

21. Chrysler, "This Is Redstone," II-2 to II-10, IV-21, RATL; Bullard, *Redstone*, 61–67; Stuhlinger and Ordway, *Crusader*, 101; biography of William A. Mrazek in William Pickering Publications Collection, 1932–1971, folder 34, JPL. Mrazek became director of the Structures and Mechanics Laboratory of the Army Ballistics Missile Agency in April 1956.

22. Bullard, *Redstone*, 67–68; Haeussermann, "Guidance and Control," 228–29, 232; Stuhlinger and Ordway, *Crusader*, 102; F. Mueller, OHI, tape 4, side 2, and tape 5, side 1. On Buchhold and Wagner, see chapter 2 of this book.

23. F. Mueller, OHI, tape 4, side 2, and tape 5, side 1; Haeussermann, "Guidance and Control," 232; MacKenzie, *Inventing Accuracy*, 156–57.

24. F. Mueller, OHI, tape 4, side 2, and tape 5, side 1, NASM; Haeussermann, "Guidance and Control," 233–36; F. Mueller, "Inertial Guidance," 185–90; Farrior, "Inertial Guidance," 406. This account greatly simplifies the technical accounts in Haeussermann's and Mueller's printed accounts. On PIGAs, see chapter 2 of this book.

25. Bullard, *Redstone*, 67–70; Ordway and Sharpe, *Rocket Team*, 372.

26. Stuhlinger and Ordway, *Crusader*, 102; Chrysler, "This Is Redstone," IV-2, IV-4, IV-21, RATL. This account greatly simplifies the very complex workings of the ST-80. See the latter source for details.

27. Farrior, "Inertial Guidance," 406.

28. Huzel, *Peenemünde to Canaveral*, 227–30; Braun, "Redstone's First Flight"; von Braun, "Redstone, Jupiter, and Juno," 109.

29. Bullard, *Redstone*, 42–43, 55, 74–78, 82–88 (quotation, 75); Dunar and Waring, *Power to Explore*, 39–40; Medaris, *Countdown for Decision*, 99.

30. Bullard, *Redstone*, 70, 92, 100, 162–65; von Braun, "Redstone, Jupiter, and Juno," 110–11; Nicholas and Rossi, *Missile Data Book*, 3-5 to 3-9; Ordway and Wakeford, *International Missile and Spacecraft Guide*, 15.

31. M. Neufeld, "Orbiter," 231–37; Launius, preface, ix; Green and Lomask, *Vanguard*, 17–19; R. C. Hall, "Vanguard and Explorer," 101–4; R. C. Hall, "Earth Satellites"; von Braun, "Minimum Satellite Vehicle," 275. The larger tanks would increase the burning time from 110 to 132.4 seconds and the length of the midsection of the missile by 62 inches in von Braun's proposal. The standard steel warhead skin would be replaced by lightweight aluminum to reduce the weight of the first stage.

32. For details, see M. Neufeld, "Orbiter," which covers a great many nuances in the process and offers new evidence; also Bille and Lishock, *First Space Race*, 76–84. Viking and Vanguard are covered in Hunley, *Viking to Space Shuttle*, chap. 1.

33. Von Braun, "Redstone, Jupiter, and Juno," 111 (also quotations); Medaris, *Countdown for Decision*, 119; Baker, Hughes, and Bowne, *Redstone Arsenal Complex Chronology*, 2.B.1.

34. Von Braun, "Redstone, Jupiter, and Juno," 111–14; Medaris, *Countdown for Decision*, 142–43; Armacost, *Thor-Jupiter Controversy*, 144; presentation of William R. Lucas, "Redstone, Juno, and Jupiter," in "Rocketry in the 1950's," 46–58, NHRC; Nicholas and Rossi, *Missile Data Book*, 3-5, 3-6, for the speeds of the two missiles; Swenson,

Grimwood, and Alexander, *This New Ocean*, 64, for a good explanation of ablation. As G. Harry Stine reveals in *ICBM*, 107, Arthur Kantrowitz of Cornell's School of Aeronautical Engineering had been working under contracts with the Office of Naval Research when he learned of the reentry problem. He developed the test facility using shock tubes (see glossary entry); on these, see Heppenheimer, *Countdown*, 84; Simon G. Bauer, "Shock Tube," in *McGraw-Hill Encyclopedia*, 16: 431-2. Kantrowitz also worked with an Air Force contractor, AVCO, which built him a research lab. Thus he contributed to both the Army's and later the Air Force's ablative nose cones.

35. Von Braun, "Redstone, Jupiter, and Juno," 113, gives this description for Jupiter C, but according to Wolfe and Truscott, "Juno I," 43, NHRC, the Hydyne was used in Juno I but not Jupiter C.

36. Von Braun, "Redstone, Jupiter, and Juno," 113; Greever, "Juno I and Juno II," 71; von Braun, "Rundown," 32.

37. JPL, "Explorer I," 20, 22, 83–84; comments of William H. Pickering in "Rocketry in the 1950's," 70–71, NHRC; von Braun, "Redstone, Jupiter, and Juno," 113; von Braun, "Rundown," 80; Medaris, *Countdown for Decision*, 202; Robillard, "Explorer Rocket Research Program," 493; news release, JPL, no headline or date, in "Juno I Launch Vehicle," NASM. My thanks to Howard Wolko, a distinguished structural engineer who served as a volunteer with the Aeronautics Division of NASM, for explaining the effects of the resonance. All of the other sources apparently assumed it should be obvious to their readers.

38. Von Braun, "Redstone, Jupiter, and Juno," 113–14; Bille and Lishock, *First Space Race*, 116, 119; A. Wilson, "Jupiter/Juno," 12–14; Wolfe and Truscott, "Juno I," 30, 40, NHRC. I have used the last report for distances achieved by the first and third launches. According to von Braun, "This [nose cone] was the first man-made object ever recovered from outer space" (113).

39. For the data on Sputnik and the Vanguard satellite, Hastedt, "Sputnik and Technological Surprise," 401; von Braun, "Minimum Satellite Vehicle," 275 (first quotation); Medaris, *Countdown for Decision*, 152–55 (remarks to McElroy, 155). These remarks have been quoted elsewhere, e.g., McDougall, *Heavens and the Earth*, 131, but of course Medaris must have been writing them from memory and they probably aren't exactly what the two men said. On von Braun's activities publicizing space flight, see McCurdy, *Space and the American Imagination*, 37–42; Crouch, *Aiming for the Stars*, 120–21.

40. McDougall, *Heavens and the Earth*, 113–16, 131–32; Hastedt, "Sputnik and Technological Surprise," 401–2; Van Dyke, "Sputnik: A Political Symbol," 366–69.

41. A. Wilson, "Jupiter/Juno," 14; Medaris, *Countdown for Decision*, 166–70, 189–90 (quotation, 166); R. C. Hall, "Vanguard and Explorer," 110; McDougall, *Heavens and the Earth*, 150, 154.

42. Wolfe and Truscott, "Juno I," 4, 44–45, 75, 78, NHRC.

43. N. Friedman, *Seapower and Space*, 13.

44. Von Braun, "Rundown," 82–84; Robillard, "Explorer Rocket Research Program," 492– 94; A. Wilson, "Jupiter/Juno," 15; telephone interview of Stuhlinger by author, April 10, 2002, to clarify details of the procedure; Stuhlinger bio, September 1963, microfiche 1063, and Haeussermann bio, January 1963, microfiche 1068, biographical

files, MSFC/HO; Medaris, *Countdown for Decision*, 216; Bullard, *Redstone*, 166–67; four documents written by Stuhlinger—"Apex Predictor," "Apex Prediction in Missile #42," "Dry Run for Apex Determination," and "Apex Determination for Missile #27," all UP—copies of which were generously provided by Dr. Stuhlinger; on the use of the LEV-3 and the reasons for it, Stuhlinger to author, May 5, 2002 UP.

45. Medaris, *Countdown for Decision*, 222–24; Crouch, *Aiming for the Stars*, 144–47; Koppes, *JPL*, 87–88; von Braun, "Rundown," 84; Pickering, "Cluster System." There is some confusion in the sources as to the location of the West Coast tracking station. Medaris, 224, locates it in Earthquake Valley, which Koppes places near San Diego. Pickering, who should have known very well where it was, on page 156 places it at Camp Irwin, which was in the Mojave Desert considerably northeast of Los Angeles and a long way from San Diego. However, Bille and Lishock, in *First Space Race*, 132, state that it was actually located at Borrego Springs, which is in San Diego County.

46. Stuhlinger, "Army Activities in Space," 68–69; Wolfe and Truscott, "Juno I," 72–79, NHRC. Sources vary in their nomenclature, sometimes using the designation Jupiter C both for the reentry test vehicle and the launch vehicle and sometimes using Juno I for both. I have used Jupiter C only for the reentry test vehicle and Juno I only for the launch vehicle for the Explorer satellites. It is beyond the scope of this book to discuss most of the satellites launched by various launch vehicles, but for preliminary treatments of the initial Explorer satellites, see JPL, "Explorer I"; Robillard, "Explorer Rocket Research Program."

47. Hansen, *Engineer in Charge*, 1, 385; Saturn/Apollo Systems Office, "Mercury-Redstone," 1-9, NHRC. The other research lab was the Lewis Flight Propulsion Laboratory in Cleveland, Ohio. Langley was located at what became the Air Force's Langley Field in the Tidewater area of Virginia, near the city of Hampton.

48. Hansen, *Engineer in Charge*, 385. To get responsibility for the project, NACA had to compete with related projects from all three services—the Air Force's Man-in-Space-Soonest, the Army's Man Very High (renamed Project Adam), and the Navy's Manned Earth Reconnaissance proposals—but President Eisenhower supported a space-for-peace policy that favored a civilian agency. There also wasn't any clear military mission for placing a human in orbit at this time. Hence the NACA got the mission while it awaited the pending formation of the National Aeronautics and Space Administration. See Swenson, Grimwood, and Alexander, *This New Ocean*, 91, 92, 96, 99–102.

49. Galloway, "Organizing"; Glennan, *Birth of NASA*, 1–2, 9, 13. On the creation of NASA, see also Schoettle, "Establishment of NASA"; Griffith, *National Aeronautics and Space Act*.

50. Swenson, Grimwood, and Alexander, *This New Ocean*, 122; Saturn/Apollo Systems Office, "Mercury-Redstone," 1-9, NHRC; Baker, Hughes, and Bowne, *Redstone Arsenal Complex Chronology*, 2.B.Introduction.

51. Saturn/Apollo Systems Office, "Mercury-Redstone," 1-9, NHRC; Swenson, Grimwood, and Alexander, *This New Ocean*, 114; Bullard, *Redstone*, 156–57.

52. Saturn/Apollo Systems Office, "Mercury-Redstone," 2-3, NHRC; Swenson, Grimwood, and Alexander, *This New Ocean*, 171.

53. Kuettner, "Mercury-Redstone," 69; Saturn/Apollo Systems Office, "Mercury-Redstone," 3-1, NHRC.

54. Kuettner, "Mercury-Redstone," 70; Liquid Propellant Information Agency, *Liquid Propellant Safety Manual*, chap. 8, p. 2. Kuettner does not say what the higher-quality vanes were made of, and other sources do not mention this change.

55. Kuettner, "Mercury-Redstone," 70; NASA Office of Congressional Relations, *Mercury-Redstone III*, 5-1; Sharpe and Burkhalter, "Mercury-Redstone," 369.

56. Kuettner, "Mercury-Redstone," 71–72; Saturn/Apollo Systems Office, "Mercury-Redstone," 5-1 to 5-37, 8-5, NHRC; C. Walker, *Atlas*, 233.

57. Kuettner, "Mercury-Redstone," 73–74; Saturn/Apollo Systems Office, "Mercury-Redstone," 1-5, NHRC.

58. Kuettner, "Mercury-Redstone," 71; Swenson, Grimwood, and Alexander, *This New Ocean*, 181; NASA Office of Congressional Relations, *Mercury-Redstone III*, 5-1.

59. The story of the transfer is complex and interesting but does not belong in this history. See Medaris, *Countdown for Decision*, 257–69; Dunar and Waring, *Power to Explore*, 25–30; Glennan, *Birth of NASA*, 9–11, 22–23; M. Neufeld, "End of the Army Space Program."

60. Kuettner, "Mercury-Redstone," 77; Saturn/Apollo Systems Office, "Mercury-Redstone," 8-1 to 8-5, NHRC; Swenson, Grimwood, and Alexander, *This New Ocean*, 293–97.

61. Saturn/Apollo Systems Office, "Mercury-Redstone," 8-1, 8-7, NHRC; Swenson, Grimwood, and Alexander, *This New Ocean*, 297–301.

62. Saturn/Apollo Systems Office, "Mercury-Redstone," 1-5, 8-9 to 8-11, NHRC; Kuettner, "Mercury-Redstone," 77–78; Swenson, Grimwood, and Alexander, *This New Ocean*, 314–18.

63. Kuettner, "Mercury-Redstone," 78–79; Swenson, Grimwood, and Alexander, *This New Ocean*, 328–30; Ezell, *NASA Historical Data Book* 2: 143–44.

64. Swenson, Grimwood, and Alexander, *This New Ocean*, 332–35; Heppenheimer, *Countdown*, 189–92.

65. See Saturn/Apollo Systems Office, "Mercury-Redstone," 9-1 to 9-9, NHRC. For coverage of the MX-774B, see chapter 7 of this book and chapter 1 of Hunley, *Viking to Space Shuttle*; the latter chapter also covers the Viking sounding rocket.

Chapter 7. The Atlas, Thor, and Jupiter Missiles, 1954–1959

1. Rosen, "Big Rockets," 67. For information on Soviet missiles, see Zaloga, *The Kremlin's Nuclear Sword*, which, among other things, discusses a *Soviet* missile gap with the United States. Thanks to Dwayne Day for bringing Zaloga's book to my attention.

2. On Vanguard, see Hunley, *Viking to Space Shuttle*, chap. 1.

3. For 1957 figures, Lonnquest, "Face of Atlas," 208, 220 (thanks to John for providing a copy); for 1960 figures, Chapman, *Atlas*, 8. T. Hughes, *Rescuing Prometheus*, 70, gives the same number of contractors and subcontractors and 70,000 workers in 1957. By then the Air Force and contractor personnel overseeing Atlas were responsible for Thor and Titan as well.

4. For these points, see Beard, *Developing the ICBM*, esp. 8, 12, 29, 72, 142, 154, 224,

228; cf. Lonnquest, "Face of Atlas," 14–16, 21, 23, 28, 34, 37, 41. Lonnquest presents the more nuanced view. See also Lonnquest and Winkler, *To Defend and Deter*, 3–4; J. Neufeld, *Ballistic Missiles*, 48.

5. Lonnquest, "Face of Atlas," 29–30; J. Neufeld, *Ballistic Missiles*, 68–70.

6. Beard, *Developing the ICBM*, 140–41.

7. Lonnquest and Winkler, *To Defend and Deter*, 36; Pike, "Atlas," 94; Beard, *Developing the ICBM*, 11, 141–43, 165–66; Lonnquest, "Face of Atlas," 33–34, 65–71, 77–82. Already in December 1952 an ad hoc committee chaired by Caltech's Clark Millikan had recommended, based on development toward the so-called H-bomb in the "Mike" test of November 1952, that Atlas's CEP be changed to 1 mile and its payload reduced to 3,000 pounds.

8. "Trevor Gardner," 86; Lonnquest, "Face of Atlas," 43; T. Hughes, *Rescuing Prometheus*, 81, 83–84; York, *Race to Oblivion*, 18–22, 84 (first quotation, 84); Beard, *Developing the ICBM*, 166 (string of quotations, which Beard presents in that fashion); Lonnquest and Winkler, *To Defend and Deter*, 34–43.

9. Lonnquest, "Face of Atlas," 71–76 (first quotation, 71) for an excellent biographical discussion of von Neumann; Gorn, *Universal Man*, 8–9, for discussion of Minta Gymnasium; von Kármán, *Wind and Beyond*, 106–7 (last quotation, 107); Ceruzzi, *Modern Computing*, 21, 23–24.

10. Lonnquest, "Face of Atlas," 74–76; Lonnquest and Winkler, *To Defend and Deter*, 33–37. On von Kármán, see his *Wind and Beyond* and Gorn's *Universal Man*, both passim.

11. For entire report with Gardner's covering memo, see J. Neufeld, *Ballistic Missiles*, 249–65 (quotations, 255, 259, 252); on Quarles, "Donald A. Quarles" and "Biography: Donald A. Quarles." Quarles became secretary of the Air Force himself from August 15, 1955, to April 30, 1957.

12. Lonnquest and Winkler, *To Defend and Deter*, 39; T. Hughes, *Rescuing Prometheus*, 93, 102; Terhune, OHI; "BMO Chronology," 18–19, UP. My thanks to Raymond L. Puffer, formerly BMO chief historian, for allowing me to copy the chronology.

13. J. Neufeld, "Schriever," 281–90; Texas A&M, "History and Development" for its previous name; Lonnquest, "Face of Atlas," 222.

14. J. Neufeld, "Schriever," 290; Lonnquest, "Face of Atlas," 73, 204, 220–23; "Trevor Gardner"; Terhune, OHI, 16, 19; Glasser, OHI, 54–55. One staff officer who had a much lower opinion of Schriever was Col. Ed Hall, who is discussed in this and subsequent chapters. Incidentally, so frequently was Schriever late for meetings (as well as sometimes early, causing the attendees to be hastily assembled) that his staff spoke of SST, Schriever Standard Time (plus or minus an hour); see Glasser, OHI, 179.

15. ICBMs, by most definitions, had a range of at least 5,000 miles, whereas intermediate range ballistic missiles (IRBMs) like Thor and Jupiter had ranges of at least 1,500 miles.

16. Lonnquest and Winkler, *To Defend and Deter*, 43; T. Hughes, *Rescuing Prometheus*, 103–5; Beard, *Developing the ICBM*, 190–93; "BMO Chronology," 25–26, UP. For much more on these developments, see Lonnquest, "Face of Atlas," 193–99, 274–75. In July 1960 the Air Force essentially abolished the Gillette Procedures. J. Neufeld,

Ballistic Missiles, 267–321, has reproduced the Gillette committee report and related documents.

17. The members of these committees did not think Convair was capable of serving in the traditional role of prime contractor for Atlas.

18. Dyer, *TRW*, 168–85, 198, 207; T. Hughes, *Rescuing Prometheus*, 86, 89–92, 97–103, 110–15, 132–35; Beard, *Developing the ICBM*, 171–77; Lonnquest, "Face of Atlas," 144–59; "BMO Chronology," 18–22, UP.

19. Lonnquest, "Face of Atlas," 226–31; T. Hughes, *Rescuing Prometheus*, 119–26; "BMO Chronology," 19, 28, UP. The official name of Black Saturday was the Commander's Monthly Internal Management Conference, and it lasted all day. It had begun as a weekly Friday meeting, which proved to be more often than necessary.

20. Lonnquest, "Face of Atlas," 232–44; T. Hughes, *Rescuing Prometheus*, 107–8; Lonnquest and Winkler, *To Defend and Deter*, 44.

21. T. Hughes, *Rescuing Prometheus*, 108; Glasser, OHI, 44–45.

22. Pike, "Atlas," 92; Martin, "Steel Balloon," 1–2; Chapman, *Atlas*, 28–35, 86–88; Martin, "Brief History," 54–55; C. Walker, *Atlas*, 22; J. Neufeld, *Ballistic Missiles*, 36–37.

23. Chapman, *Atlas*, 28, 87 (first quotation, 28); Martin, "Steel Balloon," 2; C. Walker, *Atlas*, 23, 285.

24. Chapman, *Atlas*, 50–52, 58–59; Pike, "Atlas," 93–94. On Standley, see C. Walker, *Atlas*, 16.

25. J. Neufeld, *Ballistic Missiles*, 68–77; Chapman, *Atlas*, 60–65; Pike, "Atlas," 93–94; C. Walker, *Atlas*, 35–36. Patterson served as both engineer and marketer, although his education was in mathematics (B.A.) and physics (M.A.); see Walker, 296.

26. Martin, "Steel Balloon," 1, 4 (quotations, 4); Chapman, *Atlas*, 88. Martin felt obliged to identify Reynolds Wrap as a brand of metal foil used to wrap food. As discussed below, Rocketdyne provided the engines for Atlas as well as Thor and Jupiter.

27. J. Neufeld, *Ballistic Missiles*, 116–17, 125; "BMO Chronology," 20–21, UP; Lonnquest, "Face of Atlas," 139–40.

28. Martin, "Steel Balloon," 6–10; Chapman, *Atlas*, 88–89, 102. Martin qualified his comment about Miles by saying, "My recollection is that John Miles came forward with some simplifying assumptions" (9). For more on Dempsey, see C. Walker, *Atlas*, 8–9, 29–30, 39–40, 143, 251, 260; see Walker also on thickness of stainless steel (49) and tank pressurization (53).

29. Myers, "Navaho," 129; J. Simmons, "Navaho Lineage," 18–19; Gibson, *Navaho Missile Project*, 40–41; Chapman, *Atlas*, 65.

30. Kraemer, *Rocketdyne*, 58–62; Lonnquest and Winkler, *To Defend and Deter*, 4.

31. Jenkins, "Stage-and-a-Half," 100; "BMO Chronology," 21–22, UP; Smith and Minami, "Atlas Engines," 54–55.

32. Smith and Minami, "Atlas Engines," 54–57, 63; Jenkins, "Stage-and-a-Half," 100–101, which gives different figures for the engines' thrust (I have relied on Smith and Minami, who do not say whether the sustainer engine's specific impulse was measured at sea level, but that is likely); J. Simmons, "Navaho Lineage," 18. Presumably the sustainer's specific impulse was much higher at altitude, for which its nozzle was designed.

33. Smith and Minami, "Atlas Engines," 55, 63–65; Hartt, *Mighty Thor*, 80; J. Simmons, "Navaho Lineage," 20–21. Simmons says the baffles appeared in the MA-3, but Smith and Minami seem to suggest they appeared in the MA-2, referring to "resolving potential combustion instability" during 1955–59 in the MA-1/MA-2 engine systems. Although the Hartt book is about a different missile, it used substantially the same propulsion system, and Hartt is obviously describing the same problem and providing additional details about the solution.

34. Smith and Minami, "Atlas Engines," 55, 60–61, 64–65; *CPIA/M5*, unit 176, Atlas MA-3 Booster, and unit 178, Atlas MA-3 Sustainer; "Experimental Engines Group," 25–26; C. Walker, *Atlas*, 52.

35. "Experimental Engines Group," 21–27 (quotations, 21, 22, 25, 26); Ezell and Mitchell, "Engine One," 57.

36. Hujsak, "Bird That Did Not Want to Fly." Clarence William "Bill" Schnare told the author on May 23, 2002, in a telephone conversation about his OHI at AFHRA, that the Non-Rotating Engine Branch routinely used product-improvement clauses in its contracts after the early 1950s. Schnare, together with Kenny Buchtel, was instrumental in instituting the system. I thank him for sharing his memories and permitting me to cite them and his interview as corrected by our conversations.

37. "BMO Chronology," 22, UP; MacKenzie, *Inventing Accuracy*, 120–21; Draper, *Inertial Guidance*, 22.

38. MacKenzie, *Inventing Accuracy*, 22, 64–65, 80; Draper, Wrigley, and Hvorka, "Origins of Inertial Navigation," 449–60; Draper, "On Course to Modern Guidance." For Minneapolis Honeywell's crediting of MIT under Draper's direction, see "Vanguard Guidance Sidebar," NHRC; on the complexity of invention and the limited extent to which Draper can be called the inventor of liquid-floated gyroscopes, MacKenzie, 91–92.

39. For most of the three paragraphs above, and first three and last four quoted words and phrases, Battin, "Space Guidance Evolution," 97–99, 102; for middle three quoted words or phrases, MacKenzie, *Inventing Accuracy*, 318–20. MacKenzie is also the source for the fact (as distinguished from what Battin had heard) that Atlas used delta guidance. Battin spells Schwidetzky's name with an *ei*, but Lethbridge, "History of Rocketry," spelled it with no *e* before the *i* and is the source for the fact that Schwidetzky was one of six people accompanying von Braun to this country in the first installment of Germans coming from Peenemünde. Battin identifies him as having worked there with von Braun. Chapman, *Atlas*, 66, also spells the name "Schwidetzky."

40. MacKenzie, *Inventing Accuracy*, 119–22; Pike, "Atlas," 96; Lonnquest, "Face of Atlas," 39, 212–13; Klass, "How Command Guidance Controls Atlas."

41. "Searchlights to Space Systems," 4–11; "Arma Inertial Guidance," 5–11 (second quotation, 6); "BMO Chronology," 23 (also first quotation), UP. According to Stumpf, *Regulus*, 165, Draper was the one who originally discovered flurolube (which Stumpf spells "fluorolube") as the fluid for use in his fluid-floated gyros.

42. Babcock, *Magnificent Mavericks*, chap. 12, pp. 19, 21, 30–33. I am deeply in Liz's debt for providing her forthcoming history in draft and allowing me to cite it. It is interesting that WDD decided to test the system at NOTS, since the Air Force had tracks of its own at Edwards AFB and Holloman AFB. But the NOTS track's rails were

continuously welded, while Holloman's, at least, were not, so perhaps that was a factor. See Weals, Pryor, and Di Pol, OHI.

43. Chandler and Bankston, "USAF Projects WS-107A and WS-315A," 1, 3–7, 17–18, 47, NA. Incidentally, NOTS also used SNORT to test the Army Ballistic Missile Agency's LEV-3 autopilot in linear accelerations up to 20 Gs during 1956. The autopilot "remained stable to the desired accuracy throughout each run" ("Technical Program Review, 1956," 223, NA).

44. "Arma Inertial Guidance," 7.

45. Ibid., 5, which reprints the news release; "Searchlights to Space Systems," 12; MacKenzie, *Inventing Accuracy*, 119n72, 122. Contrary to what MacKenzie says on page 122 about the greater accuracy of the radio-inertial system, on 428 he shows a CEP for Atlas D, which normally flew with a radio-inertial system, of 1.8 nautical miles, whereas that for Atlas E and F with all-inertial systems was 1.0, far more accurate. Hence my "according to some accounts." The figures on his pages 122 and 428 may be for different periods or from different sets of data.

46. Muenger, *Searching the Horizon*, 67–68; Hartman, *Adventures in Research*, 217–18; J. Neufeld, *Ballistic Missiles*, 79n.

47. "BMO Chronology," 22–23, UP; T. Hughes, *Rescuing Prometheus*, 126–29; Pike, "Atlas," 96; Lonnquest and Winkler, *To Defend and Deter*, 210–11; Hartt, *Mighty Thor*, 144, 166, 193; U.S. Congress, "Missile Programs," H. Rep. 1121, 108–10; Fuhrman, "Polaris to Trident," 11, which describes the X-17; Powell, "Thor-Able and Atlas-Able," 219–21; Air Force Ballistic Missile Division, news release REL-59-23, May 14, 1959, copy in NHRC. On Army development, see chapter 6 of this book as well as the congressional report cited here. The speed of the Atlas is given by Hartt, 166, as 17,250 mph and by Ballistic Missile Division, "Thor," NASM, as 10,000 mph; Nicholas and Rossi, *Missile Data Book*, 3-5, 3-7, gives only the Mach numbers. For the Able upper stage, and fuller coverage of the Thor-Able flights, see Hunley, *Viking to Space Shuttle*, chap. 2. A Thor-Able nose cone with AVCO's Avcoite ablative material was recovered from the South Atlantic on April 8, 1959.

48. Pike, "Atlas," 175, 178; Grassly, "Summary," I-1-1-1, AFHRA.

49. Nicholas and Rossi, *Missile Data Book*, 3-2, 3-8; Lonnquest and Winkler, *To Defend and Deter*, 216. As Hunley, *Viking to Space Shuttle*, discusses, many of the operational missiles later were used as launch vehicles.

50. Armacost, *Thor-Jupiter Controversy*, esp. 50–55; Lonnquest and Winkler, *To Defend and Deter*, 40–41, 47–49; J. Neufeld, *Ballistic Missiles*, 132–33.

51. U.S. Congress, "Missile Programs," H. Rep. 1121, 101 (also quotations). "Thor-Jupiter Controversy," the subtitle of Armacost's book, is also the heading of the section of Report 1121 that begins on page 101; ensuing pages give further details.

52. "BMO Chronology," 23, UP; Truax, autobiography (in draft, kindly provided by Capt. Truax), chap. 18, 240–41, and chap. 19, 242–44, UP; A. Thiel, OHI extract, 3–4; Hoelzer, OHI, tape 2, side 2, for Thiel's background on the V-2. Truax remembered incorrectly that Gardner was with Hycon when the young naval officer knew him, but Truax was at Aerojet as the Bureau of Aeronautics' representative from March to September 1944, according to his officer biographical files at the NHC. At that time, Gard-

ner worked for General Tire and Rubber. My thanks to Ernst Stuhlinger for a copy of the Thiel interview, in which Thiel recalled, "I put Commander Truax to work with me" (4). Truax, probably more accurately, said, "I formed a team from the Ramo-Wooldridge Corporation to assist on the technical side," a team that included Thiel.

53. Truax, autobiography, chap. 19, 244–45, 249, and chap. 20, 254, UP; Hartt, *Mighty Thor*, 41–43, 69, 72.

54. Hartt, *Mighty Thor*, 45–59; J. Neufeld, *Ballistic Missiles*, 146–47; "BMO Chronology," 27, UP. Hartt puts the signing date on December 28. The two Air Force sources date it from December 27, although the chronology says "awarded." Arms, "Thor," 2-5, SMC/HO, also gives the date of December 28.

55. T.O. 21-SM75-1, 1-1 to 1-2, 4-1 to 4-6, UP; W.T.G., "Thor," 864; "Thor Fact Sheet," NASM. Thanks to John Lonnquest for providing a copy of T.O. 21-SM75-1.

56. Battin, "Space Guidance Evolution," 99–103; MacKenzie, *Inventing Accuracy*, 120–21, 319; Klass, "Thor Guidance," 38–39.

57. Klass, "Thor Guidance," 39, 41; Geschelin, "Thor's Brains," 27; W.T.G., "Thor," 866; Hartt, *Mighty Thor*, 189–91; Arms, "Thor," 3-5, SMC/HO. For an overview of the Mace and Regulus II, see Werrell, *Cruise Missile*, 111, 117, 119.

58. Draper, Wrigley, and Havorka, *Inertial Guidance*, 18–19.

59. Klass, "Thor Guidance," 39, 41; Geschelin, "Thor's Brains"; MacKenzie, *Inventing Accuracy*, 120n; W.T.G., "Thor," 866.

60. W.T.G., "Thor," 864; Fuller and Minami, "Thor/Delta Booster Engines," 44–45.

61. Herzberg, "Rocket Propulsion Laboratory," 1: xvi–xxii, I-3 to I-4, AFFTC/HO; Rockefeller, "History of Thor, 1955–1959," 16, AFHRA; fact sheet, "Air Force Astronautics Laboratory," February 1988, and brochure, "Directorate of Rocket Propulsion," undated, in "Units, RPL," AFFTC/HO. On the Non-Rotating Engine Branch, see also E. N. Hall, OHI by Hunley. Systems Command was the successor organization to Air Research and Development Command, effective April 1, 1961. For further information about the Atlas test stands and testing at the Edwards rocket site, see C. Walker, *Atlas*, 76, 130, 134–35, 137, 217.

62. Arms, "Thor," 2-9 to 3-4, B-2, B-3, SMC/HO; cf. the less complete comments in Hartt, *Mighty Thor*, 103, 108, 112–13, 116–17, 128, 131–32. On the nose cone, see General Electric, "Heat Sink Nose Cone," NASM; Butz, "GE System Stabilizes."

63. Grassly, "Summary," I-4-1-1, I-4-3-1, AFHRA; Arms, "Thor," SMC/HO, 3-4 to 3-5, B-3 to B-5, SMC/HO; Nicholas and Rossi, *Missile Data Book*, 3-4.

64. U.S. Congress, "Missile Programs," H. Rep. 1121, 103–5; A. Thiel, OHI extract, 4 (quotations); Perry, "Atlas, Thor, Titan, and Minuteman," 153.

65. Hartt, *Mighty Thor*, 129–30, 134–35; Arms, "Thor," B-3, SMC/HO; U.S. Congress, "Missile Programs," H. Rep. 1121, 105–6; U.S. Congress, "Missile Programs," Hearings, 339, 381.

66. U.S. Congress, "Missile Programs," H. Rep. 1121, 106–7 (Schriever quotation, 107); A. Thiel, OHI extract, 5, for information and quotation from Thiel.

67. Hartt, *Mighty Thor*, 164–69; U.S. Congress, "Missile Programs," H. Rep. 1121, 107; Arms, "Thor," B-5 to B-6, SMC/HO.

68. Lonnquest and Winkler, *To Defend and Deter*, 52, 268; "Thor Fact Sheet," NASM;

Ballistic Missile Division, "Thor," NASM. There is discussion of Thor, Atlas, and Jupiter as launch vehicles in Hunley, *Viking to Space Shuttle*.

69. Grimwood and Strowd, "Jupiter," 2–13, AMC; Armacost, *Thor-Jupiter Controversy*, 44–49, 92–93. The Navy's comparatively brief participation in the Jupiter program will be covered more fully in chapter 9 of this book. Here it is treated only for the effects it had on Jupiter.

70. Grimwood and Strowd, "Jupiter," 8, 30, 32, 35, 37, 41, 43, 5-1, AMC; Armacost, *Thor-Jupiter Controversy*, 119; Lonnquest and Winkler, *To Defend and Deter*, 49, 52.

71. Grimwood and Strowd, "Jupiter," 14–16, 19, AMC.

72. Ibid., 17, 25–28, 31; U.S. Congress, "Missile Programs," H. Rep. 1121, 50, 103–5; U.S. Congress, "Missile Programs," Hearings, 380–81.

73. Grimwood and Strowd, "Jupiter," 67–71, 5-2, 5-3, AMC; Standard Missile Characteristics, SM-78 Jupiter, January 1962, copied by John Lonnquest at the Air Force Museum at Wright-Patterson AFB. On the Jupiter, see also McCool and Chandler, "Liquid Propellant Engines," 294. Thor data are from Fuller and Minami, "Thor/Delta Booster Engines," 44–45.

74. Grimwood and Strowd, "Jupiter," 58–59, 69, AMC; *CPIA/M5*, unit 47, LR101-NA-1000 lb Thrust, Vernier for Atlas MA-5 and Thor MB-3-2 . . . ; Standard Missile Characteristics, SM-78 Jupiter (see note 73). Thiokol developed the vernier motor for the Jupiter. See Thiokol, *Rocket Propulsion Data*, item TX-83-4, which gives the average thrust as 570 pounds.

75. F. Mueller, OHI, tape 5, side 1; F. Mueller, "Inertial Guidance," 190; Haeussermann, "Guidance and Control," 233–36; Standard Missile Characteristics, SM-78 Jupiter (see note 73); Lonnquest and Winkler, *To Defend and Deter*, 259, 269; von Braun, "Redstone, Jupiter and Juno," 117–18; Grimwood and Strowd, "Jupiter," 58–59, 65–66, AMC; MacKenzie, *Inventing Accuracy*, 131n110, which cites Gen. Nathan Twining of the Air Force as giving the Jupiter CEP at half a mile and Thor's as "less than 2 nautical miles." Lonnquest and Winkler give Thor's as 2 miles and Jupiter's as 1,500 meters (equal to the 0.8 nautical miles provided in the official Air Force Standard Missile Characteristics). Because the official Air Force publication on Jupiter concurs with Lonnquest and Winkler, I have used 0.8 nautical miles.

76. Grimwood and Strowd, "Jupiter," 59–63, AMC; Standard Missile Characteristics, SM-78 Jupiter (see note 73); Thiokol, *Rocket Propulsion Data*, item TX-148; General Electric, "Heat Sink Nose Cone," NASM.

77. Grimwood and Strowd, "Jupiter," 71–73, 157, AMC; Medaris, *Countdown to Decision*, 137–39. Medaris called the devices "big 'beer cans' [made] out of wire mesh" (138), but Grimwood and Strowd not only describe them differently, they provide photos showing that they hardly resembled beer cans.

78. On the successful launch and Medaris's view of the cause of sloshing, see Grimwood and Strowd, "Jupiter," 157, AMC; Medaris, *Countdown to Decision*, 137–38; Medaris, " The Anatomy of Program Management," in Kast and Rosenzweig, eds., *Science, Technology, and Management*, 114. On the first successful Thor launch, see above in this chapter. A copy of E. N. Hall, "Comment," UP, was kindly provided by Michael Neufeld. Hall later told me, "I don't recall making this specific comment, but indeed

it was a tragic thing. In reality, there never was a Jupiter missile. It flew with a Thor engine which Medaris called 'ground support equipment.' The Jupiter program was a . . . nuisance to me. To provide the Army with engines for this misfit, I had continually to distort my development, test, and production schedules" (letter, UP).

79. Grimwood and Strowd, "Jupiter," 82, 157–58, AMC.

80. Lonnquest and Winkler, *To Defend and Deter*, 259, 264; Nicholas and Rossi, *Missile Data Book*, 3-2.

81. Lonnquest and Winkler, *To Defend and Deter*, 259, 268; Nicholas and Rossi, *Missile Data Book*, 3-5, 3-7, 3-8, 3-10; T.O. 21-SM75-1, 1-1, UP; Grimwood and Strowd, "Jupiter," 5-1, AMC; Standard Missile Characteristics, SM-78 Jupiter (see note 73); "Thor Fact Sheet," NASM; North American Rockwell, "Jupiter Propulsion System," NHRC.

82. U.S. Congress, "Missile Programs," Hearings, 379.

83. See Hunley, *Viking to Space Shuttle*, chap. 1.

Chapter 8. The Titan I and II, 1955–1966

1. Titans III and IV are covered in Hunley, *Viking to Space Shuttle*, chap. 6.

2. Adams, "Titan I to Titan II," 203, states that Titan I "represented the first application of altitude ignition of a large liquid rocket motor in an essentially zero g environment."

3. Beard, *Developing the ICBM*, 210.

4. Greene, "SM-68 Titan," 1: 11–14 (quotations, 11–12), SMC/HO.

5. Ibid., 1: 16–18, 20–21; "BMO Chronology," 21, 25, UP; Stumpf, *Titan II*, 13–15; Harwood, *Raise Heaven and Earth*, 305.

6. Chulick et al., "Titan," 20–22; Stumpf, *Titan II*, 18.

7. Chulick et al., "Titan," 22–23; Gordon et al., *Aerojet*, III-98; Adams, "Titan I to Titan II," 205; Aerojet, *Liquid Rocket Engines*, 20.

8. Chulick et al., "Titan," 23–24.

9. Ross, "Life at Aerojet-General University," 47, 72 (quotation, 72), UP. Many of Ross's recollections from this document reappeared in Gordon et al., *Aerojet*, but I have not found this one there.

10. Gordon et al., *Aerojet*, III-95, III-98 (quotation, III-95); Chulick et al., "Titan," 20–21; Aerojet, *Liquid Rocket Engines*, 4, 20. I have followed this last source for the engine thrust—actually expressed in terms of lbf (pounds force) rather than just pounds—and specific impulse. The other sources give slightly lower figures.

11. Adams, "Titan I to Titan II," 205–6; Stumpf, *Titan II*, 17; Harwood, *Raise Heaven and Earth*, 302, 307–8.

12. Adams, "Titan I to Titan II," 204; Hoselton, "Titan I Guidance System," 4–6; Greene, "SM-68 Titan," 1: 22, SMC/HO; Aerospace Corporation, "Final Report," II.C-49, II.C-55, AFHRA. The "apparently" stems from the fact that on page II.C-49 of the Gemini report there is a description of the three-axis reference system (TARS) and on II.C-55 there is the statement "The basic TARS package was used on the Titan I program." I infer from these two separate items that the "basic TARS package" used on Titan I included the "three strapped-down integrating rate gyros" described on II.C-49. Incidentally, Gary A. Hoselton was a guidance system mechanic on the Titan I system

for the Air Force, so his description, which is more detailed than the others, presumably is reliable.

13. Hoselton, "Titan I Guidance System," 4–5; Greene, "SM-68 Titan," 1: 112, SMC/HO.

14. Adams, "Titan I to Titan II," 204.

15. Hoselton, "Titan I Guidance System," 7; Greene, "SM-68 Titan," 1: 97, 113, SMC/HO. Greene does not say, but presumably the 0.8-nautical-mile figure for accuracy represented the missile's CEP. MacKenzie, *Inventing Accuracy*, 428, 433–34, assigns it a CEP of 0.65 nautical miles, but his comments make it clear this is not an entirely reliable figure, although he calls it "plausible." Stumpf, *Titan II*, 22, does give 0.8 nautical miles as the CEP, citing Greene for the figure.

16. Greene, "SM-68 Titan," 1: 110–11, SMC/HO; Stumpf, *Titan II*, 19, 58–60. It was AVCO, incidentally, that proposed the change in terminology from "nose cone" to "reentry vehicle," in March 1957. At first objecting that the shift would produce confusion, the Air Force eventually acceded. See Greene, 1: 111n.

17. Martin Company, "Flight Test Report, Lot A Summary," Report SR-59-38, October 1959, in Greene, "SM-68 Titan," 3: supporting documents, pp. 291–308, SMC/HO; Greene, "SM-68 Titan," 1: 91–92, SMC/HO; Stumpf, *Titan II*, 19, 275.

18. R. Greenspun and D. Mikelson, "Flight Test Report, Lot B Summary," June 1960, Martin Report CR-60-20, item 4.1, in Greene, "SM-68 Titan," 3: supporting document 121; Greene, "SM-68 Titan," 1: 92–93, SMC/HO; Stumpf, *Titan II*, 19, 275.

19. Greene, "SM-68 Titan," 1: 93–94, 126–28, SMC/HO; "BMO Chronology," 58, UP. Carrying significant experience from the Viking program to Titan were, among others, Bill Purdy, Titan chief engineer, and John Youngquist; see Harwood, *Raise Heaven and Earth*, 256, 305.

20. Greene, "SM-68 Titan," 1: 94, SMC/HO; "Titan Again Fails, Exploding on Pad," *New York Times*, December 13, 1959, in Greene, 3: supporting document 91.

21. Priority TWX, AFBMD (Air Force Ballistic Missile Division) to COFS (Chief of Staff), USAF and ARDC (Air Research and Development Command), December 24, 1959, in Greene, "SM-68 Titan," 3: supporting document 59, SMC/HO.

22. Greene, "SM-68 Titan," 1: 94–95, SMC/HO; Harwood, *Raise Heaven and Earth*, 317 (anecdote).

23. Schriever to General T. D. White, January 7, 1960, in Greene, "SM-68 Titan," 3: supporting document 107, SMC/HO; see also supporting documents 106, "Washington Briefing—6 Jan 60," and 101, "Summary Statement to be made by General Ritland prior to giving Recommendations," December 31, 1959. Incidentally, the briefing revealed that industrial contractors employed 20,000 people on Titan alone.

24. Greenspun and Mikelson, "Flight Test Report, Lot B Summary" (see note 18), items 3.1, 3.2, 4.2; Greene, "SM-68 Titan," 1: 95, SMC/HO.

25. D. Mikelson, "Flight Test Report, Lot C Summary," September 1960, Martin Report CR-60-41, summary and item 3.4, in Greene, "SM-68 Titan," 3: supporting document 124, SMC/HO; Stumpf, *Titan II*, 275–76, which I am using here only for the specific dates of the flights, the author's indications of the outcomes of the flights (successful or unsuccessful) not according with other evidence, although generally Stumpf

has done a great deal of research and is reliable. As noted, the -3 engines began flying in May of 1960, which was in the midst of the lot-G testing.

26. Mikelson, "Flight Test Report, Lot C Summary" (see note 25); Greene, "SM-68 Titan," 1: 95, SMC/HO; Stumpf, *Titan II*, 275–76.

27. Greene, "SM-68 Titan," 1: 95, SMC/HO; Stumpf, *Titan II*, 275–76. There were no lots D, E, F, or H.

28. Greene, "SM-68 Titan," 1: 96–99, SMC/HO; Stumpf, *Titan II*, 275–77; Grassly, "Summary," I-2a-3-1 to I-2a-3-2, AFHRA. The count in the narrative does not include ten operational, demonstration-and-shakedown-operation, and special test missions. Harwood, *Raise Heaven and Earth*, 317, counts these among the R&D flights and comes up with a 66 percent success rate for all sixty-seven.

29. Stumpf, *Titan II*, 29–31; Adams, "Titan I to Titan II," 211; Lonnquest and Winkler, *To Defend and Deter*, 232–35; Nicholas and Rossi, *Missile Data Book*, 3-4.

30. See, e.g., Levine, *Missile and Space Race*, 82–84. He claims, "Titan I was probably the first ICBM to make an important contribution to Western Defense" (82), but presumably he would agree that it did so for only a short time.

31. Clark, *Ignition!* 42–45 (and see Hunley, *Viking to Space Shuttle*, chap. 1, for the early history of UDMH development); Gordon et al., *Aerojet*, III-100; Lonnquest and Winkler, *To Defend and Deter*, 231; Stine, *ICBM*, 229, on Demaret and colleagues making the proposal to BMD.

32. Stumpf, *Titan II*, 34–35, 29.

33. J. Neufeld, *Ballistic Missiles*, 191–92.

34. Stumpf, *Titan II*, 33–35; Lonnquest and Winkler, *To Defend and Deter*, 233; Greene, "SM-68 Titan," 1: 74, SMC/HO; Adams, "Titan I to Titan II," 217.

35. Stumpf, *Titan II*, 36; Lonnquest and Winkler, *To Defend and Deter*, 230; Greene, "SM-68 Titan," 1: 74–75, SMC/HO; MacKenzie, *Inventing Accuracy*, 203–6. MacKenzie argues, however, that there were too few Titan IIs to serve the counterforce mission and that Minuteman II was a more effective counterforce weapon than Minuteman I.

36. Stumpf, *Titan II*, 36; "BMO Chronology," 68, UP; Greene, "SM-68 Titan," 1: 114, SMC/HO.

37. Chulick et al., "Titan," 24–27; Stumpf, *Titan II*, 37; Aerojet, *Liquid Rocket Engines*, 4, 8, 20, 22. As comparisons of diagrams and data in the last source reveal, what is sometimes said about the similarity in size of the Titan I LR87-AJ-3 and LR91-AJ-3 and the Titan II LR87-AJ-5 and LR91-AJ-5 is erroneous. For example, in Titan I the stage-1 engine was 108 inches long and 41.3 inches in diameter at the nozzle exit, whereas the Titan II equivalent was 93.3 inches long and 43.2 inches wide at the far end of the nozzle. There was an even greater disparity in the diameters of the stage-2 engines.

38. Chulick et al., "Titan," 27.

39. Ibid., 24–27; Adams, "Titan I to Titan II," 214–15; Stumpf, *Titan II*, 37–40; Air Force Plant Representative, "Semi-Annual Historical Report," Tab 9, captions for photos showing operation of Titan II engine, AFHRA.

40. Chulick et al., "Titan," 25–29; Harwood, *Raise Heaven and Earth*, 320; Aerojet, *Liquid Rocket Engines*, 4, 8, 20, 22. I have followed this last source for total thrust and specific impulse where there are differences with the other sources.

41. Aerojet, "Aerojet Engineers behind Propulsion," NASM.

42. Adams, "Titan I to Titan II," 214; Stumpf, *Titan II*, 64. MacKenzie, *Inventing Accuracy*, 157, states that the gyros in Titan II used ball bearings, but Stumpf, 64, talks about problems involving "air bubbles in the flotation fluid" (see below in the narrative), strongly suggesting that the gyros were fluid-floated like those on the Thor. J. Bonin and E. J. Loper, "Performance of Titan II Accelerometer on Centrifuge, Sled and Missile Flight Test," Inertial Guidance Test Symposium, vol. 2, October 16–17, 1962, V-1 to V-13, microfilm roll 32,265, AFHRA, beginning on frame 615, describes rocket sled tests at Holloman AFB, flight tests at Cape Canaveral, and centrifuge tests at AC Spark Plug of pendulous integrating gyro accelerometers for the Titan II, but the other literature does not make it clear that this was the type of accelerometer used on the actual missile. The tests did "indicate adequacy" (V-13) of the PIGA design, however.

43. Stumpf, *Titan II*, 64, 72, 182, 184–85. As with Titan I, MacKenzie, *Inventing Accuracy*, 428, 435, shows an accuracy of 0.65 nautical miles CEP, with figures he has found in the literature ranging from 0.5 to 0.8 nautical miles. I am more convinced by Stumpf's figures, which show a CEP slightly under 0.8 and specific figures varying from 0.58 to 0.97 nautical miles, 0.78 being the midpoint.

44. Stumpf, *Titan II*, 61–62. Cf. Lonnquest and Winkler, *To Defend and Deter*, 230–31.

45. Stumpf, *Titan II*, 44–47.

46. Ibid., 48–50; Adams, "Titan I to Titan II," 214–16.

47. Stumpf, *Titan II*, 44–47; Adams, "Titan I to Titan II," 221.

48. Stumpf, *Titan II*, 71–73, 86; Adams, "Titan I to Titan II," 220–21; Hacker and Grimwood, *Shoulders of Titans*, 104–5, 140; Harwood, *Raise Heaven and Earth*, 320; Grassly, "Summary," I-2b-1-1, AFHRA. Adams and Harwood agree on 25 complete successes among 33 launches. Stumpf also agrees, but the February 16, 1963, silo launch that he counts as successful because the Titan II cleared the silo intact can hardly be counted as an overall success, making for only 24. Grassly has 10 failures and 23 successes without any explanation, with Hacker and Grimwood also talking about 10 full or partial failures.

49. Hacker and Grimwood, *Shoulders of Titans*, esp. 76–78, 104–5; Space Division, "Space and Missile Systems Organization: A Chronology," 4, 95, SMC/HO. Launius, *NASA*, 81–82, provides a convenient, brief synopsis of much material about Project Gemini that is scattered through the early pages of *Shoulders of Titans*. My account does not mention that Marshall Space Flight Center was initially involved with the Agena part of Project Gemini since it had an Agena Project Office. This responsibility later shifted to SSD. The treatment here is not intended as an operational history of Gemini, which Hacker and Grimwood have already provided, but simply a recounting of the parts of that history relevant to launch-vehicle development.

50. Hacker and Grimwood, *Shoulders of Titans*, 77, 87, 105; Dyer, *TRW*, 231–32; Stumpf, *Titan II*, 72.

51. Hacker and Grimwood, *Shoulders of Titans*, 125–26, 135; Stumpf, *Titan II*, 72, 78.

52. Hacker and Grimwood, *Shoulders of Titans*, 134–37, 142–44, 166–68; Stumpf,

Titan II, 72, 79; "Gemini-Titan II," 6-13 to 6-16, SMC/HO; Huzel and Huang, *Modern Engineering*, 366, which has much better illustrations of the standpipe and piston than the other sources.

53. Chulick et al., "Titan," 31; Stumpf, *Titan II*, 40–42; Hacker and Grimwood, *Shoulders of Titans*, 168–69.

54. Stumpf, *Titan II*, 42–44, 72–73, 75–79. Stumpf bases this account on interviews. I have seen no other account of this problem and its resolution. Hacker and Grimwood, *Shoulders of Titans*, 141, mentions that the SSD commander briefed Gilruth, NASA's Manned Spacecraft Center chief, on August 8, 1963, that the clogging of the gas generator seemed to have been solved, but the two authors discuss the problem only briefly on page 137 and don't go into its technical solution. The Space Task Group became the Manned Spacecraft Center in 1961.

55. Stumpf, *Titan II*, 73, 97, 254–65, 281; Nicholas and Rossi, *Missile Data Book*, 3-2; Lonnquest and Winkler, *To Defend and Deter*, 234–35.

56. Stumpf, *Titan II*, 31, 36, 49, 281; Nicholas and Rossi, *Missile Data Book*, 3-7, 3-10; Lonnquest and Winkler, *To Defend and Deter*, 230–31; Aerojet, *Liquid Rocket Engines*, 4, 8, 20, 22. These sources differ considerably among themselves. I have followed Stumpf, 36, for the range, which is with the Mark 6 reentry vehicles shown in the table. With the Mark 4 reentry vehicle, the Titan II had a range of 8,400 nautical miles, further than the 9,000 (statute?) miles shown by Lonnquest and Winkler and way further than the 7,250 nautical miles shown by Nicholas and Rossi. Nicholas and Rossi give 6,300 nautical miles as the range for Titan I; so do Lonnquest and Winkler, although they don't specify "nautical." I have followed Aerojet for the engine thrust levels, since it built the engines and published its book for potential users. Stumpf gives the lengths of the respective Titans as 97.4 and 103.4 feet, based on actual measurement of missiles in storage for the latter figure. I rounded up to the more usual dimensions. He says various Air Force documents range from 108 to 114 for Titan II. He gives a total weight for Titan I of 223,211 pounds including the reentry vehicle and 327,119 (maximum 341,000) pounds for Titan II. Again, I've gone with the more conventional figures in Lonnquest and Winkler, probably without reentry vehicle, since the weight could vary, as Stumpf indicates.

57. Swenson, Grimwood, and Alexander, *This New Ocean*, 174–76, 187–89, 255. Culbertson had transferred to NASA in 1965 after being chief project engineer for the Atlas launch vehicle at General Dynamics. Among other high-level technical and policy jobs he held at NASA, he was on the space shuttle design and development team. He later headed the U.S. space station program and retired in January 1988 as associate administrator for policy and planning. See news release 87-188, "Culbertson to Retire from NASA," in "Culbertson, Philip E.," NASM.

58. Swenson, Grimwood, and Alexander, *This New Ocean*, 272–78; Grimwood, *Project Mercury*, 105–7.

59. Swenson, Grimwood, and Alexander, *This New Ocean*, 277–78, 299–300, 307–8, 318–22; Grimwood, *Project Mercury*, 111–12, 123–24; Convair, "Flight Test," AFHRA.

60. Swenson, Grimwood, and Alexander, *This New Ocean*, 336–37, 382–83; Grimwood, *Project Mercury*, 132.

61. Swenson, Grimwood, and Alexander, *This New Ocean*, 377, 383–84, 389; Grimwood, *Project Mercury*, 148–49. Crouch, *Aiming for the Stars*, 180, has a convenient table comparing the Soviet Vostok and the American Project Mercury missions and results.

62. Swenson, Grimwood, and Alexander, *This New Ocean*, esp. 397–400, 461–62, 505–11, 640–41; Grimwood, *Project Mercury*, 158–60, 164–65, 172–74, 191–93.

63. Hacker and Grimwood, *Shoulders of Titans*, 79, 299; "Statement of Dr. George E. Mueller, Associate Administrator for Manned Space Flight, NASA, before the Subcommittee on Manned Space Flight, Committee on Science and Astronautics, House of Representatives," February 17, 1964, 31–32, in G. Mueller, "Speeches," LC/MD. See Hunley, *Viking to Space Shuttle*, chaps. 2–3, for the development of Agena as an upper stage and for the conversion of the Atlas missile to launch-vehicle use. See Hacker and Grimwood, 88–89, 112–14, 157–61, 297–304, 330–32, for details of the development of Agena for Project Gemini.

64. Aerospace Corporation, "Final Report," II.B-1, II.B-22, II.B-23, AFHRA; Stumpf, *Titan II*, 73; Hacker and Grimwood, *Shoulders of Titans*, 523; quotation from "Statement of Dr. George E. Mueller" (see note 63), 29.

65. "Gemini-Titan II," 3-6, 6-18, SMC/HO; Aerospace Corporation, "Final Report," II.C-1, AFHRA.

66. "Gemini-Titan II," 6-18 to 6-20, SMC/HO; Aerospace Corporation, "Final Report," II.C-1, II.C-46 to II.C-55, AFHRA.

67. "Gemini-Titan II," 6-9, 6-16 to 6-21, SMC/HO; Aerospace Corporation, "Final Report," II.C-1, II.C-58, II.C-59, AFHRA.

68. Aerospace Corporation, "Final Report," II.C-1, AFHRA.

69. Chulick et al., "Titan," 30–31; Aerojet, *Liquid Rocket Engines*, 8, 10, 22, 24; *CPIA/M5*, unit 80, YLR91-AJ-7, and unit 93, YLR87-AJ-7; "Gemini-Titan II," iii, SMC/HO. The LR stood for "liquid rocket" and was sometimes used as shorthand for XLR and YLR. The X stood for experimental and the Y for operational engines, but the two terms were often interchanged in practical use.

70. Hacker and Grimwood, *Shoulders of Titans*, 197–99, 523.

71. Ibid., 209, 523.

72. Ibid., 524–29; Crouch, *Aiming for the Stars*, 191–92; Heppenheimer, *Countdown*, 218–19; on the Soviet feat, esp. Siddiqi, *Challenge to Apollo*, 454–59.

73. Hacker and Grimwood, *Shoulders of Titans*, 524–25; Crouch, *Aiming for the Stars*, 194–95; Heppenheimer, *Countdown*, 220.

74. Hacker and Grimwood, *Shoulders of Titans*, 268–69, 298–303.

75. Ibid., 526; Crouch, *Aiming for the Stars*, 194–96; Heppenheimer, *Countdown*, 221–22.

76. Hacker and Grimwood, *Shoulders of Titans*, 330–31, 383–89, 523–26; Aerospace Corporation, "Final Report," I.A-1, AFHRA; Crouch, *Aiming for the Stars*, 196–98; Heppenheimer, *Countdown*, 222–24; Launius, *NASA*, 81–82; Walters, "Plant Representative—Detachment 23," 17, AFHRA. For a detailed analysis of the failure of the Atlas on *Gemini 9*, see Convair, "Atlas SLV-3," NASM.

77. Stumpf, *Titan II*, 265–71; Memorandum for Record, Capt. Jeffrey J. Finan, Titan

II 23G-1 Launch Controller, subj: Titan II 23G-1 Processing History, November 21, 1988, supporting document III-9 in Geiger and Clear, "History," AFHRA. The story of Titan launch vehicles is continued in Hunley, *Viking to Space Shuttle*, chap. 6.

78. Herzberg, "Rocket Propulsion Laboratory," II-123 to II-133 (quotations, II-125, II-133), AFFTC/HO; bio of Richard R. Weiss, in "Units, RPL," AFFTC/HO.

79. Herzberg, "Rocket Propulsion Laboratory," II-137 to II-139, AFFTC/HO; CPIA, "History of CPIA." The Liquid Propellant Information Agency was formed only in 1958. As reported below in chapter 9, its solid-propellant counterpart dated from 1948, with a precursor known as the Rocket Propellant Intelligence Agency founded in 1946. For both organizations, the Navy had been the founding service.

80. Herzberg, "Rocket Propulsion Laboratory," II-148 to II-155, AFFTC/HO, with quotation from "Report of the Ad Hoc Working Group on Liquid Propellant Combustion Instability," CPIA publication 47, March 1964, 6.

81. Huzel and Huang, *Design*, 143–49.

82. Combustion instability, in particular, remained a problem in 1995. Yang and Anderson in the preface to *Liquid Rocket Engine Combustion Instability* noted, "In spite of the relatively long history of combustion instability, including a number of spectacular failures, and the resultant large body of work that has been devoted to the area, there is still a lack of fundamental knowledge regarding its prediction and the specific physical mechanisms that bring upon [about] its onset" (n.p.); see also 8–9, 15–16, 35.

Chapter 9. Polaris and Minuteman: The Solid-Propellant Breakthrough, 1955–1970

1. The dates when the missiles were phased out, along with summary information about their development and technical specifications, are conveniently provided in Lonnquest and Winkler, *To Defend and Deter*, 209–15, 227–34, 259–64, 268–73. Initial operational capabilities are shown in Nicholas and Rossi, *Missile Data Book*, 1-5, 3-2 to 3-4.

2. For earlier developments in solid-propellant rocketry, see chapter 5. Among the many sources on nuclear strategy, see, e.g., Spinardi, *Polaris to Trident*, 4–6, 31–34, 99–100, 148–53, 162, 187, which cites other sources.

3. For the technical problems, compare. Fuhrman, "Polaris to Trident," 6 with Sapolsky, *Polaris System Development*, 26–29, 47.

4. The best account of this complex series of developments is Armacost, *Thor-Jupiter Controversy*, 50–71, but see also Sapolsky, *Polaris System Development*, 18–23. For the fullest discussion of the advocacy of a fleet ballistic missile before Burke became CNO, see Davis, *Politics of Innovation*, 22–26. Both Freitag and Merrill had backgrounds in guided missiles. Freitag served from 1955 to 1957 in the Office of the Chief of Naval Operations, where he was project officer for Jupiter and Polaris. Merrill became the first technical director for both projects. See "Freitag, Robert F.," NHRC; Merrill, OHI by Hunley.

5. Baar and Howard, *Polaris!* 14, 63; Spinardi, *Polaris to Trident*, 27.

6. Sapolsky, *Polaris System Development*, 23–24.

7. Spinardi, *Polaris to Trident*, 25; A. Burke, OHI, 11, 17, 19, 36; bio, transcript of naval service, article from *Navy Times*, June 10, 1961, 59, and obituary from *New York*

Times, March 14, 1990, all in "Raborn, William F., Jr.," NHC; Baar and Howard, *Polaris!* 76–77, 96; Merrill, OHI by Hunley, 6.

8. Merrill, OHI by Hunley, 5, 11; Merrill, OHI by Stillwell, 183, 200, copy generously furnished by Capt. Merrill; Sapolsky, *Polaris System Development*, 88; Raborn, "Management," 144. Thanks to Paul Stillwell of the Naval Institute for permission to cite OHIs in its files.

9. Sapolsky, *Polaris System Development*, 8, 11, 26–28, 35–36, 45–47, 49–51, 55–56; Baar and Howard, *Polaris!* 15; Spinardi, *Polaris to Trident*, 25–26, 28; Umholz, "Solid Rocket Propulsion and Aerojet," fifth page; Merrill, OHI by Hunley, 6; "Smith, Levering," NHC. Babcock, *Magnificent Mavericks*, chap. 16, 7–9, reveals that Big Stoop was a huge two-stage rocket funded by the Atomic Energy Commission as a means of launching a nuclear device. It used two Bumblebee motors developed by the Allegany Ballistics Laboratory and had three 20-mile test flights in 1951. The project ended in 1952 but gave Smith the confidence to use two stages in Polaris.

10. This paragraph is based on Hunley, "Evolution," 26, and sources cited there (including correspondence and telephone interviews with Karl Klager); Sparks and Friedlander, "Atlantic Research Corporation," 1, 8; letter, Charles B. Henderson to author, October 16, 1997; Henderson, OHI. For a different but not incompatible discussion of Rumbel and Henderson's discovery, see Reily, *Rocket Scientists*, 43–46. NOTS apparently also had a role in this discovery, although Henderson did not mention it. According to Fuhrman, "Polaris to Trident," 12, "Tests at the Naval Ordnance Test Station in 1955, which were confirmed by the Atlantic Research Corporation in 1956, demonstrated a significant increase in specific impulse obtained by the addition of finely divided aluminum to the propellant." My research has not yielded details of this important discovery at NOTS.

11. Gordon et al., *Aerojet*, IV-98 to IV-103 (quotations from Dr. Marvin Gold, IV-99); short biographies of Dr. Klager that he generously provided; Hunley, "Evolution," 27, and sources cited there; *CPIA/M1*, unit 233, July 1960; Fuhrman, "Polaris to Trident," 11–16. For his work in developing both the grain and the propellant, Klager received the Navy's Distinguished Public Service Award in 1958, and Dr. Gold, also of Aerojet, earned the Navy Civilian Meritorious Public Service Citation for his leadership of the research, development, and scale-up of the nitrochemistry employed in the Polaris missile.

12. Above two paragraphs are based on Spinardi, *Polaris to Trident*, 29–31; Bothwell, "Birth of the Polaris Missile," which also contains other interesting information on the concept of minimum deterrence that should be given more consideration in U.S. strategic thinking than it has received; e-mail, Franklin H. Knemeyer to author, April 9, 2001; Knemeyer, OHI; Babcock, *Magnificent Mavericks*, chap. 21. My thanks to Leroy Doig, China Lake historian, for the web address of Bothwell's study and to Liz Babcock for her suggestion to consult Franklin Knemeyer on Bothwell's study, in which he participated, leading to his e-mail, which added information not in my original draft. Incidentally, Liz Babcock has Levering Smith reporting to the SPO in late March 1956, while his bio says April. I have followed the bio, but the difference is only a matter of days.

13. Tangentially relevant to these developments, in an interesting document, "Notes on Conference in Gen. Medaris' Office, 2 July 1956," kindly sent by Michael Neufeld from the Medaris Papers at the Florida Institute of Technology, Medaris discussed his objections to the Navy's developing a separate guidance system for Jupiter S.

14. This section is based on Spinardi, *Polaris to Trident*, 42–45; MacKenzie, *Inventing Accuracy*, 16–18, 22–25, 120–23, 147–49; Fuhrman, "Polaris to Trident," 15; Sapolsky, *Polaris System Development*, 134; Eldon Hall, *Journey to the Moon*, 14–18, 37–43, which provides much more detail on the design of the computer and on miniaturization than presented here; Eldon Hall, "From the Farm," 24–27, which provides the date of Polaris's first guided flight. See chapter 7 above for the date of Atlas's first guided flight, more on Q-guidance, and the weights of the Jupiter and Thor guidance and control systems.

15. Spinardi, *Polaris to Trident*, 54–55; Sapolsky, *Polaris System Development*, 135–36; Wertheim, OHI; Fuhrman, "Polaris to Trident," 7; MacKenzie, *Inventing Accuracy*, 261–62.

16. Spinardi, *Polaris to Trident*, 53–54; Wertheim, OHI; Fuhrman, "Polaris to Trident," 11–12.

17. Table 9.1 below for Polaris; above in the text for Jupiter S; Fuhrman, "Polaris to Trident," 4, shows Jupiter S's length as being only 41.3 feet, still very much larger than Polaris.

18. Klager, "Early Polaris and Minuteman," 6–8, UP; Fuhrman, "Polaris to Trident," 12; Spinardi, *Polaris to Trident*, 52. Specifications for the cases are taken from Klager. See also letter, Commander, U.S. Naval Air Missile Test Center [NAMTC], to Commandant, Eleventh Naval District, December 6, 1957, in RG 181, box 35, file A12, NA (Laguna Niguel), which talks about a jetavator as a "jet thrust deflector for directing thrust forces in a missile flight. It is presently used for correcting JATO thrust misalignment. It was first developed at NAMTC in 1951." Fiedler had arrived at Point Mugu in 1946 and moved to Lockheed in 1956. See Menken, "Pacific Missile Range . . . July 1959 to 30 June 1960," 11, NA, which gives the date of arrival as 1946, and Pace, "Willy A. Fiedler," which places Fiedler's arrival (erroneously?) in 1948 and his move to Lockheed in 1956. Fiedler died January 17, 1998.

19. Fuhrman, "Polaris to Trident," 13; Spinardi, *Polaris to Trident*, 53; Klager, "Early Polaris and Minuteman," 7–8, UP. Incidentally, Aerojet claimed credit for the thrust termination system in "Polaris Solid Triumph for Aerojet," 2, where it stated, "The problem has been solved by an Aerojet patented invention which is still classified."

20. Sapolsky, *Polaris System Development*, 59, 104, 118–21, 126–27, 133–39; Raborn, "Management," 148–52; Spinardi, *Polaris to Trident*, 36; Watson, OHI, 19; Pehrson, OHI, 31–33; Merrill, OHI by Stillwell, 189, 193–94; for the overall scope of Polaris, "Polaris Fleet Ballistic Missile Weapon System Fact Sheet," NASM, but see also Raborn, "Management," 140–44.

21. See Hunley, *Viking to Space Shuttle*, chap. 5.

22. Fuhrman, "Polaris to Trident," 14–15 (quotation); Spinardi, *Polaris to Trident*, 52. As Ed Price emphasized in comments to the author on August 4, 2000, despite the interservice cooperation so crucial to Polaris's success, principal credit belonged to the Navy and the SPO.

23. R. Miller, OHI; unclassified excerpts from unspecified China Lake documents from 1957–59, NAWCD, provided by China Lake historian Leroy Doig III, which also discuss NOTS's pop-up program at San Clemente Island to test the undersea launch and tests at its main facilities in the desert where high explosives were detonated atop a sample of Polaris's composite propellant to ensure that this would not detonate also; comments of George C. Schnitzer on April 23, 2002, about his work for Lockheed on the IBM 709; Lockheed, "Special Polaris Report," NASM.

24. See Hunley, "Evolution," 32, and sources cited there; letter, Edward W. Price to author, October 12, 1999; Smith, OHI; Price, "Combustion Instability," NAWCWD.

25. Spinardi, *Polaris to Trident*, 42–45; Fuhrman, "Polaris to Trident," 15 (quotation).

26. A. Burke, OHI, 18, 27, 32, 35–36; Watson, OHI, 58, 62, USNI; [Watson], *Adventures in Partnership*, 11, for Burke's quotation. The book bears no date, but Watson dates it (OHI, 58) as coming out soon after Lyndon Johnson became president, with the proofs in final form when John F. Kennedy was assassinated. Raborn's comment was quoted in Watson, OHI, 62.

27. "Smith, Levering," NHC; Merrill, OHI by Hunley, 6, 11 (quotation, 11); Wertheim, OHI, 25 (quotations).

28. "Smith, Levering," NHC, including his bio and "Kennedy Flouts Routine on Polaris Promotion," *Washington Star*, August 12, 1961. For other quotations, see Masterson, "Reminiscences," 301, NHC. Masterson called Smith the "technical brains" behind the entire Polaris system, not just the missile. Masterson had been an executive member of the Navy Ballistics Missile Committee that played a major role in supporting the development of Polaris. See also "Polaris Insurance," *Washington News*, November 13, 1967.

29. Fuhrman, "Polaris to Trident," 16; Hunley, "Evolution," 28; Dyer and Sicilia, *Modern Hercules*, 9, 318–20; "Lightweight Pressure Vessels," in Ritchey, "Technical Memoir," UP. On the explosive nature of double-base propellants, see also Gruntman, *Blazing the Trail*, 173. Gruntman, incidentally, discusses the early rockets of Russia/USSR, France, Great Britain, Japan, India, and Israel as well as those of the United States.

30. Hunley, "Evolution," 28; Dyer and Sicilia, *Modern Hercules*, 322; *CPIA/M1*, unit 410, Polaris Model A2 Stage 2.

31. Moore, "Solid Rocket Development at Allegany," 5–7; Hunley, "Evolution," 27–28; Warren and Herty, "U.S. Double-Base Solid Propellant Tactical Rockets" (generously sent by Ed Price); e-mails, Ronald L. Simmons to author, July 9 and 12, 2002.

32. Henderson, "Composite-Modified Double Base Propellants," UP; three e-mails, Ronald L. Simmons to author, July 15, 2002.

33. Fuhrman, "Polaris to Trident," 16–17; Spinardi, *Polaris to Trident*, 63–65; *CPIA/M1*, unit 410, 59-DS-32,750, Polaris Model A2 Stage 2; "Modifications Raise Polaris' Range," 36; "Smith, Levering," NHC.

34. Fuhrman, "Polaris to Trident," 17–18; Hunley, "Evolution," 28; "Polaris A3 Firing Tests Under Way"; Judge, "Aerojet Moves into Basic Glass," 34; *CPIA/M1*, unit 376, 64-KS-74,500, Polaris Model A3 Mod 0, Stage 1, with which cf. unit 375 for Model A2, Stage 1.

35. Fuhrman, "Polaris to Trident," 18; Spinardi, *Polaris to Trident*, 69; "Problems May Cut Polaris A3 Range Goal"; *CPIA/M1*, units 375 and 376.

36. Spinardi, *Polaris to Trident*, 69; *CPIA/M1*, unit 411, 78-DS-31,935, Rocket Motor Mark 6 Mod 0, Polaris Model A3 Stage 2, and unit 410, 59-DS-32,750, Rocket Motor Mark 4 Mod 0, Polaris Model A2 Stage 2; A. Oberth, *Solid Propellant Development*, 2–13; Fuhrman, "Polaris to Trident," 18.

37. Spinardi, *Polaris to Trident*, 63, 67–68; Yaffee, "Pyrolitic Graphite Studied," 26–27; MacKenzie, *Inventing Accuracy*, 261–62. Yaffee, "Ablation Wins Missile Performance Gain," provides considerable background information on nose cone development. Note that the warheads on Polaris A3 were not independently targeted but flew in a pattern that created a triangle or "claw" upon landing.

38. Eldon Hall, *Journey to the Moon*, 43–46; Spinardi, *Polaris to Trident*, 65, 69–71; Fuhrman, "Polaris to Trident," 18; MacKenzie, *Inventing Accuracy*, 263–64. As Spinardi relates (71), the Polaris team developed a strapdown guidance system for Polaris but did not use it. Because such a system was fixed to the missile rather than stabilized in space, it was simpler, hence probably cheaper and more reliable, but it required greater computer power to replace (through computation) the physical reference provided by a stable platform. At that time, small digital computers were still not fast or powerful enough to allow strapdown systems to compete with stable platforms for missile guidance, although they were used in upper-stage launch-vehicle applications.

39. Fuhrman, "Polaris to Trident," 18; Spinardi, *Polaris to Trident*, 72.

40. Fuhrman, "Polaris to Trident," 40, 65, 69; Saxton, "Introduction to PMS (A3P)," 312, NASM; Nicholas and Rossi, *Missile Data Book*, 3-6.

41. Nicholas and Rossi, *Missile Data Book*, 3-9, 3-10.

42. This paragraph is pieced together from Neal, *Ace in the Hole*, 49, 52–62; Edward N. Hall, "Air Force Missile Experience," 23; Hall, "USAF Engineer in Wonderland" (typescript generously provided by Col. Hall), 44–45, UP; Hall, OHI by Hunley, 3–4; Hall, *Art of Destructive Management*, 31–38 (quotation, 38). On these matters, see also Schnare, OHI, 142–49.

43. Klager, "Early Polaris and Minuteman," 14, UP.

44. Lockheed bought Grand Central in 1960–61; Rocketdyne eventually took over Phillips Petroleum's rocket operations.

45. On the AFLRP, see esp. Gordon et al., *Aerojet*, IV-15 to IV-16, but also Neal, *Ace in the Hole*, 73; Klager and Dekker, "Early Solid Composite Rockets," 22, UP; Edward Hall, letter, UP; Hunley "Evolution," 25.

46. ICBM System Program Office, "Minuteman Weapon System," 17, OO/ALC, which gives the 1958 date but states that the roadmap leading to Minuteman began in 1956. Edward Hall wrote to the author that Secretary of Defense McElroy approved Polaris and Minuteman the same day, but this must have been some sort of supplementary approval for Polaris, because the literature on that program states clearly that it won approval in December 1956 from McElroy's predecessor as secretary of defense, Charles E. Wilson. See, e.g., Sapolsky, *Polaris System Development*, 33–34, and Fuhrman, "Polaris to Trident," 4, which gives the precise date December 8, 1956.

47. Sapolsky, *Polaris System Development*, 21–23, 31, 35–37, 39, 45–47, 49–51,

55–56, 64–65, 74, 97–99, 149–50, 157–59, quotation on 39; Fuhrman, "Polaris to Trident," 2–3; Lonnquest and Winkler, *To Defend and Deter*, 43–44; Beard, *Developing the ICBM*, 188–93. Many Navy admirals opposed Polaris because it was expensive and cut into funds available for other vital programs. The reasons why they did not fight it included simple loyalty to superiors and, probably, Raborn's salesmanship and the confidence he engendered. On this, see, e.g., Raborn, OHI, 21–23; A. Burke, OHI, 29; Watson, OHI, 8–9, 22–23. Support for Sapolsky's argument comes from Ritland, OHI, 200: "I can remember LeMay was sitting behind the desk and [AF chief of staff Nathan] Twining was sitting on the edge of the chair and saying, 'Curt, we've got to support this program because . . . we've got to have a good solid-propellant missile to compete with the Polaris program.'"

48. Biographical synopsis of Adelman provided by Karen Schaffer of Chemical Systems Division, one of the names of the organization Adelman helped to found after he left Ramo-Wooldridge; Adelman, OHI. On Ramo-Wooldridge and its role, see Kennedy, Kovacic, and Rea, "Solid Rocket History," 1–13, which recognizes that Hall and Adelman were "the primary proponents of solid-propellant ICBMs and led a small group of engineers to monitor and direct the progress of the technology" (13). On Ramo-Wooldridge and TRW see also, besides Dyer, *TRW*, Hall's less flattering descriptions in his interviews and writings, including *The Art of Destructive Management* and "USAF Engineer in Wonderland," UP.

49. Kennedy, Kovacic, and Rea, "Solid Rocket History," 13; Neal, *Ace in the Hole*, 82. Both Adelman and Hall emphasized in OHIs the unreliability of Neal's study, but it provides many details not available elsewhere and, if compared carefully with other sources, helps to piece together a complex and poorly documented history.

50. Cf. Kennedy, Kovacic, and Rea, "Solid Rocket History," 13; Neal, *Ace in the Hole*, 58; Adelman, OHI; Hall, OHI by Hunley; Hall, "USAF Engineer in Wonderland," 61, UP. For the changing of WDD's name, in addition to the sources cited in chapter 7, see R. Anderson, "Minuteman to MX," 34: "SAMSO: A Look Back," reprinted from *Air Force Magazine*, August 1979.

51. Schnare, OHI, 145–47, 166; Schnare, comments, PI. Schnare, who was Fagan's boss, could not remember the source of the information on use of aluminum, but he was in close touch with Aerojet.

52. Kennedy, Kovacic, and Rea, "Solid Rocket History," 13–14; Neal, *Ace in the Hole*, 85–86; R. Anderson, "Minuteman to MX," 33–36. For the proposal to make Minuteman mobile, see "Mobility Designed into Minuteman." Minuteman's advantage in number of people required to launch a missile shrank to twentyfold over Titan II, which had a crew of four. See Stumpf, *Titan II*, 152.

53. Neal, *Ace in the Hole*, 87–91; Hall, "USAF Engineer in Wonderland," 61–62, UP; Hall, OHI by Neufeld, 20–21; Adelman, OHI; Memorandum of Record, Col. Ray E. Soper, Chief, Ballistic Division, Office of the Assistant Chief of Staff for Guided Missiles, subj: Secretary of Defense Review of Air Force Ballistic Missile Proposals, February 10, 1958, in DeHaven, "Aerospace," Document 212, AFHRA (copy kindly provided by Archangelo Difante); Terhune, OHI, 21. As indicated, these sources disagree about details but concur on the overall picture. Neal, for example, places this meeting in

October 1957, but Soper's memo has the February 8, 1958, date. Since the memo was written on February 10, 1958, and Neal is demonstrably unreliable about details, I have accepted the memo as obviously the more reliable source.

54. Neal, *Ace in the Hole*, 92–100; Hall, "USAF Engineer in Wonderland," 62–63, UP; Hall, OHI by Neufeld, 21–23; Adelman, OHI. Neal does not report Ramo-Wooldridge/STL's opposition, but his account is otherwise generally consistent with those of Adelman and Hall, who both emphasize the opposition. Adelman's account concurs with Neal's in pointing to Schriever's role in funding subsequent to the initial $50 million, which Hall claims as being the result of his advocacy. In his letter to Hunley, UP, Hall stated that Adelman was not involved in this sequence of events and that McElroy granted the $50 million during Hall's briefing. He says that STL, which he refers to as TRW (Thompson Ramo Wooldridge), "continued to sabotage the Minuteman program. . . . At this period, I had no support from any part of TRW. The effect of this constant noisy opposition was to have Mr. [William] Holaday, the missile czar, terminate the program. It took a great deal of effort on my part to have it resuscitated." Adelman, on the other hand, remembers being involved in advocating Minuteman, and Neal supports his view.

55. Hall, "USAF Engineer in Wonderland," 63, UP; Hall, OHI by Hunley; Adelman, OHI; Hunley, "Evolution," 30–31; Kennedy, Kovacic, and Rea, "Solid Rocket History," 10; Neal, *Ace in the Hole*, 123; R. Anderson, "Minuteman to MX," 34, 49, which gives Hall's tenure as first director of the Minuteman program as 1957–58. For his contributions to missile development, Hall was inducted into the Air Force Space and Missile Hall of Fame in 1999; see Hoffman, "Three Inducted."

56. A. Thiel, OHI extract, 5; Truax, autobiography, chap. 18, 249, UP.

57. Glasser, OHI, 67, 71, 83, quotation from 83; Terhune, OHI, 21; Gen. Samuel C. Phillips, "Minuteman," in "Rocketry in the 1950's," 80, NHRC.

58. In a very long endnote to *Rescuing Prometheus*, 330–32n159, Thomas P. Hughes discusses Hall's criticisms of Ramo-Wooldridge and seems largely to discount them, in the process implying that Hall was not a significant figure. Hughes makes such comments as "Schriever remembers Hall as a malcontent who had to be relieved as project manager." Certainly Hall's criticism of Ramo-Wooldridge was one-sided, but Ramo-Wooldridge management clearly did oppose Minuteman at first, despite Adelman's role in it, and Hall was an important figure who played a very significant role in the social construction of Minuteman.

59. "Launch on December 63rd"; Klager, "Early Polaris and Minuteman," 14–15, UP; Kennedy, Kovacic, and Rea, "Solid Rocket History," 10, 15; McGuire, "Minuteman Third-Stage Award Nears," 36–37; Leavitt, "Minuteman"; Lonnquest and Winkler, *To Defend and Deter*, 242.

60. Kennedy, Kovacic, and Rea, "Solid Rocket History," 14, 17; Klager, "Early Polaris and Minuteman," 17–18, UP; Neal, *Ace in the Hole*, 136–37. Each of these sources provides part of the explanation but, unfortunately, none of them addresses the important question of exactly how the participants arrived at the solution and who among them provided it. Moreover, the specifics of the solutions come from Klager, who described them for the second stage. Presumably, however, they also applied to stage 1. This is

suggested by Kennedy, Kovacic, and Rea, who state with regard to the throat inserts, "The basic failure mechanisms of the early nozzle designs were identified and the need for tungsten throat inserts and nozzle entrance flow straightening was established" (17). They make no mention of specific stages in this regard. Note the engineer's typical use of the passive voice, which obscures who did the identifying and establishing.

61. Kennedy, Kovacic, and Rea, "Solid Rocket History," 15; Neal, *Ace in the Hole,* 76, 83, 125, 137–38.

62. "Launch on December 63rd."

63. Ibid.; Neal, *Ace in the Hole,* 147–51; Leavitt, "Minuteman"; U.S. Air Force, Astronautics Laboratory, fact sheet, "Air Force Astronautics Laboratory," February 1988, 2–3, on the rocket lab and its history, AFFTC/HO. On the early critics, see esp. Hall, OHI by Neufeld, 22.

64. Glasser, OHI, 75–77, quotations on 76; Kennedy, Kovacic, and Rea, "Solid Rocket History," 15. Lonnquest and Winkler, *To Defend and Deter,* 246n, gives a variant of this account in which, as Glasser told Lonnquest, Douglas "got the [operational] dates [for Minuteman] confused and told Congress 1962," a year earlier than planned, leading to the all-up testing. Glasser said, "It was a 'miracle' that the missile performed so well."

65. "Launch on December 63rd"; Leavitt, "Minuteman."

66. ICBM System Program Office, "Minuteman Weapon System," 14, 17, OO/ALC.

67. E. Sutton, "Polymers to Propellants," 13; E. Sutton, "Polysulfides to CTPB Binders," 6–7; Mastriola and Klager, *Solid Propellants Based on Polybutadiene Binders,* 122–23; A. Oberth, *Solid Propellant Development,* 2–8. My thanks to Dr. Klager and the CPIA for copies of the last two publications and to Ernie Sutton for the first two.

68. Sutton, "Polysulfides to CTPB Binders," 6–8; Stone, "Minuteman Propulsion," 57; Propulsion Characteristics Summary, TU-122, Wing I, February 1, 1963, AFMC/HO; Thiokol, *Rocket Propulsion Data,* item TU-122. Theoretical specific impulse provides a basis of comparison for propellants' performance without regard to the particular conditions of employment, such as altitude and air density, temperature, and the like. The Propulsion Characteristics Summaries were generously provided by Bill Elliott of AFMC/HO.

69. Stone, "Minuteman Propulsion," 54–62; Propulsion Characteristics Summary, TU-122 (see note 68); Thiokol, *Rocket Propulsion Data,* items TU-122 and TX-12. The Thiokol figures differ somewhat from the AF data in the Propulsion Characteristics Summary, perhaps because of a difference in dates of applicability. I have used the data from the Propulsion Characteristics Summary.

70. Klager, "Early Polaris and Minuteman," 16, UP; Stone, "Aerojet Second Stage," 71; Propulsion Characteristics Summary, M-56 Wing I, February 3, 1964, AFMC/HO; *CPIA/M1,* unit 362, M56A1. The last source is the only one for the composition of the propellants. The data it provides are for Minuteman stage 2 Wing II, but the designations of the propellants, ANP-2864 and ANP-2862, are the same as in the Propulsion Characteristics Summary for Wing I. The two sources differ slightly on specific impulse.

71. Klager, "Early Polaris and Minuteman," 15, UP; Stone, "Aerojet Second Stage," 68–69; Propulsion Characteristics Summary, M-56 Wing I (see note 70).

72. See Hunley, "Evolution," esp. 27–28.

73. Ibid.; Dyer and Sicilia, *Modern Hercules*, 322; Stone, "Glass Fiber Case," 162–65; "Minuteman: The Mobilization Programme"; A. Oberth, *Solid Propellant Development*, 2–13; Propulsion Characteristics Summary, 65-DS-15,520, September 29, 1961, AFMC/HO. Ronald L. Simmons worked on tailoring the Hercules CMDB formulation for Minuteman stage-3 propellant and other upper-stage applications, as he related by e-mail, July 9, 2002. He also worked on the cross-linked "slurry-cast CMDB" used on later Fleet Ballistic Missiles, including the Trident D5. My profound thanks to him for explaining to me many of the intricacies of the CMDB process.

74. Propulsion Characteristics Summary, 65-DS-15,520 (see note 73); Stone, "Glass Fiber Case," 167.

75. R. Anderson, "Minuteman to MX," 39; Wuerth, "Minuteman Guidance and Control," 64, 67–70; Propulsion Characteristics Summary, 65-DS-15,520 (see note 73); Propulsion Characteristics Summary, M-56AJ-3, Wing I, April 10, 1962, AFMC/HO; Propulsion Characteristics Summary, TU-122 (see note 68); Ogden Air Logistics Center, "Minuteman Weapon System," 27, 33, OO/ALC; Lonnquest and Winkler, *To Defend and Deter*, 242, 246. For the CIGTF and sled testing, see Air Force Missile Development Center, "Central Inertial Guidance Test Facility," n.d., microfilm 32,302, frame 1128, AFHRA; Capt. N. D. LaVally, Instructor in Astronautics, Air Force Academy, "Report on Visit to Inertial Guidance Test Facility, AFMDC, Holloman AFB, NM, January 13, 1960," microfilm 31,729, frames 19–26, AFHRA; "Special Study of Track Test Techniques," extract from a report by the Historical Division AFMDC, n.d., microfilm 31,729, frames 27–31, AFHRA; Bushnell, "Guidance System Testing," 37–40, AFHRA; Draft CIGTF Summary Report, Phase III Minuteman (WS-133A) Guidance Sled Tests, October 1961, microfilm 32,270, frames 46–47, AFHRA; R. J. Boyles and A. R. Nock, Space Technology Laboratory, "The Evaluation of Inertial Guidance System Errors from Flight Test Data," October 16–17, 1962, IV-1 to IV-40, microfilm 32,265, frames 568ff., AFHRA. Bushnell pointed out that the sled tests revealed some problems with the guidance equipment, which presumably North American's engineers corrected. From his perspective, the maiden flight of Minuteman I with its "all-inertial guidance system installed and functioning . . . could scarcely have been achieved except for the input already obtained from track testing" (39). Boyles and Nock for STL pointed out that because tracking of actual missile flights was hampered in its accuracy by attenuation of signals from the missile as a result of the exhaust flames and products of solid-propellant combustion, data from the Azusa tracking system were "virtually useless for evaluating guidance system accuracy" (frame 597). In general, the authors concluded, "individual gyro and accelerometer error sources cannot be identified [from flight testing data] to the accuracy required for current ballistic missile inertial guidance systems" (IV-40). These statements highlighted the importance of sled testing for pinpointing and correcting guidance-and-control system errors.

76. Propulsion Characteristics Summary, M-56AJ-3, Wing I (see note 75); Propulsion Characteristics Summary, M-56A1, Wing II, November 9, 1962, AFMC/HO; Kennedy, Kovacic, and Rea, "Solid Rocket History," 16; Stone, "Aerojet Second Stage," 68; Klager, "Early Polaris and Minuteman," 16, UP; R. Anderson, "Minuteman to MX," 43. I

have taken the weight reduction from the first two sources rather than Stone, since the former are official Air Force documents and likely to be more accurate. Stone placed the weight reduction at 220 pounds for the case and 25 for the nozzles, for a total of 245 pounds. The Propulsion Characteristics Summaries give the loaded weight of the Wing I second stage as 11,858 and of Wing II as 11,558.9. Despite the increase in propellant weight, the mass fraction improved from 0.865 to 0.897. Klager talked about the shift from steel to titanium as occurring in Minuteman II, but he apparently confused Wing II and Minuteman II. The other sources all indicate that the change occurred in Wing II. Lonnquest and Winkler, *To Defend and Deter*, 241, and Nicholas and Rossi, *Missile Data Book*, agree on the range. For some reason, incidentally, Grassly's "Ballistic Missile and Space Launch Summary," AFHRA, does not give success and failure rates for Minuteman launches. Apparently sources for these had not been declassified when she wrote.

77. Kennedy, Kovacic, and Rea, "Solid Rocket History," 17–18; Propulsion Characteristics Summary, M-56A1, Wing II (see note 76); Propulsion Characteristics Summary, SRJ19-AJ-1, January 22, 1966, AFMC/HO.

78. Sparks and Friedlander, "Atlantic Research Corporation," 8; E. Sutton, "Polysulfides to CTPB," 8 (also quotation).

79. Sparks and Friedlander, "Atlantic Research Corporation," 8.

80. Gordon et al., *Aerojet*, IV-84.

81. Kennedy, Kovacic, and Rea, "Solid Rocket History," 18.

82. Ibid.; Propulsion Characteristics Summaries M-56A1, Wing II (see note 76) and SRJ19-AJ-1 (see note 77); Hunley, "Evolution," 29 and sources cited there.

83. Hunley, "Solid-Propellant Rocketry," 7; E. Sutton, "Polysulfides to CTPB," 8; E. Sutton, "Polymers to Propellants," 14; Corley, "Career Chronology" and "History Questions," UP, generously sent by Robert C. Corley via e-mail. For HTPB's uses in the electrical industry, as a sealant for glass windows and other purposes in the construction industry, and as an adhesive, information came from "Functional Polymers Products and Research." I am grateful to Daniel Lednicer, a chemist who volunteered at NASM, for finding this information for me. The solid rocket motor for the Titan IVB that was developed as an upgrade to the original Titan IV motor used HTPB, but the first use of Titan IVB took place after the period covered by this history. In a biography he sent to me before his death, Karl Klager called himself "developer of the HTPB propellant formulations, now used in many parts of the world," but evidently he was one of many such developers.

84. Propulsion Characteristics Summaries M-56A1, Wing II (see note 76) and SRJ19-AJ-1 (see note 77); Klager, "Early Polaris and Minuteman," 18, UP; Kennedy, Kovacic, and Rea, "Solid Rocket History," 18; Fuhrman, "Polaris to Trident," 18; Ogden Air Logistics Center, "Minuteman Weapon System," 50, OO/ALC; MacKenzie, *Inventing Accuracy*, 150, where he talks about the Air Force's seeing Polaris as a threat. MacKenzie, 136, also quotes Admiral Burke as comparing the Air Force's techniques with those of the Communists, showing just how intense interservice rivalry had become.

85. MacKenzie, *Inventing Accuracy*, 206–13; Wuerth, "Minuteman Guidance and Control," 64–65, 67–68, 70, 74–75; R. Anderson, "Minuteman to MX," 44; B. Miller,

"Microcircuits Boost Minuteman," esp. 77. Apparently, selection of the PIGA, produced by AC Spark Plug, resulted from sled tests at Holloman. In September 1961 the AF Missile Development Center was preparing to test three different accelerometers for the Minuteman, a Bosch-Arma "vibrating string accelerometer," the AC Spark Plug PIGA, and two models of the Autonetics VM-4. See letter, Col. Blake W. Lambert, DCS/Operations, to BSD (BSL), September 27, 1961, microfilm 32,270, frames 1187–88, AFHRA.

86. Lonnquest and Winkler, *To Defend and Deter*, 247; Nicholas and Rossi, *Missile Data Book*, 3-2.

87. Hunley, "Evolution," 32, and sources cited there; comments of Edward W. Price on a draft of one of the author's manuscripts, August 4, 2000; Schnare, OHI, 131–32; CPIAC, "History of CPIAC."

88. Ogden Air Logistics Center, "Minuteman Weapon System," 54, OO/ALC; Kennedy, Kovacic, and Rea, "Solid Rocket History," 20; B. Wilson, "Composite Motor Case Design"; Gordon et al., *Aerojet: The Creative Company*, IV-104 to IV-105; Propulsion Characteristics Summary, SRJ19-AJ-1 (see note 77); Propulsion Characteristics Summary, M57A1, January 4, 1966, AFMC/HO; Propulsion Characteristics Summary, SR73-AJ-1, Januray 1970, AFMC/HO; "Stage III for USAF's Minuteman III"; *CPIA/M1*, unit 457, Minuteman III, 3rd Stage. The last source gives different percentages for PBCT (as it calls CTPB) and aluminum than the Air Force's Propulsion Characteristics Summary, but Robert Geisler wrote in an e-mail on June 29, 2006, that while looking at documents at Aerojet, he found the different percentages of CTPB and aluminum shown in the Propulsion Characteristics Summaries for stages 2 and 3 to be in error. A submerged nozzle is one that, instead of extending from the rear of the combustion chamber is partly or wholly embedded in it. It displaces a small amount of propellant but also shortens the missile or rocket. For a launch vehicle, it shortens the gantry supporting the vehicle for launch and also reduces the length of the ground elevator, wiring, etc., thus reducing costs. On the debit side of the ledger, a submerged nozzle usually requires insulation on both sides of the submerged portion, adding to weight. Also, in rockets using aluminum as fuel, there is a performance loss because of entrapment of aluminum oxide in the nozzle cavity. I am indebted to Wilbur Andrepont for explaining these intricacies to me in e-mails of August 9, 1999, and July 22, 2000.

89. Ogden Air Logistics Center, "Minuteman Weapon System," 54, OO/ALC; Kennedy, Kovacic, and Rea, "Solid Rocket History," 20; *CPIA/M1*, unit 412, Minuteman Stage 3, Wing 1. Of course, the third stage was still traveling forward at a high rate of speed, but the momentary negative thrust lowered its speed relative to that of the post boost vehicle.

90. Lonnquest and Winkler, *To Defend and Deter*, 241–43; Nicholas and Rossi, *Missile Data Book*, 3-5, 3-8, which gives costs only for the A model of Minuteman I; MacKenzie, *Inventing Accuracy*, 428; ICBM System Program Office, "Minuteman Weapon System," 14, OO/ALC; R. Anderson, "Minuteman to MX," 33. For the missiles' lengths, I have followed Lonnquest and Winkler, 243, which agrees with the ICBM System Program Office figures. I have used the figures in Nicholas and Rossi, 3-5, for the launch weights but Lonnquest and Winkler, 241, for the ranges. I used MacKenzie's figures for

the accuracy. Lonnquest and Winkler, 241, show CEPs of 1.5 miles for MM I, 1.0 miles for MM II, and 800 feet for MM III. The last figure equals 0.15 mile.

91. Glasser, OHI, 73.

92. Cf. *CPIA/M1*, unit 411, Rocket Motor Mark 6 Mod 0, Polaris Model A3 Stage 2, and unit 415, XM 94, Altair II. The Altair II motor bore the designation X258B1, whereas Polaris A3 stage 2 had the designation X260, indicating that the Bureau of Naval Weapons probably initiated the Altair II contract earlier than the Polaris A3 stage 2 contract, but both completed their development in 1963, although Altair II finished five and a half months earlier than Polaris A3. The X248 motor apparently had its beginnings as the Altair I, a propulsion unit for the third stage of the Delta and the fourth stage of the Scout launch vehicles. Hercules developed this motor under contract to the U.S. Naval Research Laboratory and NASA, completing development in October 1958. But the propellant did not include HMX in its ingredients. See *CPIA/M1*, unit 427, Altair I. Since Hercules developed both the X258 and X260 motors, presumably their technology was shared to some degree.

Conclusions

1. Cf. the comment in Bille and Lishock, *First Space Race*, 54: "History is not chemistry: it is not possible to say with any certainty that 'Factor X contributed 46% to the decision and Factor Y 54%.'"

Glossary of Terms, Acronyms, and Abbreviations

The definitions of terms below are intended primarily for the general reader. As such, they are intentionally couched in simple, lay language and will not satisfy the rigor that the scientific and engineering communities expect. Unfortunately, many more precise definitions are not comprehensible to nontechnical audiences. These definitions thus constitute a compromise between exactness and comprehensibility. The technical community will probably find this glossary useful mostly for the spelling out of acronyms and abbreviations that may be unfamiliar or forgotten.

A – *Aggregat* (German: assembly, as in A-4)
AAF – Army Air Forces, predecessor of the USAF
AAS – American Astronautical Society
ABL – Allegany Ballistics Laboratory
ablation – The vaporizing of a substance to dissipate or carry away heat—used for nose cones of reentry bodies and nozzles or nozzle throats where heating could be intense
ABMA – Army Ballistic Missile Agency
accelerometer – An instrument for measuring acceleration
AFB – Air Force Base
AFHRA – Air Force Historical Research Agency
AFLRP – Air Force Large Solid Rocket Feasibility Program
AFOSR – Air Force Office of Scientific Research
AFSC – Air Force Systems Command
AFSWC – Air Force Special Weapons Center
AG – *Aktiengesellschaft* (German: joint stock company)
AIAA – American Institute of Aeronautics and Astronautics
AMC – Air Materiel Command; Army Missile Command
AOMC – Army Ordnance Missile Command
apogee – The farthest point from Earth's center of a satellite orbiting Earth
ARC – Atlantic Research Corporation
ARPA – Advanced Research Projects Agency
ASIS – Abort Sensing and Implementation System, used on Atlas-Gemini
attitude – The position of a rocket or spacecraft in relation to its axes and an external data point such as the horizon or a star

axial – Circling around a cylinder or other elongated body in a direction essentially perpendicular to the length of that body. *Cf.* helical

baffle – A device to prevent sloshing in liquid-propellant tanks or combustion instability in liquid-propellant combustion chambers

BMD – Air Force Ballistic Missile Division (formerly WDD)

BMO – Air Force Ballistic Missile Organization

booster – A somewhat ambiguous term referring to any kind of rocket that adds to or provides lift. Most specifically, it refers to (1) a strap-on or other rocket motor or engine, especially solid, that augments lift capability during launch and the early part of flight. More generally, the term can mean (2) an entire launch vehicle; (3) the first stage of a multistage launch vehicle; or even (4) a rocket that lifts a satellite from one orbit to a higher one, as in the phrase "Agena booster," which refers to an Agena upper stage used in this way.

boundary layer – In aerodynamics, the thin layer of air next to the surface of a body passing through the atmosphere in which viscous forces influence the motion of the fluid (air), causing a transition from smooth (laminar) to turbulent airflow

BSD – Air Force] Ballistic Systems Division, one of two divisions that split from BMD on April 1, 1961

C – Centigrade

Caltech – California Institute of Technology

castable – Of a fluid including a binder, oxidizer, and fuel, the ability to be poured into a case and cured into a solid propellant

cavitation – The formation of bubbles in the "plumbing" of a liquid-propellant rocket engine, interfering with the flow of propellants or other liquids

CEP – Circular error probable, a measure of missile accuracy in which 50 percent of warheads from a missile land in a circle with that radius; thus a CEP of 2,000 feet means that 50 percent of warheads must strike within a circle with a 2,000-foot radius

cigarette burning – Of a solid propellant, burning that is restricted to one end of a propellant grain

CIT – California Institute of Technology

CMDB – Composite-modified double base

combustion instability – Oscillations in an operating combustion chamber, whether liquid- or solid-propellant, that can be so great as to destroy the engine or motor

composite – A type of solid propellant consisting of separate particles of oxidizer, possibly a separate fuel, and other substances dispersed in an elastic matrix that serves as a binder and also a fuel

control – Providing commands to rocket or missile actuators that will cause the vehicle to follow a desired trajectory, usually with provision for feedback to a guidance and control system

Convair – Consolidated Vultee Aircraft Corporation

CPIA – Chemical Propulsion Information Agency

cross-range – Of deviation, lateral; crosswise to the intended trajectory of a rocket or missile in the yaw axis

cryogenic – Extremely cold

CTPB – Carboxy-terminated polybutadiene, a solid-propellant binder used somewhat sparingly because of its high cost compared with PBAN

cut-and-try engineering – Cutting or constructing parts and trying them out to fix undiagnosed problems; if they worked, the fabricator still might not understand fully the nature of the underlying problem, but their success or failure added to the engineering data base for future design; in this book, roughly synonymous with "trial-and-error" and "empirical"

delta guidance – A system of guidance in which a rocket's velocity and position at the end of powered flight are calculated in advance but achieved in a flexible manner, not simply returned to a preplanned trajectory whenever a deviation occurs

DoD – Department of Defense

Doppler effect – A change in the frequency with which (radio) waves from a moving object reach a receiver, depending on speed; used in control devices for missiles and rockets to track trajectory and for related purposes

drag – A retarding force from the atmosphere operating on a body passing through that atmosphere; sometimes likened to friction

DU-1 – [JATO] Droppable Unit-1

elastomer – A rubberlike substance

exhaust velocity – The speed of the gases expelled from a rocket nozzle, a measure of effectiveness

extrusion – A method of producing a propellant grain by forcing it through dies, either with the propellants suspended in a solvent or in a dry (solventless) process

F – Fahrenheit

FBI – Federal Bureau of Investigation

Fiberite – Any of a variety of products, made initially from cotton fiber waste coated with a resin to produce a composite material, one of which, apparently a resin molded at high pressure and loaded with silica and graphite cloth, was used for the exit cones of Minuteman I

film cooling – Introducing a flow (film) of fuel down the inside wall of a liquid-propellant combustion chamber and/or exhaust nozzle to protect it/them from the heat of combustion

form drag – Drag due to pressure minus drag associated with lift

G – Acceleration equal to the force of gravity at sea level

GALCIT – Guggenheim Aeronautical Laboratory, Caltech

gas generator – A device for starting operation of turbopumps that uses the same propellants as the combustion chamber, avoiding the use in earlier liquid-propellant rockets of a separate system with its own propellants

GE – General Electric Company

GEMSIP – Gemini Stability Improvement Program

gimbal – A device that permits a body (such as an engine, nozzle, or guidance and control device) to rotate freely in any direction or that suspends it so that it remains level when the larger structure, such as a rocket, changes its attitude; as a verb, to "steer" by rotating an engine or nozzle using a gimbal

GmbH – *Gesellschaft mit beschränkter Haftung* (German: company with limited liability)

grain – A mass of propellant, usually configured to provide a predetermined thrust time curve (a graph of the amount of thrust over time as a propellant burns)

guidance – Determining a trajectory and velocity for a missile or rocket to reach a desired position from another position or location

gyroscope – A device that rotates about an axis like a children's top. It responds to a disturbing angular force by moving slowly (precessing) in a direction at right angles to that of the force. The precession is predictable, so by mounting gyroscopes in such a way that they respond in only one or two directions, they can be used to indicate acceleration or angular velocity. These indications together with electrical pickoff devices can then be used by a guidance and control system to direct servomechanisms to vanes, gimbals, or other devices to adjust the attitude of a rocket and thus, in effect, steer it.

helical – Coiling in a spiral fashion

heterogeneous engineering – "Engineering" that involves winning support for a project or goal as distinguished from designing or manipulating objects in the physical world

HMX – A high explosive
HTPB – Hydroxy-terminated polybutadiene, a solid-propellant binder that is superior to CTPB and also cheaper but not as low-cost as PBAN
Hydyne – Unsymmetrical dimethylhydrazine and diethylene triamine, a liquid fuel used on the Jupiter C
hypergolic – Of propellants, igniting upon contact with each other without need of an igniter
IAA – International Academy of Astronautics
IAF – International Astronautics Federation
ICBM – Intercontinental ballistic missile, usually defined as having a range of at least 5,000 miles
IGY – International Geophysical Year, July 1, 1957 to December 31, 1958
inertial guidance – Determining a trajectory and velocity for a missile or rocket so it will reach a desired position from another position or location using self-contained, automatic devices such as gyroscopes and accelerometers that respond to inertial forces (changes of direction and/or speed) and feed information to a computer
injector – A device in a liquid-propellant engine that atomizes and mixes propellants as it introduces them to the combustion chamber for ignition
internal cavity – A hollow area in a propellant grain where the burning of propellants occurs
IRBM – Intermediate range ballistic missile, usually defined as having a range of at least 1,500 miles but less than 5,000 miles
IWFNA – Inhibited white fuming nitric acid, a liquid propellant
JATO – Jet-assisted takeoff
jetavator – A thrust vector control mechanism used on Polaris, featuring rings that rotated into the exhaust stream from the missile's nozzles to provide control in pitch, yaw, and roll
JNWC – Joint New Weapons Committee
JPL – Jet Propulsion Laboratory, Caltech
lbf-sec/lbm – Pounds of thrust per pound of propellant burned per second, a measure of specific impulse
like-on-like – A type of propellant injection in which fuel impinges on fuel and oxidizer on oxidizer, instead of fuel on oxidizer
longeron – A lengthwise framing member of an airplane or rocket structure
LR – Liquid rocket, as in engine designations such as in LR 87-AJ-3 (Titan I stage 1 engine)

MA – Mercury-Atlas, as in MA-1, MA-2, which were also designations of Atlas engines
Mach number – Speed in relation to that of sound
MAPO – An imine used as a curing agent for a solid propellant
mass fraction – The mass of the propellant in a rocket stage divided by the total mass of the stage; the higher the mass fraction, the more efficient the stage, other parameters being equal
MDS – Malfunction detection system, used on Gemini-Titan
MIT – Massachusetts Institute of Technology
MSFC – Marshall Space Flight Center (NASA)
NA – National Archives
NAA – North American Aviation
NACA – National Advisory Committee for Aeronautics, predecessor of NASA
NAMTC – Naval Air Missile Test Center
NASA – National Aeronautics and Space Administration
NASM – National Air and Space Museum, Smithsonian Institution
NDRC – National Defense Research Committee
net centric warfare – Information-oriented naval warfare, heavily dependent upon satellites and computers
nominal – Functioning as designed
NOTS – Naval Ordnance Test Station
nozzle – A device at the end (initially, the bottom) of a rocket engine or motor that accelerates the expanding gases from the combustion chamber to increase thrust
NRL – Naval Research Laboratory
OHI – Oral history interview
ORDCIT – Ordnance–California Institute of Technology, as in ORDCIT contract
OV – Orbiting vehicle
oxidizer – A substance used together with a rocket fuel to supply oxygen that enables the fuel to burn at high altitudes and in space where atmospheric oxygen is sparse or unavailable
PBAA – Polybutadiene-acrylic acid, a solid-propellant material consisting of an elastomeric (rubberlike) copolymer of butadiene and acrylic acid
PBAN – Polybutadiene-acrylic acid-acrylonitrile, a solid-propellant material, successor to PBAA, with greater tear strength
PBY – A patrol bomber made by Consolidated Aircraft Company
perigee – The nearest point to Earth's center of a satellite orbiting Earth

PERT – Program Evaluation and Review Technique, a management system developed by the Navy for Polaris

phenolic – A type of resin made from a crystalline acidic compound (phenol) and used for coatings, such as for nozzles and nozzle throats

PIGA – Pendulous integrating gyro accelerometer, a gyroscope mounted as a pendulum to measure acceleration

PIPA – Pulsed integrating pendulous accelerometer, a type of pendulum, held in place by electronic pulses, which measures acceleration, though less accurately than the most sensitive PIGAs

pinion gear – A small wheel with few teeth that meshes with a larger gear or gears

pitch – The up-down movement of the nose of a rocket as it flies more or less horizontally

polybutadiene – A rubbery solid binder material used in propellants like CTPB, HTPB, PBAA, and PBAN

polymer – A compound consisting of many repeated, linked, simple molecules, with a chemical structure that makes it rubbery, so that it retains its shape while resisting cracking and still permits fuel and oxidizers to be loaded within it before it cures—all useful characteristics for a solid-propellant binder

polyurethane – A class of polymers containing urethane links

propellants – Fuels and oxidizers that burn in a combustion chamber to produce expanding gases and thus supply thrust

psi – pounds per square inch

PTS – Payload Transfer System

Q-guidance – A system that permitted much of the computation for guidance (the Q-matrix) to be performed long before launch, leaving little calculation to be done by the computer on the missile or other rocket

ramjet – A simple jet engine in which the air for combustion is compressed in a tube by the forward motion of the vehicle through the atmosphere instead of by the complex compressor devices used in turbojets

rate gyro – A gyroscope used to detect angular deviations

reentry body – An item such as the nose cone of a missile that reenters the atmosphere from space

regenerative cooling – Cooling of a liquid-propellant combustion chamber by circulating a propellant around the outside wall

RFNA – Red fuming nitric acid, a liquid oxidizer

RG – National Archives Record Group

rocket engine – The propulsion device of a rocket, usually one burning liquid propellants
rocket motor – A propulsion device in a rocket, usually one burning solid propellants
roll – The rotation of a rocket about its longitudinal axis
RP-1 – A kerosene rocket fuel
RPL – Rocket Propulsion Laboratory (Air Force)
rpm – Revolutions per minute
SAMSO – Air Force Space and Missile Systems Organization
scaling up – Increasing size and performance
servo – An automatic mechanism used for control, as in a servomechanism to actuate flight-control surfaces
shock tube – As used by Arthur Kantrowitz, test chamber containing explosive gases separated from a vacuum chamber by a thin diaphragm. When testers detonated the gases, the diaphragm burst, sending gas over a model in the (former) vacuum chamber to simulate the conditions of reentry. The shock lasted only a small fraction of a second, but measurements taken by the tube's instruments could be extrapolated to approximate the temperatures and speeds a nose cone would face.
shroud tip – A device linking the tips of turbine blades (as in a turbopump) to restrict flutter and prolong blade life
SNORT – Supersonic Naval Ordnance Research Track
specific impulse – A measure of performance for a propellant combination or propulsion system, expressed as a measurement of thrust per amount of propellant burned per unit of time (in this book, pounds of thrust per pound of propellant burned per second, expressed as lbf-sec/lbm); I_{sp} is the symbol
SPO – Special Projects Office
SSD – Air Force Space Systems Division, one of two divisions that split from BMD on April 1, 1961
stabilized platform – A guidance and control device that maintains a fixed orientation in space by use of gyroscopes and gimbals
stage – A component of a rocket or missile containing its own propulsion system and structure. Typically, when one stage has expended its propellants, the next higher stage ignites and the lower stage drops away, reducing the weight to be accelerated to the design speed for the final stage of the vehicle.
static testing – Testing of a rocket or rocket system on the ground instead of in flight or on a rocket sled

steel balloon – A propellant tank of the Atlas missile and space-launch vehicle, which had a very thin skin and provided structural support through being inflated with helium, much like a balloon

STG – Space Task Group (later Manned Spacecraft Center, then Johnson Space Center, Houston)

STL – Space Technology Laboratories, successor to the Guided Missile Research Division in the Ramo-Wooldridge Corporation, which merged with Thompson Products to become TRW

strapdown guidance – Guidance provided by a system fixed to the missile or launch-vehicle structure, rather than rotating to maintain a fixed orientation in space like a stable platform. Strapdown guidance systems require additional computer power to replace the physical reference provided by a stable platform.

stringer – A longitudinal element to reinforce the skin of an aircraft or rocket structure

submerged nozzle – A nozzle that, instead of extending in its entirety from the rear of the combustion chamber, is partly or wholly embedded in it. While such a nozzle displaces a small amount of propellant, it also shortens the rocket or missile, enabling, for example a solid-propellant missile to fit in a smaller silo or in the confines of a submarine.

sustainer – An engine that, after one or more booster engines have dropped off, stays with a missile or launch vehicle to carry it and the payload to the designed speed.

SVS – Stage Vehicle System

systems engineering – In designing a product or procedure, the integration of all component systems and other elements, including the people developing or operating it, so as to achieve the desired goal most efficiently and effectively

TARS – Three-axis reference system

theoretical specific impulse – A measure of performance for a propellant or propulsion unit that provides a basis of comparison with other propellants without regard to the particular conditions of employment, such as altitude

thrust – The force imparted by a rocket engine or motor that impels the rocket in a forward direction

thrust-to-weight ratio – The amount of thrust in relation to the weight of a given rocket or stage, usually expressed as a decimal

thrust vector control – Control of the direction of a rocket's or stage's thrust for purposes of steering

trajectory – The path of a rocket's or missile's flight
TRW – Thompson Ramo Wooldridge
TT – Technical translation
TWX – TeletypeWriter eXchange; a message sent by teletype
UCLA – University of California at Los Angeles
UDMH – Unsymmetrical dimethyl hydrazine
USAF – United States Air Force
USN – United States Navy
vector – A quantity that has both magnitude and direction; as used here, primarily the direction
VfR – *Verein für Raumschiffahrt* (Society for Space Travel)
viscous drag – Drag due to the stickiness of the atmosphere, i.e., its resistance to objects flying through it, also called skin friction
WAC – Women's Army Corps; Without Attitude Control
WDD – Air Force Western Development Division, which became BMD on June 1, 1957
XLR – Experimental liquid rocket (engine)
yaw – The left-right or side-to-side directional motion of the nose of a rocket flying more or less horizontally
YLR – Operational liquid rocket (engine)

Sources

Archival and Private Sources

Abbreviations

AFFTC/HO	Air Force Flight Test Center History Office, Edwards AFB, Calif.
AFHRA	Air Force Historical Research Agency, Maxwell AFB, Ala.
AFMC/HO	Air Force Materiel Command History Office, Wright-Patterson AFB, Ohio
AMC	Army Missile Command History Office, Redstone Arsenal, Ala.
Caltech	California Institute of Technology Archives, Pasadena, Calif.
CU	Clark University Archives and Special Collections, Worcester, Mass.
FBI	Federal Bureau of Investigation, Washington, D.C.
JPL	Jet Propulsion Laboratory Archives, Pasadena, Calif.
LC/MD	Library of Congress, Manuscripts Division, Washington, D.C.
MSFC/HO	Marshall Space Flight Center History Office, Huntsville, Ala.
NA	National Archives, Washington, D.C.; College Park, Md.; Laguna Niguel, Calif.
NASM	Smithsonian Institution, National Air and Space Museum Archives, Washington, D.C., and Silver Hill, Md.
NAWCWD	Naval Air Warfare Center, Weapons Division, China Lake, Calif., History Office
NHC	Naval Historical Center, Washington Navy Yard, Washington, D.C.
NHRC	NASA Historical Reference Collection, NASA Headquarters, Washington, D.C.
OHI	Oral history interview
OO/ALC	Ogden Air Logistics Center History Office, Hill AFB, Utah
PI	Private interviews by author or others (to be donated to NHRC)
RATL	Redstone Arsenal Technical Library, Huntsville, Ala.
RG	Record Group (in National Archives)
SMC/HO	(U.S. Air Force) Space and Missiles Systems Center History Office, Los Angeles AFB, Calif.
UP	Unpublished papers (to be donated to NHRC)
USNI	U.S. Naval Institute, Annapolis, Md.
USSARC	U.S. Space and Rocket Center, Huntsville, Ala.

Sources

"A-4 Missile—Illustrated Synopsis." Von Braun Collection. USSARC.

"Accession List." *See* U.S. Army Ordnance Research and Development Translation Center.

Adelman, Barnet R., OHI by J. D. Hunley. 2000. PI.

Aerojet-General Corporation. "Aerojet Engineers behind Propulsion for the Air Force Titan II." News release. March 8, 1965. General Collection, "Titan II," folder OT-490020-16. NASM.

Aerospace Corporation. "Gemini Program Launch Systems Final Report." Aerospace report TOR-1001 (2126-80)-3. January 1967. File K243.0473-7. AFHRA.

Air Force Missile Test Center, Patrick AFB, Fla. "Thor Fact Sheet." News release. April 1961. General Collection, "Thor Missile, General," folder OT-380000-01. NASM.

Air Force Plant Representative, Aerojet-General Corporation, Sacramento, Calif. "Semi-Annual Historical Report, July 1, 1966, to Dec. 31, 1966." File R243.0708-35. AFHRA.

Arms, W. M. "Thor: The Workhorse of Space—A Narrative History." McDonnell Douglas Astronautics Company, Huntington Beach, Calif. 1972. Filed under "Standard Launch Vehicles." SMC/HO.

Ballistic Missile Division, USAF. "Thor." Fact sheet. June 25, 1959. General Collection, "Thor IRBM," folder OT-380025-01. NASM.

Ballistic Missile Organization (BMO), History Office Staff. "Chronology of the Ballistic Missile Organization, 1945–1990." August 1993. From private collection of Raymond L. Puffer of Air Force Flight Test Center History Office, Edwards AFB, Calif. UP. Cited as "BMO Chronology."

Bartley, Charles. OHI by John Bluth. 1995. JPL.

Biographical files. History Office Master Collection, microfiches 1063 and 1068. MSFC/HO.

"BMO Chronology." *See* Ballistic Missile Organization.

"Bossart, Karel Jan." General Collection, folder CB-467-000-01. NASM.

Bradshaw, E. W., Jr., and M. M. Mills. "Development and Characteristics of the WAC Corporal Booster Rocket." JPL project note 4-30. 1948. JPL.

Bragg, James W. "Development of the Corporal: The Embryo of the Army Missile Program." 2 vols. Reports and Historical Branch, U.S. Army Ordnance Missile Command, Redstone Arsenal, Ala. 1961. AMC.

British Ministry of Supply. "Report on Operation 'BACKFIRE.'" Vol.2. 1946. RATL.

"Bumper 8: 50th Anniversary of the First Launch on Cape Canaveral." Group OHI by Roger Launius and Lori Walters. Kennedy Space Center. July 24, 2000. NHRC.

Burke, Arleigh. OHI by John T. Mason Jr. 1972. USNI.

Burke, James D., OHI by P. Thomas Carroll and James H. Wilson. 1972. History Collection, HF-3-577. JPL.

Bushnell, David. "Guidance System Testing at the Air Force Missile Development Center, 1960." File K280.10-57B. AFHRA.

[Cagle, Mary T.] "History of the Sergeant Weapon System." U.S. Army Missile Command, Redstone Arsenal, Ala. (1963). AMC.

Carroll, P. Thomas. Notes of telephone conversations between Carroll, JPL, and Dr. Richard W. Porter, General Electric Company, August 7, 1972. History Collection, folder 3-574. JPL.
Chandler, Frank S., and Jesse O. Bankston. "USAF Projects WS-107A and WS-315A, Third Quarterly Progress Report." NOTS 1624, TPR-172, Supersonic Track Division, Test Department, U.S. Naval Ordnance Test Station, China Lake, Calif. November 2, 1956. RG 181 Technical Reports, 1955–1958, box 1. NA (Laguna Niguel).
Chrysler Corporation, Missile Division. "This Is Redstone." Undated, received April 29, 1966. RATL.
"Comparative History of Research and Development Policies Affecting Air Materiel, 1915–1944." Prepared for the Scientific Advisory Group, Office of the Chief of the Air Staff, by Historical Division, Assistant Chief of Air Staff, Intelligence. June 1945. RG 18, (Army Air Forces) Air Adjutant General, Bulky Decimal File 353.41-360.2. NA.
"Conference Minutes, ORDCIT Project, June 1944–June 1945." Frank J. Malina Collection, folder 4.4. LC/MD.
Convair Division, General Dynamics Corporation. "Atlas SLV-3 Space Launch Vehicle Flight Evaluation Report, SLV-3 5303." GDC/BKF66-029. 1966. Bellcom Collection, box 12, folder 2. NASM.
———. "Flight Test Evaluation Report, Missile 67C." March 7, 1961. File K243.0473-1. AFHRA.
Corley, Robert C. "Career Chronology." 2002. UP.
———. "History Questions." 2002. UP.
"Correspondence with the American Rocket Society, 1944–1952." Von Kármán collection, folder 38.4. NASM.
"Correspondence with Clark Millikan, 1938–1961." Von Kármán collection, folder 20.26. NASM.
"Correspondence with Martin Summerfield, 1941–1960." Von Kármán collection, folder 29.2. NASM.
"Culbertson, Philip E." Folder 000408. NHRC.
DeHaven, Ethel M. "Aerospace—The Evolution of USAF Weapons Acquisition Policy, 1945–1961." Vol. 3. June 1962. AFSC Historical Publications Series 62-24-8. AFHRA.
Denison, Frank G., Jr., and Chauncey J. Hamlin Jr. "The Design of the Axial-Cooled Rocket Motor for the Corporal E." ORDCIT project, progress report 4-112. 1949. History Collection, folder 3-999. JPL.
Denison, F[rank] G., and Larry Thackwell. "Progress of Sergeant Program." 1949. JPL Report no. 5, folder 53. JPL.
"Description of the Experiment Station of the ACJP." Frank J. Malina Collection, folder 9.9. LC/MD.
"Dornberger, Walter R. (Gen.)." General Collection, folder CD-614000-01. NASM.
Dunn, L. G., and M. M. Mills. "The Status and Future Program for Research and Development of Solid Propellants." JPL memorandum 4-5. 1945. JPL.

"Facilities and Equipment of the Air Corps Jet Propulsion Research Project." May 28, 1943. Frank J. Malina Collection, folder 10.1. LC/MD.

Fahrney, R. Adm. D. S. "The History of Pilotless Aircraft and Guided Missiles." Typescript. World War II Command File, Shore Establishment, Aeronautics, Bureau of Aeronautics, boxes 401–2. NHC. (Also at NHRC, filed as a book under UG1242. D7.F35 1958.)

FBI. File on John Whiteside (Jack) Parsons, file 65-59589. FBI.

"Freitag, Robert F." Biographical Data. July 25, 1966. NHRC.

Friedman, Henry. "A Summary Report on A-4 Control and Stability." Report F-SU-2152-ND, prepared at Headquarters Air Materiel Command, Wright Field, Dayton, Ohio. June 1947. File "V-2," Lektriever 4. NHRC.

Geiger, Jeffrey, and Kirk W. Clear. "History of Space and Missile Test Organization and Western Space and Missile Center." October 1, 1987–September 30, 1988. File K241.011. AFHRA.

"Gemini-Titan II Launch Vehicle Fact Sheet." n.d. Photo storage area. SMC/HO.

General Electric. "Progress Report on Bumper Vehicle." 1950. History Collection, folder 3-350. JPL.

General Electric, Missile and Space Vehicle Department. "Summary: Heat Sink Nose Cone." January 1959. General Collection, "Thor IRBM," folder OT-380013-01. NASM.

Georgia Institute of Technology, Engineering Experiment Station. "Missile Catalog: A Compendium of Guided Missile and Seeker Information." Report A169/T1. April 1, 1956. NHRC.

Glasser, Otto J., OHI by John J. Allen. 1984. File K239.0512-1566. AFHRA.

Goddard, Esther C. "Excerpts from Diary of Robert H. Goddard 1898–1945." Typescript. CU.

Goddard, Robert H. "1919 and 1936 Rocket Essays." Folder 000832. NHRC.

———. "*NY Times* (Jan. 18, 1920) Editorial." Folder 000829. NHRC.

Goddard Papers. Boxes 4, 14, 18. CU. (Not to be confused with printed *Goddard Papers*.)

"Goddard Patent Infringement." Folders 000837 and 000838. NHRC.

"Goddard/NACA Jet Propulsion 1941." Folder 000835. NHRC.

Goldberg, S. J. "Field Preparations, Firing Procedure, and Field Results for the WAC Corporal." JPL report 4-22. 1945. JPL.

———. "Firing Tests of 'Private A' at Leach Spring, Camp Irwin, California." JPL report 4-3. 1945. JPL.

Grassly, Sarah A. "Ballistic Missile and Space Launch Summary as of 30 June 1969." SAMSO Historical Office. March 1970. File K243.012, 67/07/01–69/06/30. AFHRA.

Greene, Warren E. "The Development of the SM-68 Titan." Three volumes. AFSC Historical Publications Series 62-23-1. August 1962. SMC/HO.

Grimwood, James M., and Frances Strowd. "History of the Jupiter Missile System." U.S. Army Ordnance Missile Command, Redstone Arsenal, Ala. 1962. AMC.

Guzzo, A. T. "Progress Report: Development of Standard Operating Procedures for

Manufacture of Polysulfide-Perchlorate Propellants." Thiokol Corporation, Redstone Division report 26-51. November 1951. UP.

Haley, Andrew G. Letter to Capt. R. C. Schulte, Asst. Chief of Air Staff/MM and D, June 25, 1943. RG 18, (Army Air Forces) Air Adjutant General, box 630, file 360.2. NA.

Hall, E[dward] N. "Comment by Col. E. N. Hall (BMD-ARDC) Regarding Successful JUPITER launch." ABMA/BMD memo for file. June 12, 1957. Copy provided by Michael Neufeld from Medaris Papers, Florida Institute of Technology. UP.

———. Letter to J. D. Hunley, July 15, 2000. UP.

———. OHI by J. D. Hunley. By telephone. 1998. PI.

———. OHI by Jacob Neufeld. 1989. File K239.0512-1820. AFHRA.

———. "USAF Engineer in Wonderland, Including the Missile Down the Rabbit Hole." Undated typescript. UP.

Henderson, C[harles] B. "The Development of Composite-Modified Double Base Propellants at the Atlantic Research Corporation." With Keith Rumbel. 1998. UP.

———. OHI by J. D. Hunley. By telephone. 1997. PI.

"Hermann, Rudolf." General Collection, folder CH-335500-01. NASM.

Hermann, Rudolf. OHI by Sandy Sherman. 1988. Folder CH-335500-01. NASM.

Herzberg, Louis F. "History of the Air Force Rocket Propulsion Laboratory, 1 January–30 June 1964." In "Units, RPL" (q.v.). AFFTC/HO.

Hoelzer, Helmut. OHI by Michael Neufeld. 1989. Space History Division, NASM.

"Hoffman, Samuel Kurtz." General Collection, folder CH-480000-01. NASM.

Intercontinental Ballistic Missile (ICBM) System Program Office, USAF. "Minuteman Weapon System History and Description." Hill AFB, Utah May 1996. OO/ALC.

"Jet Propulsion Laboratory Conference Minutes, 1944–1945." Von Kármán collection, folder 73.16. NASM.

JPL. Organization charts. JPL 119, Jack James Collection, 1945–86, folder 6. JPL.

"JPL/Monthly Summaries, 1942–1944." Frank J. Malina Collection, folder 8.1. LC/MD.

"Juno I Launch Vehicle." General Collection, folder OJ-800001-01. NASM.

"Juno Launch Vehicles." Folder 012072. NHRC.

Klager, Karl. "Early Polaris and Minuteman Solid Rocket History." n.d. UP.

Klager, Karl, and Albert O. Dekker. "Early Solid Composite Rockets." Unpublished paper. October 1972. UP.

Knemeyer, Franklin H., OHI by Elizabeth Babcock, Leroy L. Doig III, and Mark Pakuta. 1991. NAWCWD.

Lockheed Missiles and Space Division. "Special Polaris Report." *Trajectory*, Winter 1960–61 (special issues of house organ). "Polaris Missiles," folder OP-701075-01, NASM.

Malina, Frank J. "Biographical Information." May 1, 1968. Folder 001418, "Malina, Frank J." NHRC.

———. "Development and Flight Performance of a High Altitude Sounding Rocket the 'WAC Corporal.'" JPL report 4-18. 1946. JPL.

———. OHI by R. Cargill Hall. 1968. JPL.

———. OHI by James H. Wilson. 1973. JPL.
———. "Report on Jet Propulsion for the National Academy of Sciences Committee on Air Corps Research." December 21, 1938. Frank J. Malina Collection, folder 9.1. LC/MD.
———. "A Review of Developments in Liquid Propellant Jet (Rocket) Propulsion at the ACJP Project and the Aerojet Engineering Corporation." 1944. Frank J. Malina Collection, folder 9.2. LC/MD.
Malina, F[rank] J., et al. "The Jet Propulsion Laboratory, GALCIT." June 25, 1945. Frank J. Malina Collection, folder 10.6. LC/MD.
Malina, Frank J., John W. Parsons, and Edward S. Forman. "Final Report for 1939–40." GALCIT project 1, report 3. JPL.
Masterson, Kleber S. "The Reminiscences of Vice Admiral Kleber S. Masterson." Based on interviews by John T. Mason Jr., Arlington, Va., 1972–73. NHC.
Meeks, Paul J., David Altman, and John I. Shafer. "Summary of Solid-Propellant Activities at the Jet Propulsion Laboratory California Institute of Technology." Memorandum JPL-11. 1951. History Collection, folder 3-981. JPL.
Menken, Arthur. "History of the Pacific Missile Range, 30 June 1959." RG 181, box 32, "Command Histories, 1959–1974," folder "History of the PMR, 1959." NA (Laguna Niguel).
———. "History of the Pacific Missile Range, An Historical Report Covering the Period 1 July 1959 to 30 June 1960." RG 181, box 33, "Command Histories of PMR, 1959–1974." NA (Laguna Niguel).
Merrill, Grayson. OHI by J. D. Hunley. 2002. PI.
———. OHI by Paul Stillwell. 1997. USNI.
Microfilm rolls 31,729; 32,265; 32,270; 32,302. AFHRA.
Miles, R. C. "The History of the ORDCIT Project up to 30 June 1946." n.d. JPL.
Miller, Ray. OHI by J. D. Hunley. 2002. PI.
Millikan, Clark. "The Diaries of Clark Millikan." Box 30, no. 6, 1945. Caltech.
Mueller, Fritz. OHI by Michael Neufeld. 1989. Space History Division, NASM.
Mueller, George E. "Speeches." George E. Mueller Collection, folder 13. LC/MD.
Newell, Homer E., Jr. "Guided Missile Kinematics." Naval Research Laboratory. May 22, 1945. RG 218, Joint Chiefs of Staff, Joint New Weapons Committee (JNWC), subject file May 1942–1945, box 48, folder 401.1. NA.
Nichols, Peter L., Jr., Robert J. Parks, and James D. Burke. "Solid-Propellant Development at the Jet Propulsion Laboratory." JPL, Caltech publication 105, presented to Ad Hoc Committee on Large Solid-Propellant Rocket Engines, July 17, 1957. History Collection, folder 3-525. JPL.
North American Rockwell Corporation, "Data Sheet, Jupiter Propulsion System." December 15, 1967. NHRC.
Oberth, Hermann. OHI by Martin Harwit and Frank Winter. 1978. Space History Division, NASM.
Ogden Air Logistics Center. "Minuteman Weapon System: History and Description." OO-ALC/MMG, Hill AFB, Utah. 1990. OO/ALC.
Parsons, J. "A Consideration of the Practicality of Various Substances as Fuels for Jet Propulsion." 1937. JPL.

Parsons, J. W., and M. M. Mills. "The Development of an Asphalt Base Solid Propellant." GALCIT project 1, report 15. 1942. JPL.
Peenemünde Document Collection. General Collection, 3 folders. NASM.
"Peenemunde East Through the Eyes of 500 Detained at Garmisch." Folder 002685. NHRC.
Peenemünde Guided Missile (PGM) microfilms. Identified by roll number and either Archiv or FE (Fort Eustis) number. NASM.
Pehrson, Gordon O., OHI by John T. Mason Jr. 1974. USNI.
PGM. *See* Peenemünde Guided Missile.
Pickering, William H., OHI by Michael Q. Hooks. 1989. JPL.
"Polaris Fleet Ballistic Missile Weapon System Fact Sheet." May 1, 1964. "Polaris Missiles," folder OP-701200-01. NASM.
Porter, Dr. Richard. OHI by David DeVorkin. 1984. Space History Division, NASM.
Price, E. W. "Combustion Instability." NOTS technical article 3, TP 2400. February 1960. NAWCWD.
"Private A." Undated report. RG 218, box 47, folder 354.4. NA.
Propulsion Characteristics Summaries. Extracts from the "Gray Books," formerly classified but now for Minuteman declassified. AFMC/HO.
Puckett, Allen E. "Performance of the 12-Inch Wind Tunnel." Memorandum 4-52. 1949. Folder 3-809. JPL.
Raborn, William F., OHI by John T. Mason Jr. 1972. USNI.
"Raborn, William F. Jr." Officer Biographical Files. NHC.
Ritchey, H. W. "Technical Memoir." c. 1980. UP.
Ritland, Osmond J., OHI by Lyn R. Officer. 1974. File K239.0512-722. AFHRA.
Rockefeller, Alfred, Jr. "History of Thor, 1955–1959." File K243.012-27, 1955–1959. AFHRA.
"Rocketry in the 1950's, Transcript of AIAA Panel Discussions." NASA Historical Report 36. October 28, 1971. NHRC.
"Rockets . . . V-2." Folder. NHRC.
Ross, Chandler C. "Life at Aerojet-General University: A Memoir." Aerojet History Group. 1981. UP.
Rudolph, Arthur. OHI by Michael Neufeld. 1989. Space History Division, NASM.
Sandberg, W. A., and W. B. Berry. "Design and Fabrication of the WAC Corporal Missile, Booster, Launcher, and Handling Facilities." JPL report 4-21. 1946. JPL.
Saturn/Apollo Systems Office, NASA Marshall Space Flight Center. "The Mercury-Redstone Project." December 1964. Folder 012091, "History of the Redstone." NHRC.
Saxton, L. "Introduction to PMS (A3P)." Lockheed Missiles and Space Company, Sunnyvale, Calif. March 16, 1964. "Polaris Missiles," folder OP-701075-22. NASM.
Schnare, Clarence William. Comments to J. D. Hunley. By telephone. May 23, 2002. PI.
———. OHI by James C. Hasdorf. 1975. File K239.0512-863. AFHRA.
Schulze, H. A. "Technical Data on the Development of the A4 (V-2)." Historical Office, George C. Marshall Space Flight Center. February 1965. Schulze Collection. USSARC.

Seifert, Howard S. "History of Ordnance Research at the Jet Propulsion Laboratory, 1945–1953." 1953. History Collection, folder 3-97. JPL.
———. OHI by James H. Wilson. 1971. JPL.
Shafer, John I., OHI by P. Thomas Carroll. 1970. JPL.
Simmons, Ronald L. Biography, résumé, and e-mails. 2002. UP.
"Smith, Levering." Officer Biographical Files. NHC.
Smith, Levering. OHI by Leroy Doig III and Elizabeth Babcock. 1989. NAWCWD.
Space Division, Air Force Systems Command. "Space and Missile Systems Organization: A Chronology, 1954–1979." SMC/HO.
Stadhalter, W. R. (Program Manager, Rocketdyne). "The Redstone: Built-In Producibility." n.d. General Collection, "Redstone Missile," folder OR-180000-01. NASM.
[Stanton, Roger.] "Research and Development at the Jet Propulsion Laboratory, GALCIT." June 1946. Frank J. Malina Collection, folder 12.12. LC/MD.
"Stewart, Homer Joe." Folder 002216. NHRC.
Stewart, Homer Joe. OHI by John L. Greenberg. 1982. Caltech.
———. "Static and Dynamic Stability Estimates for the Corporal (Missile No. 2, XF30L20,000)." Progress report 4-4. 1944. JPL.
"Stiff, Ray C., Jr." General Collection, folder CS-882500-01. NASM.
Stuhlinger, Ernst. "Apex Determination for Missile #27." RPO-D technical memo no. 4. August 31, 1956. UP.
———. "Apex Prediction in Missile #42." Disposition form. December 17, 1957. UP.
———. "Apex Predictor." Memo. September 1998. UP.
———. "Dry Run for Apex Determination." Disposition form. July 5, 1957. UP.
———. Letters to J. D. Hunley, March 3, 1995, and May 5, 2002. UP.
———. OHI by J. D. Hunley. 1994. PI.
Summerfield, Martin. OHI by J. D. Hunley. 1994. PI.
———. "Questions by Dr. Martin Summerfield, Jet Propulsion Laboratory, California Institute of Technology, Answers by Professor W. von Braun with the collaboration of Mr. J. Paul and Professor C. Wagner." Edited by T/Sgt. E. Wormser from notes taken by Summerfield, April 19, 1946. Technical report 10, Headquarters, Res & Dev Sv Sub-Office (Rocket), Fort Bliss, Texas. Schulze Collection. USSARC.
Summerfield, M., J. I. Shafer, H. L. Thackwell Jr., and C. E. Bartley. "The Applicability of Solid Propellants to High-Performance Rocket Vehicles." JPL ORDCIT project memorandum 4-17. October 1, 1947. JPL.
"Technical Program Review, 1956." NOTS 1627, China Lake, Calif. January 1, 1957. RG 181, box 1, Technical Reports, 1955–1958. NA (Laguna Niguel).
Terhune, Charles H., OHI by Robert Mulcahy. 2001. SMC/HO.
Thackwell, H. L., Jr., and J. I. Shafer. "The Applicability of Solid Propellants to Rocket Vehicles of V-2 Size and Performance." JPL ORDCIT project memorandum 4-25. July 21, 1948. Used in a redacted version provided to the author by the NASA Management Office at JPL.
Thiel, Adolf K., OHI (extract only) by F. I. Ordway. 1988. PI.
"Thiel, Walter." General Collection, folder CT-168000-01. NASM.
"Thor Fact Sheet." *See* Air Force Missile Test Center.

T.O. 21-SM75-1. Technical manual, USAF Model SM-75 Missile Weapon System. October 20, 1961. UP.

Toftoy, Col. H. N. "A Brief History of the Hermes II Project." General Collection, "Hermes II Missile," folder OH-033005-01. NASM.

Truax, Robert C. Autobiography. Draft, 2002. Computer disk. UP.

"Truax, Robert C." Officer Biographical Files. NHC.

"Tsien, H. S." Folder 002375. NHRC.

Tsien, Hsue-shen, ed. *Jet Propulsion*. Lectures for CalTech/JPL graduate course, 1943–1948. Frank J. Malina Collection, folder 7.4. LC/MD.

"Units, RPL." Material on variously named organizations that became the Air Force Rocket Propulsion Laboratory. Four unnumbered boxes. AFFTC/HO.

U.S. Army Missile Command. "History of the Sergeant Weapon System." c. 1965. RATL.

U.S. Army Ordnance Corps. "The Corporal: A Surface-to-Surface Guided Ballistic Missile." Report 20-100. 1958. History Collection, folder 3-355. JPL.

U.S. Army Ordnance Corps/General Electric Company. "Hermes Guided Missile Research and Development Project, 1944–1954." Unclassified condensation, September 25, 1959. Folder 012069, "US Army Hermes Rocket." NHRC. (Also in folder OH-033000-01, "Hermes A-1 Missile Documents," NASM.)

———. "Hermes Guided Missile Systems, Inception through 30 June 1955." Vol. 10 of "Ordnance Guided Missile and Rocket Programs." Technical report. Filed as a book. NASM.

U.S. Army Ordnance Research and Development Translation Center, Ft. Eustis, Va. "Accession List of German Documents Pertaining to Guided Missiles." 3 vols. 1946. RATL.

———. "History of German Guided Missiles." USSARC.

"V-2 (A-4) Missile (US, Post-War), White Sands Articles." General Collection, folders OV-003003-01 and OV-003003-02. NASM.

"Vanguard Guidance Sidebar." Folder 006640, "Vanguard." NHRC.

"von Braun, Dissertation." Folder 002558. NHRC.

von Braun, Wernher. "Behind the Scenes of Rocket Development in Germany, 1928 through 1945." Von Braun Collection, "Manuscript: Behind . . ." USSARC.

———. "The Development of German Rocketry Prior to 1945." Von Braun Collection, "Ft. Bliss Period." USSARC.

"von Braun, Wernher." General Collection, folder CV-644000-01. NASM.

Walters, A. J. "History of the Air Force Plant Representative—Detachment 23." January 1, 1967–June 30, 1967. File K243.0707-23. AFHRA.

Watson, Clement Hayes. OHI by John T. Mason Jr. 1972. USNI.

Weals, Frederick, James Pryor, and John Di Pol. OHI by J. D. Hunley. 2002. PI.

Wertheim, Robert. OHI by J. D. Hunley. By telephone. 2002. PI.

White, L. D. "Final Report, Project Hermes V-2 Missile Program." General Electric report R52A0510. September 1952. Folder 012069, "US Army Hermes Rocket." NHRC.

White, Maxwell. "An Interpretive History, The Pacific Missile Test Center, Point Mugu, California: The Genesis, 1936–1946." 1989 UP.
Wolfe, Allen E., and William J. Truscott. "Juno I: Re-entry Test Vehicles and Explorer Satellites." Vol. 1 of "Juno Final Report." JPL technical report 32-31. September 6, 1960. Folder 012071, "Jupiter C (Juno I)." NHRC.
Wunderman, Liljan Malina. OHI by Benjamin Zibit. 1996. PI.
"Wyld, James Hart." General Collection, folder CW-94100-02. NASM.

Published Sources

Besides books and articles, this section includes items "published" on the Internet and papers presented at meetings, which the engineering community counts as publications.

Adams, Laurence J. "The Evolution of the Titan Rocket—Titan I to Titan II." In Hunley, *Rocketry and Astronautics* (24th Symposium), 201–23.
Aerojet Liquid Rocket Company. *Liquid Rocket Engines*. Sacramento, Calif.: Aerojet, 1975.
Anderson, Margaret Lavinia. *Practicing Democracy: Elections and Political Culture in Imperial Germany*. Princeton, N.J.: Princeton University Press, 2000.
Anderson, Robert C. "Minuteman to MX: The ICBM Evolution." *TRW/DSSG/Quest* (Autumn 1979): 31–49.
"Arma Inertial Guidance." *Arma Engineering* 3, no. 2 (February–March 1960): 5–11.
Armacost, Michael H. *The Politics of Weapons Innovation: The Thor-Jupiter Controversy*. New York: Columbia University Press, 1969.
Baar, James, and William E. Howard. *Polaris!* New York: Harcourt, Brace, 1960.
Babcock, Elizabeth. *Magnificent Mavericks: Evolution of the Naval Ordnance Test Station from Rockets to Guided Missiles and Underwater Ordnance*. Vol. 3 of *History of the Navy at China Lake, California*. Washington, D.C.: China Lake Museum Foundation, in press.
Baedeker, Karl, ed. *The United States, with an Excursion into Mexico: A Handbook for Travelers, 1893*. Reprint, New York: Da Capo, 1971.
Bainbridge, William Sims. *The Spaceflight Revolution: A Sociological Study*. New York: Wiley, 1976.
Baker, David. *The Rocket: The History and Development of Rocket and Missile Technology*. New York: Crown, 1978.
Baker, Michael E., Kaylene Hughes, and James D. Bowne. *Redstone Arsenal Complex Chronology*, part 2, *Nerve Center of Army Missilery, 1950–62*, section B, *The ABMA/AOMC Era (1956–62)*. Redstone Arsenal, Ala.: U.S. Army Missile Command Historical Division, 1994.
Baranowski, Shelley. *The Sanctity of Rural Life: Nobility, Protestantism, and Nazism in Weimar Prussia*. Oxford: Oxford University Press, 1995.
Barth, Hans, ed. *Hermann Oberth: Briefwechsel*. 2 vols. Bucharest: Kriterion, 1979–84.
———. *Hermann Oberth: "Vater der Raumfahrt."* Munich: Bechtle, 1991. Cited as Barth, *Oberth*.

Battin, Richard H. "Space Guidance Evolution—A Personal Narrative." *Journal of Guidance and Control* 5, no. 2 (March–April 1982): 97–102.

Beard, Edmund. *Developing the ICBM: A Study in Bureaucratic Politics*. New York: Columbia University Press, 1976.

Becklake, John, ed. *History of Rocketry and Astronautics, Proceedings of the Twenty-Second and Twenty-Third History Symposia of the International Academy of Astronautics*. AAS History Series 17. San Diego: Univelt, 1995.

Benecke, Theodor, and A. W. Quick, eds. *History of German Guided Missile Development: AGARD First Guided Missile Seminar, Munich, Germany, April 1956*. Braunschweig: E. Appelhans, 1957.

Benson, Charles D., and William Barnaby Faherty. *Moonport: A History of Apollo Launch Facilities and Operations*. SP-4204. Washington, D.C.: NASA, 1978.

Béon, Yves. *Planet Dora: A Memoir of the Holocaust and the Birth of the Space Age*. Translated by Yves Béon and Richard L. Fague. Boulder, Colo.: Westview Press, 1997.

Bergaust, Erik. *Wernher von Braun*. Washington, D.C.: National Space Institute, 1976.

Bijker, Wiebe E., Thomas P. Hughes, and Trevor J. Pinch, eds. *The Social Construction of Technological Systems: New Directions in the Sociology and History of Technology*. Cambridge, Mass.: MIT Press, 1987.

Bille, Matt, and Erika Lishock. *The First Space Race: Launching the World's First Satellites*. College Station: Texas A&M University Press, 2004.

Bilstein, Roger E. *Stages to Saturn: A Technological History of the Apollo/Saturn Launch Vehicles*. SP-4206. Washington, D.C.: NASA, 1980.

———. *Testing Aircraft, Exploring Space: An Illustrated History of NACA and NASA*. Baltimore: Johns Hopkins University Press, 2003.

"Biography: Donald A. Quarles." <www.af.mil/bios/bio.asp?bioID=6830>.

Bornemann, Manfred. *Geheimprojekt Mittelbau: Die Geschichte der deutschen V-Waffen-Werke*. Munich: Lehmanns, 1971.

Bothwell, Frank. "Birth of the Polaris Missile and the Concept of Minimum Deterrence." <www.armscontrolsite.com/PolarisIntro.html>, accessed April 9, 2001.

Bower, Tom. *The Paperclip Conspiracy: The Hunt for the Nazi Scientists*. Boston: Little, Brown, 1987.

Bowman, John. "Steinhoff Dreams of Flying Reached Back into Boyhood." *Alamogordo Daily News*, October 3, 1976, 20.

Bracher, Karl Dietrich. *The German Dictatorship: The Origins, Structure, and Effects of National Socialism*. Translated by Jean Steinberg. New York: Praeger, 1970.

Braun, Julius H. "The Legacy of Hermes." In Hunley, *Rocketry and Astronautics* (24th Symposium), 135–42.

———. "Redstone's First Flight, Success or Failure." Paper 96-IAA.2.1.01, presented at the 47th Congress of the International Astronautical Federation, Beijing, October 7–11, 1996.

Bromberg, Joan Lisa. *NASA and the Space Industry*. Baltimore: Johns Hopkins University Press, 1999.

Brügel, Werner, ed. *Männer der Rakete*. Leipzig: Hachmeister und Thal, 1933.

Bullard, John W. *History of the Redstone Missile System*. Historical monograph AMC 23 M. Redstone Arsenal, Ala.: Army Missile Command, 1965.

Burchard, John E., ed. *Rockets, Guns and Targets*. Boston: Little, Brown, 1948.

Busemann, A[dolph]. "Ludwig Prandtl, 1875–1953." *Biographical Memoirs of Fellows of the Royal Society* 5: 193–205. London: Royal Society, 1960.

Butz, J. S., Jr. "GE System Stabilizes Thor Nose Cone." *Aviation Week*, August 10, 1959, 33.

Carroll, P. Thomas. "Historical Origins of the Sergeant Missile Powerplant." In Lattu, *Rocketry and Astronautics* (7th–8th Symposia), 121–46.

Carter, Paul A. *Politics, Religion, and Rockets: Essays in Twentieth-Century American History*. Tucson: University of Arizona Press, 1991.

Cassidy, David C. *Uncertainty: The Life and Science of Werner Heisenberg*. New York: W. H. Freeman, 1991.

Ceruzzi, Paul E. *A History of Modern Computing*. Cambridge, Mass.: MIT Press, 1998.

Chang, Iris. *Thread of the Silkworm*. New York: Basic Books, 1995.

Chapman, John L. *Atlas: The Story of a Missile*. New York: Harper, 1960.

Chemical Propulsion Information Agency. *CPIA/M1 Rocket Motor Manual*. Vol. 1. Laurel, Md.: CPIA, 1994. Cited as *CPIA/M1*.

———. *CPIA/M5 Liquid Propellant Engine Manual*. Laurel, Md.: CPIA, 1994. Cited as *CPIA/M5*.

———. "History of CPIA—Over 50 Years of Service to the Propulsion Community." <cpia.jhu.edu/About/cpiaat50.htm>, accessed July 18, 2002.

Chemical Propulsion Information Analysis Center. "History of CPIAC [Chemical Propulsion Information Analysis Center]." <cpia.jhu.edu/templates/cpiacTemplate/about/index.php?action=history>.

Christman, Albert B. *Sailors, Scientists, and Rockets: Origins of the Navy Rocket Program and of the Naval Ordnance Test Station, Inyokern*, vol. 1 of *History of the Naval Weapons Center, China Lake, California*. Washington, D.C.: Naval History Division, 1971.

Chulick, M. J., L. C. Meland, F. C. Thompson, and H. W. Williams. "History of the Titan Liquid Rocket Engines." In Doyle, *Liquid Rocket Engine Development*, 19–35.

Clark, John D. *Ignition! An Informal History of Liquid Rocket Propellants*. New Brunswick, N.J.: Rutgers University Press, 1972.

Clarke, Arthur C., ed. *The Coming of the Space Age*. New York: Meredith Press, 1967.

Clary, David A. *Rocket Man: Robert H. Goddard and the Birth of the Space Age*. New York: Hyperion, 2003.

Corum, James S. *The Luftwaffe: Creating the Operational Air War, 1918–1940*. Lawrence: University Press of Kansas, 1997.

CPIAC. *See* Chemical Propulsion Information Analysis Center.

Crouch, Tom D. *Aiming for the Stars: The Dreamers and Doers of the Space Age*. Washington, D.C.: Smithsonian Institution Press, 1999.

———. "'To Fly to the World in the Moon': Cosmic Voyaging in Fact and Fiction from Lucian to Sputnik." In *Science Fiction and Space Futures: Past and Present*, edited by Eugene M. Emme, 1–17. AAS History Series 5. San Diego: Univelt, 1982.

Dannenberg, Konrad K. "From Vahrenwald via the Moon to Dresden." In Hunley, *Rocketry and Astronautics* (24th Symposium), 119–34.

———. "Hermann Oberth—Half a Century Ahead." <www.meaus.com/centuryAhead.html>.

Davis, Vincent. *The Politics of Innovation: Patterns in Navy Cases.* Denver: University of Denver, 1967.

Day, Lance, and Ian McNeil, eds. *Biographical Dictionary of the History of Technology.* London: Routledge, 1996.

Debus, Allen G., ed. *World Who's Who in Science.* Chicago: Marquis–Who's Who, 1968.

De Maeseneer, Guido. *Peenemünde: The Extraordinary Story of Hitler's Secret Weapons, V-1 and V-2.* Vancouver: AJ Publishing, 2001.

Derdak, Thomas, ed. *International Directory of Company Histories.* Vol. 1. Chicago: St. James Press, 1988.

DeVorkin, David H. *Science with a Vengeance: How the Military Created the US Space Sciences after World War II.* New York: Springer, 1992.

"Donald A. Quarles." <www.af.mil/history/person.asp?dec=&pid=123006469>.

Dooling, Dave. "Thiokol: Firm Celebrates 30 Years of Space Technology." *Huntsville Times*, April 22, 1979.

Dornberger, Walter. *V-2.* Translated by James Cleugh and Geoffrey Halliday. New York: Viking, 1954.

Doyle, Stephen E., ed. *History of Liquid Rocket Engine Development in the United States, 1955–1980.* AAS History Series 13. San Diego: Univelt, 1992.

Dragomir, Sylviu. *The Ethnical Minorities in Transylvania.* Geneva: Sonor Printing, 1927.

Draper, Charles S[tark]. "On Course to Modern Guidance." *Astronautics and Aeronautics* 18, no. 2 (February 1980): 56–61.

———. "Origins of Inertial Navigation." *Journal of Guidance and Control* 4, no. 5 (September–October 1981): 449–63.

Dunar, Andrew J., and Stephen P. Waring. *Power to Explore: A History of Marshall Space Flight Center, 1960–1990.* SP-4313. Washington, D.C.: NASA, 1999.

Durant, Frederick C., III. "Robert H. Goddard: Accomplishments of the Roswell Years, 1930–1941." In Lattu, *Rocketry and Astronautics* (7th–8th Symposia), 317–41.

———. "Robert H. Goddard and the Smithsonian Institution." In Durant and James, *First Steps Toward Space*, 57–69.

Durant, Frederick C., III, and George S. James, eds. *First Steps Toward Space: Proceedings of the First and Second History Symposia of the International Academy of Astronautics.* Smithsonian Annals of Flight 10. Washington, D.C.: Smithsonian Institution Press, 1974.

Dyer, Davis. *TRW: Pioneering Technology and Innovation since 1900.* Boston: Harvard Business School Press, 1998.

Dyer, Davis, and David B. Sicilia. *Labors of a Modern Hercules: The Evolution of a Chemical Company.* Boston: Harvard Business School Press, 1990.

Ehricke, Krafft A. "The Peenemünde Rocket Center." *Rocketscience* 4, no. 3 (September 1950): 17–22, 31–34, 57–63.

Elder, John. "The Experience of Hermann Oberth." In Hunley, *Rocketry and Astronautics* (25th Symposium), 277–318.

Emme, Eugene M., ed. *The History of Rocket Technology: Essays on Research, Development, and Utility*. Detroit: Wayne State University Press, 1964.

Essers, I[lse]. *Max Valier—A Pioneer of Space Travel*. Translated by Agence Tunisienne de Public-Relations. TT F-664. Washington, D.C.: NASA, 1976.

———. *Max Valier: Ein Vorkämpfer der Weltraumfahrt, 1895–1930*. Düsseldorf: VDI, 1968.

Evans, Richard J. *Rereading German History: From Unification to Reunification, 1800–1996*. London: Routledge, 1997.

"The Experimental Engines Group." Interview with W. F. "Bill" Ezell, C. A. "Cliff" Hauenstein, J. O. "Jim" Bates, G. S. "Stan" Bell, and R. "Dick" Schwarz. *Threshold: An Engineering Journal of Power Technology* (Rocketdyne), no. 4 (Spring 1989): 21–27.

Ezell, Linda Neuman. *NASA Historical Data Book*. Vol. 2, *Programs and Projects, 1958–1968*. SP-4012. Washington, D.C.: NASA, 1988.

Ezell, William F., and J. K. Mitchell. "Engine One." *Threshold: An Engineering Journal of Power Technology* (Rocketdyne), no. 7 (Summer 1991): 52–63.

Farrior, J. S. "Inertial Guidance: Its Evolution and Future Potential." In Stuhlinger et al., *From Peenemünde to Outer Space*, 402–9.

Ferguson, Eugene S. *Engineering and the Mind's Eye*. Cambridge, Mass.: MIT Press, 1992.

Franklin, Thomas. *An American in Exile: The Story of Arthur Rudolph*. Huntsville, Ala.: Christopher Kaylor, 1987.

Friedman, Norman. *Seapower and Space: From the Dawn of the Missile Age to Net-Centric Warfare*. Annapolis, Md.: Naval Institute Press, 2000.

Friedman, S. Morgan. "The Inflation Calculator." <www.westegg.com/inflation/infl.cgi>.

"From the Desert to the Sea: A Brief Overview of the History of China Lake." <www.nawcwpns.navy.mil/clmf/hist.html>.

Fuhrman, R. A. "Fleet Ballistic Missile System: Polaris to Trident." Von Kármán Lecture for 1978 at the 14th Annual Meeting of the AIAA, Washington, D.C., February 1978. Paper AiAA-1978-355. Reprinted in *Journal of Spacecraft and Rockets* 15, no. 5 (September–October 1978: 265–86).

Fuller, Paul N., and Henry M. Minami. "History of the Thor/Delta Booster Engines." In Doyle, *Liquid Rocket Engine Development*, 39–51.

"Functional Polymers Products and Research." <www.elf-atochem.com>, accessed April 10, 2002.

Gai, Eli. "Guidance, Navigation, and Control from Instrumentation to Information Management." *Journal of Guidance, Control, and Dynamics* 19, no. 1 (January–February 1996): 10.

Galloway, Eilene. "Organizing the United States Government for Outer Space, 1957–1958." In Launius, Logsdon, and Smith, *Reconsidering Sputnik*, 309–26.

Gartmann, Heinz. *The Men Behind the Space Rockets.* Translated by Eustace Wareing and Michael Glenny. New York: David McKay, 1956.

GE Challenge. See General Electric Company.

"GE Reveals Hermes Missile Milestones." *Aviation Week,* March 8, 1954, 26, 30–32.

Geiger, Roger L. *To Advance Knowledge: The Growth of American Research Universities, 1900–1940.* Oxford: Oxford University Press, 1986.

General Electric Company, Missile and Space Division. *Challenge.* 10th anniversary issue, Spring 1965. Cited as *GE Challenge.*

Gerard, E[mily]. *The Land Beyond the Forest: Facts, Figures, and Fancies from Transylvania.* 2 vols. Edinburgh: Blackwood, 1888.

Geschelin, Joseph. "Thor's Brains Require Millionths-of-an-Inch Manufacturing Tolerances." *Aircraft and Missiles Manufacturing,* April 1958, 24–27.

Gibson, James N. *The Navaho Missile Project: The Story of the "Know-How" Missile of American Rocketry.* Atglen, Pa.: Schiffer, 1996.

[Glennan, T. Keith.] *The Birth of NASA: The Diary of T. Keith Glennan.* Edited by J. D. Hunley. Introduction by Roger D. Launius. SP-4104. Washington, D.C.: NASA, 1993.

Goddard, Esther C., and G. Edward Pendray, eds. *The Papers of Robert H. Goddard.* 3 vols., continuously paginated. New York: McGraw-Hill, 1970. Cited as *Goddard Papers.*

Goddard, Robert H. *Rocket Development: Liquid-Fuel Rocket Research, 1929–1941.* Edited by Esther C. Goddard and G. Edward Pendray. New York: Prentice-Hall, 1948.

Goldberg, Stanley, and Roger H. Stuewer, eds. *The Michelson Era in American Science, 1870–1930.* New York: American Institute of Physics, 1988.

Goodstein, Judith R. *Millikan's School: A History of the California Institute of Technology.* New York: Norton, 1991.

Gordon, Robert, et al. *Aerojet: The Creative Company.* Los Angeles: Stuart F. Cooper, 1995.

Gorn, Michael H. *The Universal Man: Theodore von Kármán's Life in Aeronautics.* Washington, D.C.: Smithsonian Institution Press, 1992.

Gray, Mike. *Angle of Attack: Harrison Storms and the Race to the Moon.* New York: Norton, 1992.

Green, Constance McLaughlin, and Milton Lomask. *Vanguard: A History.* Washington, D.C.: Smithsonian Institution Press, 1971.

Green, Constance McLaughlin, Harry C. Thompson, and Peter C. Roots. *The Ordnance Department: Planning Munitions for War.* Washington, D.C.: Department of the Army, 1955.

Greever, Bill B. "General Description and Design of the Configuration of the Juno I and Juno II Launching Vehicles." *IRE Transactions on Military Electronics* (Institute of Radio Engineers), vol. MIL-4, nos. 2–3 (April–July, 1960): 70–77. Seen in "Juno Launch Vehicles," folder 012072, NHRC.

Griffith, Alison. *The National Aeronautics and Space Act: A Study of the Development of Public Policy.* Washington, D.C.: Public Affairs Press, 1962.

Grimwood, James M. *Project Mercury: A Chronology*. SP-4001. Washington, D.C.: NASA, 1963.

Gruntman, Mike. *Blazing the Trail: The Early History of Spacecraft and Rocketry*. Reston, Va.: AIAA, 2004.

Hacker, Barton C. "Robert H. Goddard and the Origins of Space Flight." In *Technology in America: A History of Individuals and Ideas*, edited by Carroll W. Pursell Jr., 263–75. Cambridge, Mass.: MIT Press, 1981.

Hacker, Barton C., and James M. Grimwood. *On the Shoulders of Titans: A History of Project Gemini*. SP-4203. Washington, D.C.: NASA, 1977.

Haeussermann, Walter. "Developments in the Field of Automatic Guidance and Control of Rockets." *Journal of Guidance and Control* 4, no. 3 (May–June 1981): 225–39.

Haley, Andrew G. *Rocketry and Space Exploration*. Princeton, N.J.: Van Nostrand, 1958.

Hall, Edward N. "Air Force Missile Experience." *Air University Quarterly Review* 9 (Summer 1957): 22–33.

———. *The Art of Destructive Management: What Hath Man Wrought?* New York: Vantage, 1984.

Hall, Eldon C. "From the Farm to Pioneering with Digital Control Computers: An Autobiography." *IEEE Annals of the History of Computing*, April–June 2000, 22–31.

———. *Journey to the Moon: The History of the Apollo Guidance Computer*. Reston, Va.: AIAA, 1996.

Hall, R. Cargill. "Earth Satellites, A First Look by the United States Navy." In R. C. Hall, *Rocketry and Astronautics* (3rd–6th Symposia), 2: 253–77.

———, ed. *Essays on the History of Rocketry and Astronautics: Proceedings of the Third Through the Sixth History Symposia of the International Academy of Astronautics*. 2 vols. CP-2014. Washington, D.C.: NASA, 1977. Reprinted as *History of Rocketry and Astronautics: Proceedings of the Third through Sixth Symposia of the International Academy of Astronautics*. AAS History series 7. Parts 1 and 2. San Diego: Univelt, 1986.

———. "Origins and Development of the Vanguard and Explorer Satellite Programs." *Airpower Historian* 9, no. 1 (January 1964): 100–112.

Hallion, Richard P. "American Rocket Aircraft: Precursors to Manned Flight Beyond the Atmosphere." In Lattu, *Rocketry and Astronautics* (7th–8th Symposia), 283–89.

Hanle, Paul A. *Bringing Aerodynamics to America*. Cambridge, Mass.: MIT Press, 1982.

Hansen, James R. *Engineer in Charge: A History of the Langley Aeronautical Laboratory, 1917–1958*. SP-4305. Washington, D.C.: NASA, 1987.

Hartman, Edwin P. *Adventures in Research: A History of Ames Research Center, 1940–1965*. SP-4302. Washington, D.C.: NASA, 1970.

Hartt, Julian. *The Mighty Thor: Missile in Readiness*. New York: Duell, Sloan and Pearce, 1961.

Harwood, William B. *Raise Heaven and Earth: The Story of Martin Marietta People and Their Pioneering Achievements*. New York: Simon and Schuster, 1993.

Hastedt, Glenn P. "Sputnik and Technological Surprise." In Launius, Logsdon, and Smith, *Reconsidering Sputnik*, 401–24.

H.D.S. "Granville Stanley Hall." In *Dictionary of American Biography*, 4: 128. New York: Scribner, 1927–36.

Heppenheimer, T. A. *Countdown: A History of Space Flight*. New York: Wiley, 1997.

———. "The Navaho Program and the Main Line of American Liquid Rocketry." *Air Power History* 44 (Summer 1997): 5–17.

Hermann, Armin. "Lenard, Philipp." *Dictionary of Scientific Biography*, 8: 180–83. New York: Scribner, 1973.

Hermann, Rudolph [Rudolf]. "The Supersonic Wind Tunnel Installations at Peenemünde and Kochel and Their Contributions to the Aerodynamics of Rocket-Powered Vehicles." In Launius, *Rocketry and Astronautics* (15th–16th Symposia), 39–56.

Herring, Mack R. *Way Station to Space: A History of the John C. Stennis Space Center*. SP-4310. Washington, D.C.: NASA, 1997.

Hickman, Clarence N. "History of Rockets." *Encyclopaedia Britannica* (1959) 19: 367.

Hölsken, Heinz Dieter. *Die V-Waffen: Entstehung, Propaganda, Kriegseinsatz*. Stuttgart: Deutsche Verlags-Anstalt, 1984.

Hoelzer, Helmut. "Guidance and Control Symposium." In *The Eagle has Returned*, edited by Ernst A. Steinhoff. San Diego: Univelt, 1976, 301–16

Hoffman, Timothy. "Three Inducted into Air Force Space and Missile Hall of Fame." *Space Observer*, 8 October 1999, 3.

Holton, Gerald. "On the Hesitant Rise of Quantum Physics Research in the United States." In Goldberg and Stuewer, *Michelson Era*, 177–205.

Horrigan, Brian. "Popular Culture and the Future in Space, 1901–2001." In *New Perspectives on Technology and American Culture*, edited by Bruce Sinclair, 49–67. Philadelphia: American Philosophical Society, 1986.

Hoselton, Gary A. "The Titan I Guidance System." *AAFM* (Association of Air Force Missileers) 6, no. 1 (March 1998): 4–6.

Huggett, Clayton, C. E. Bartley, and Mark M. Mills. *Solid Propellant Rockets*. Princeton, N.J.: Princeton University Press, 1960.

Hughes, Kaylene. "Two 'Arsenals of Democracy': Huntsville's World War II Army Architectural Legacy." <www.redstone.army.mil/history/arch/index.html>.

Hughes, Thomas P. *Rescuing Prometheus*. New York: Vintage, 1998.

Hujsak, Edward J. "The Bird That Did Not Want to Fly." *Spaceflight* 34 (March 1992): 102–4.

Hunley, J. D. "Braun, Wernher von." In *Dictionary of American Biography*, supplement 10 (1976–1980), 65–68. New York: Scribner, 1995.

———. "The Enigma of Robert H. Goddard." *Technology and Culture* 36 (April 1995): 327–50.

———. "The Evolution of Large Solid Propellant Rocketry in the United States." *Quest: The History of Spaceflight Quarterly* 6, no. 1 (1998): 22–38.

———, ed. *History of Rocketry and Astronautics: Proceedings of the Twenty-Fourth Symposium of the International Academy of Astronautics*. AAS History Series 19. San Diego: Univelt, 1997.

———, ed. *History of Rocketry and Astronautics: Proceedings of the Twenty-Fifth History Symposium of the International Academy of Astronautics.* AAS History Series 20. San Diego: Univelt, 1997.

———. "The History of Solid-Propellant Rocketry: What We Do and Do Not Know." Paper AIAA-99-2925, presented at the 35th AIAA/ASME/SAE/ASEE Joint Propulsion Conference and Exhibit, Los Angeles, June 20–24, 1999.

———. "A Question of Antecedents: Peenemünde, JPL, and the Launching of U.S. Rocketry." In *Organizing for the Use of Space: Historical Perspectives on a Persistent Issue*, edited by Roger D. Launius, 1–31. AAS History Series 18. San Diego: Univelt, 1995.

———. *U.S. Space Launch Vehicle Technology: Viking to Space Shuttle.* Gainesville: University Press of Florida, 2008.

———. "Wernher von Braun, 1912–1977." In *Notable Twentieth-Century Scientists*, edited by Emily J. McMurray, 2093–96. New York: Gale Research, 1995.

Huzel, Dieter K. *Peenemünde to Canaveral.* Englewood Cliffs, N.J.: Prentice-Hall, 1962.

Huzel, Dieter K., and David H. Huang. *Design of Liquid Propellant Rocket Engines.* SP-125. Washington, D.C.: NASA, 1967.

———. *Modern Engineering for Design of Liquid-Propellant Rocket Engines.* Revised and updated by Harry Arbit et al. Washington, D.C.: AIAA, 1992.

Jekeli, Hermann. *Die Entwicklung des siebenbürgisch-sachsischen höheren Schulwesens von den Anfängen bis zur Gegenwart.* Mediaş, Romania: Reisenberger, 1930.

Jenkins, Dennis R. "Stage-and-a-Half: The Atlas Launch Vehicle." In *To Reach the High Frontier: A History of U.S. Launch Vehicles*, edited by Roger D. Launius and Dennis R. Jenkins, 70–102. Lexington: University Press of Kentucky, 2002.

Jet Propulsion Laboratory (JPL). "Explorer I." *Astronautics*, April 1958, 22, 83.

———. *Mariner-Mars 1964, Final Project Report.* SP-139. Washington, D.C.: NASA, 1967.

Johnson, Stephen B. "Craft or System? The Development of Systems Engineering at JPL." *Quest: The History of Spaceflight Quarterly* 6, no. 2 (1998): 17–31.

JPL. *See* Jet Propulsion Laboratory.

Judge, John F. "Aerojet Moves into Basic Glass." *Missiles and Rockets*, March 18, 1963, 34–35.

Karner, Stefan. "Die Steuerung der V2: Zum Anteil der Firma Siemens an der Entwicklung der ersten selbstgesteuerten Grossrakete." *Technikgeschichte* 46, no. 1 (1979): 45–66.

Karr, Erica M. "Hoffman . . . the power behind Rocketdyne." *Missiles and Rockets*, March 23, 1959, 26.

Kast, Fremont E., and James E. Rosenzweig, eds. *Science, Technology, and Management.* New York: McGraw-Hill, 1963.

Kennedy, Gregory P. *Vengeance Weapon 2: The V-2 Guided Missile.* Washington, D.C.: Smithsonian Institution Press, 1983.

Kennedy, W. S., S. M. Kovacic, and E. C. Rea. "Solid Rocket History at TRW Ballistic Missiles Division." Paper AIAA-92-3614, presented at the AIAA/SAE/ASME/ASEE 28th Joint Propulsion Conference, Nashville, July 6–8, 1992.

Kershaw, Ian. *The Nazi Dictatorship: Problems and Perspectives of Interpretation*. 2nd ed. London: Edward Arnold, 1989.

Klass, Philip J. "How Command Guidance Controls Atlas." *Aviation Week*, April 28, 1958, 74–81.

———. "Thor Guidance Goes on Production Line." *Aviation Week*, December 30, 1957, 38–39, 41–44.

Klee, Ernst, and Otto Merk. *The Birth of the Missile: The Secrets of Peenemünde*. Translated by T. Schoeters. London: Harrap, 1965.

Koelsch, William A. *Clark University, 1887–1987: A Narrative History*. Worcester, Mass.: Clark University Press, 1987.

———. "The Michelson Era at Clark, 1889–1892." In Goldberg and Stuewer, *Michelson Era*, 133–151.

Koppes, Clayton R. *JPL and the American Space Program: A History of the Jet Propulsion Laboratory*. New Haven: Yale University Press, 1982.

Kraemer, Robert S. *Rocketdyne: Powering Humans into Space*. With Vince Wheelock. Reston, Va.: AIAA, 2006.

Kuettner, Joachim P. "Mercury-Redstone Launch-Vehicle Development and Performance." In *Mercury Project Summary, Including Results of the Fourth Manned Orbital Flight, May 15 and 16, 1963*, edited by Kenneth S. Kleinknecht and W. M. Bland Jr. SP-45. Washington, D.C.: NASA, 1963.

Kurzweg, Hermann H. "The Aerodynamic Development of the V-2." In Benecke and Quick, *German Guided Missile Development*, 50–69.

Lasby, Clarence G. *Project Paperclip: German Scientists and the Cold War*. New York: Atheneum, 1971.

Latour, Bruno. *Science in Action: How to Follow Scientists and Engineers through Society*. Cambridge, Mass.: Harvard University Press, 1987.

Lattu, Kristan R., ed. *History of Rocketry and Astronautics: Proceedings of the Seventh and Eighth History Symposia of the International Academy of Astronautics*. AAS History Series 8. San Diego: Univelt, 1989.

Lattu, K[ristan], and R[ichard] Dowling. "John W. Parsons: Contributions to Rocketry, 1936–1946." Paper IAA-01–1AA.2.1.07 presented at the 52nd Congress of the International Astronautical Federation, Toulouse, October 1–5, 2001.

"Launch on December 63rd: Gen. Sam Phillips Recounts the Emergence of Solids in the Nation's Principal Deterrent." *Astronautics and Aeronautics* 10 (October 1972): 62.

Launius, Roger D., ed. *History of Rocketry and Astronautics: Proceedings of the Fifteenth and Sixteenth History Symposia of the International Academy of Astronautics*. AAS History Series 11. San Diego: Univelt, 1994.

———. *NASA: A History of the U.S. Civil Space Program*. Malabar, Fla.: Krieger, 1994.

———. Preface to Launius, Logsdon, and Smith, *Reconsidering Sputnik*, ix–xvi.

Launius, Roger D., John M. Logsdon, and Robert W. Smith, eds. *Reconsidering Sputnik: Forty Years Since the Soviet Satellite*. Amsterdam: Harwood Academic Publishers, 2000.

Law, John. "Technology and Heterogeneous Engineering: The Case of the Portuguese Expansion." In Bijker, Hughes, and Pinch, *Social Construction of Technological Systems*, 111–34.

Layton, Edwin T. "Mirror-Image Twins: The Communities of Science and Technology in 19th Century America." *Technology and Culture* 12 (1971): 562–80.

———. "Presidential Address: Through the Looking Glass, or News from Lake Mirror Image." *Technology and Culture* 28 (1987): 594–607.

———. "Technology as Knowledge." *Technology and Culture* 15 (1974): 31–41.

Leavitt, William. "Minuteman—Ten Years of Solid Performance." *Air Force Magazine* 54 (March 1971): 26.

Lehman, Milton. *Robert H. Goddard: Pioneer of Space Research*. 1963. New York: Da Capo, 1988.

Lethbridge, Cliff. "History of Rocketry, Chapter 6: 1945 to the Creation of NASA." <www.spaceline.org/history/6.html>.

Levine, Alan J. *The Missile and Space Race*. Westport, Conn.: Praeger, 1994.

Liquid Propellant Information Agency. *Liquid Propellant Safety Manual*. Silver Spring, Md.: Applied Physics Laboratory, Johns Hopkins University, 1958.

London, John R., III. "Brennschluss over the Desert: V-2 Operations at White Sands Missile Range, 1946–1952." In *History of Rocketry and Astronautics: Proceedings of the Twentieth and Twenty-First History Symposia of the International Academy of Astronautics*, edited by Lloyd H. Cornett Jr., 335–67. AAS History Series 15. San Diego: Univelt, 1993.

Lonnquest, John Clayton. "The Face of Atlas: General Bernard Schriever and the Development of the Atlas Intercontinental Ballistic Missile, 1953–1960." Ph.D. diss., Duke University, 1996.

Lonnquest, John C., and David F. Winkler. *To Defend and Deter: The Legacy of the United States Cold War Missile Program*. Rock Island, Ill.: Defense Publishing Service, 1996.

Mack, Pamela E., ed. *From Engineering Science to Big Science: The NACA and NASA Collier Trophy Research Project Winners*. SP-4219. Washington, D.C.: NASA, 1998.

MacKenzie, Donald. *Inventing Accuracy: A Historical Sociology of Nuclear Missile Guidance*. Cambridge, Mass.: MIT Press, 1990.

Maier, Charles S. *The Unmasterable Past: History, Holocaust, and German National Identity*. Cambridge, Mass.: Harvard University Press, 1988.

"Major General Holger N. Toftoy." <www.redstone.army.mil/history/toftoy/toftoy_bio.html>.

Malina, Frank J. "America's First Long-Range-Missile and Space Exploration Program: The ORDCIT Project of the Jet Propulsion Laboratory, 1943–46: A Memoir. In R. C. Hall, *Rocketry and Astronautics* (3rd–6th Symposia), 2: 339–83.

———. "Characteristics of the Rocket Motor Unit Based on the Theory of Perfect Gases." *Journal of the Franklin Institute* 230 (October 1940): 433–54.

———. "The Jet Propulsion Laboratory: Its Origins and First Decade of Work." *Spaceflight* vol. 5, nos. 5–6 (September-October 1964): 160–65, 216–23.

———. "On the GALCIT Rocket Research Project, 1936–38." In Durant and James, *First Steps Toward Space*, 113–27.

———. "Origins and First Decade of the Jet Propulsion Laboratory." In Emme, *Rocket Technology*, 46–66.

———. "Reflections of an Artist-Engineer on the Art-Science Interface." *Impact of Science on Society* 24, no. 1 (1974): 19–29.

———. "The Rocket Pioneers: Memoirs of the Infant Days of Rocketry at Caltech." *Engineering and Science* 31, no. 5 (February 1968): 9–13, 30–32.

———. "Rocketry in California." *Astronautics* 41 (July 1938): 3–6.

———. "The U.S. Army Air Corps Jet Propulsion Research Project, GALCIT Project No. 1, 1939–1946: A Memoir." In R. C. Hall, *Rocketry and Astronautics* (3rd–6th Symposia), 2: 153–201.

Mallan, Lloyd. *Men, Rockets, and Space Rats*. New York: Julian Messner, 1959.

Manuila, Sabin. *Aspects démographiques de la Transylvanie*. Bucharest: Imprimeria Nationala, 1938.

Martin, Richard E. "The Atlas and Centaur 'Steel Balloon' Tanks: A Legacy of Karel Bossart." Paper IAA-89-738, presented at the 40th Congress of the International Astronautical Federation, Málaga, October 7–13, 1989. Reprinted by General Dynamics Space Systems Division. Seen in "Bossart, Karel Jan." NASM

———. "A Brief History of the *Atlas* Rocket Vehicle." Part 1. *Quest: The History of Spaceflight Quarterly* 8, no. 2 (2000): 54–61.

Mastriola, E. J., and Karl Klager. *Solid Propellants Based on Polybutadiene Binders*. Reprinted from *Propellant Manufacture, Hazards, and Testing*, Advances in Chemistry Series 88, American Chemical Society, 1969.

Matthews, Henry. *The Saga of Bell X-2, First of the Spaceships: The Untold Story*. Beirut, Lebanon: HPM Publications, 1999.

"Maximilian F. J. C. Wolf." In Debus, *World Who's Who in Science*, 1817.

Mayr, Otto. "Lorenz, Hans." In *Dictionary of Scientific Biography*, 8: 500–501. New York: Scribner, 1973.

McCool, A. A., and Keith B. Chandler. "Development Trends of Liquid Propellant Engines." In Stuhlinger et al., *From Peenemünde to Outer Space*, 289–307.

McCurdy, Howard E. *Space and the American Imagination*. Washington, D.C.: Smithsonian Institution Press, 1997.

McDougall, Walter A. . . . *the Heavens and the Earth: A Political History of the Space Age*. New York: Basic Books, 1985.

McGraw-Hill Encyclopedia of Science and Technology. 9th ed. 20 vols. New York: McGraw-Hill, 2002.

McGuire, Frank G. "Minuteman Third-Stage Award Nears." *Missiles and Rockets*, October 17, 1960, 36–37.

Medaris, John B. *Countdown for Decision*. With Arthur Gordon. New York: Putnam, 1960.

Miller, Barry. "Microcircuits Boost Minuteman Capability." *Aviation Week and Space Technology*, October 28, 1963, 77.

"The Minuteman: The Mobilization Programme for a Thousand Minutemen." *Interavia* 1961, no. 3 : 310.

"Mobility Designed into Minuteman." *Aviation Week*, August 3, 1959, 93–94.

"Modifications Raise Polaris' Range." *Missiles and Rockets*, October 31, 1960, 36–37.

Moore, T. L. "Solid Rocket Development at Allegany Ballistics Laboratory." Paper AIAA-99-2931, presented at the 35th AIAA/ASME/SAE/ASEE Joint Propulsion Conference and Exhibit, Los Angeles, June 20–24, 1999.

Mueller, Fritz. "A History of Inertial Guidance." *Journal of the British Interplanetary Society* 38 (1985): 180–92.

Muenger, Elizabeth A. *Searching the Horizon: A History of Ames Research Center, 1940–1976*. SP-4304. Washington, D.C.: NASA, 1985.

Murray, Williamson. *The Luftwaffe, 1933–45: Strategy for Defeat.* Washington and London: Brassey's, 1996.

Myers, Dale D. "The Navaho Cruise Missile—A Burst of Technology." In Hunley, *Rocketry and Astronautics* (25th Symposium), 121–32.

NASA Education Division. *Rockets: Physical Science Teacher's Guide with Activities.* EP-291. Washington, D.C.: NASA, 1993.

NASA Office of Congressional Relations. *Mercury-Redstone III Sub-Orbital Manned Flight.* Washington, D.C.: NASA, 1961.

Neal, Roy. *Ace in the Hole*. Garden City, N.Y.: Doubleday, 1962.

Neufeld, Jacob. "Bernard A. Schriever: Challenging the Unknown." In *Makers of the United States Air Force*, edited by John L. Frisbee, 281–306. Washington, D.C.: Office of Air Force History, USAF, 1987.

———. *The Development of Ballistic Missiles in the United States Air Force, 1945–1960.* Washington, D.C.: Office of Air Force History, USAF, 1990.

Neufeld, Michael J. "The End of the Army Space Program: Interservice Rivalry and the Transfer of the von Braun Group to NASA, 1958–1959." *Journal of Military History* 69 (July 2005): 737–57.

———. "The Excluded: Hermann Oberth and Rudolf Nebel in the Third Reich." In *History of Rocketry and Astronautics: Proceedings of the Twenty-Eighth and Twenty-Ninth History Symposia of the International Academy of Astronautics*, edited by Donald C. Elder and Christophe Rothmund, 209–22. AAS History Series 23. San Diego: Univelt, 2001.

———. "The German Rocket Engineers and the Legacy of Peenemünde in Huntsville." Paper presented at the Society for the History of Technology conference, October 1993.

———. "The Guided Missile and the Third Reich: Peenemünde and the Forging of a Technological Revolution." In *Science, Technology, and National Socialism*, edited by Monika Renneberg and Mark Walker, 51–71. Cambridge: Cambridge University Press, 1994.

———. "Orbiter, Overflight, and the First Satellite: New Light on the Vanguard Decision." In Launius, Logsdon, and Smith, *Reconsidering Sputnik*, 231–57.

———. *The Rocket and the Reich: Peenemünde and the Coming of the Ballistic Missile Era*. New York: Free Press, 1995.

———. "Weimar Culture and Futuristic Technology: The Rocketry and Spaceflight Fad in Germany, 1923–1933." *Technology and Culture* 31 (1990): 725–52.

———. "Wernher von Braun, the SS, and Concentration Camp Labor: Questions of Moral, Political, and Criminal Responsibility." *German Studies Review* 25, no. 1 (February 2002): 57–78.

Newell, Homer E. *Beyond the Atmosphere: Early Years of Space Science.* SP-4211. Washington, D.C.: NASA, 1980.

Nicholas, Ted, and Rita Rossi. *U.S. Missile Data Book, 1994.* 18th ed. Fountain Valley, Calif.: Data Search Associates, November 1993.

Noyes, W. A., Jr., ed. *Chemistry: A History of the Chemistry Components of the National Defense Research Committee, 1940–1946.* Boston: Little, Brown, 1948.

Oberth, Adolf E. *Principles of Solid Propellant Development.* Publication 469. Laurel, Md.: CPIA, 1987.

Oberth, Hermann. "Autobiography." In Clarke, *Space Age,* 113–21.

———. "Ist die Weltraumfahrt möglich?" *Die Rakete: Officielles Organ des Vereins für Raumschiffahrt E.V. in Deutschland* 1, no. 11 (November 15, 1927): 144–52; no. 12 (December 15, 1927): 162–66.

———. "My Contributions to Astronautics." In Durant and James, *First Steps Toward Space,* 129–40.

———. *Die Rakete zu den Planetenräumen.* Munich: R. Oldenbourg, 1923. Reprint, Nuremberg: Uni, 1960.

———. "Raketenflug und Raumschiffahrt." *Die Rakete: Officielles Organ des Vereins für Raumschiffahrt E.V. in Deutschland* 2, no. 6 (June 15, 1928): 82–89.

———. *Ways to Spaceflight.* Translated by Agence Tunisienne de Public-Relations. TT F-622. Washington, D.C.: NASA, 1972.

Ordway, Frederick I., III. "Reaction Motors Division of Thiokol Chemical Corporation: An Operational History, 1958–1972 (Part II)." In Sloop, *Rocketry and Astronautics* (17th Symposium), 137–73.

Ordway, Frederick I., III, and Mitchell R. Sharpe. *The Rocket Team.* New York: Thomas Y. Crowell, 1979.

Ordway, Frederick I. [III], and Ronald C. Wakeford. *International Missile and Spacecraft Guide.* New York: McGraw-Hill, 1960.

Ordway, Frederick I., III, and Frank H. Winter. "Pioneering Commercial Rocketry in the United States of America, Reaction Motors Inc." Two parts. *Journal of the British Interplanetary Society* 36 (1983) and 38 (1985).

———. "Reaction Motors, Inc.: A Corporate History, 1941–1958, Part 1, Institutional Developments." In Launius, *Rocketry and Astronautics* (15th–16th Symposia), 75–100. (For part 2, *see* Winter and Ordway.)

Pace, Eric. "Willy A. Fiedler, 89, a Leading Missile Expert." *New York Times,* January 29, 1998.

Pendray, G. Edward. "Early Rocket Development of the American Rocket Society." In Durant and James, *First Steps Toward Space,* 141–55.

Perry, Robert L. "The Atlas, Thor, Titan, and Minuteman." In Emme, *Rocket Technology,* 142–61.

Peukert, Detlev J. K. *Inside Nazi Germany: Conformity, Opposition, and Racism in Everyday Life,* translated by Richard Deveson. New Haven: Yale University Press, 1982.

Pickering, William H. "Countdown to Space Exploration: A Memoir of the Jet Propulsion Laboratory, 1944–1958." With James H. Wilson. In R. C. Hall, *Rocketry and Astronautics* (3rd–6th Symposia), 2: 385–421.
———. "History of the Cluster System." In Stuhlinger et al., *From Peenemünde to Outer Space*, 141–62.
Pike, Iain. "Atlas: Pioneer ICBM and Space-Age Workhorse." *Flight International* 81 (January 1962): 89–96; 82 (February 1962): 175–79.
"Polaris A3 Firing Tests Under Way." *Missiles and Rockets*, August 13, 1962, 15.
"Polaris-Insurance." *Washington News*, November 13, 1967.
"Polaris Solid Triumph for Aerojet: Major Breakthroughs Open New Era in Sea Warfare." *Aerojet Booster* 3, no. 8-B (February 1959): 1–2.
Powell, Joel W. "Thor-Able and Atlas-Able." *Journal of the British Interplanetary Society* 37, no. 5 (May 1984): 219–25.
Powell, Joel, and Keith J. Scala. "Historic White Sands Missile Range." *Journal of the British Interplanetary Society* 47 (1994): 83–98.
Price, E. W., C. L. Horine, and C. W. Snyder. "Eaton Canyon: A History of Rocket Motor Research and Development in the Caltech-NDRC-Navy Rocket Program, 1941–1946." Paper AIAA-98-3977, presented at the 34th AIAA/ASME/SAE/ASEE Joint Propulsion Conference and Exhibit, Cleveland, July 13–15, 1998.
"Problems May Cut Polaris A3 Range Goal." *Aviation Week and Space Technology*, September 4, 1961, 31.
Public Papers of . . . John F. Kennedy, . . . 1961. Washington, D.C.: Government Printing Office, 1962.
Raborn, W. F., Jr. "Management of the Navy's Fleet Ballistic Missile Program." In Kast and Rosenzweig, *Science, Technology, and Management*, 139–52.
Rasmussen, Cecelia. "Life as Satanist Propelled Rocketeer." *Los Angeles Times*, March 19, 2000.
"Redstone." <www.redstone.army.mil/history/systems/redstone/welcome.html>.
Reed, Mack. "Rocketdyne Razing Historic Test Stand." *Los Angeles Times*, February 11, 1996.
Reily, Philip Key. *The Rocket Scientists: Achievement in Science, Technology, and Industry at Atlantic Research Corporation*. New York: Vantage, 1999.
Reisig, Gerhard H. R. *Raketenforschung in Deutschland: Wie die Menschen das All eroberten*. Berlin: Wissenschaft und Technik, 1999.
———. "Von den Peenemünder 'Aggregaten' zur amerikanischen 'Mondrakete': Die Entwicklung der Apollo-Rakete 'Saturn V' durch das Wernher-von-Braun-Team an Hand der Peenemünder Konzepte." *Astronautik* (1986): 5–9, 44–47, 73–77, 111.
Rhodes, Richard. "The Ordeal of Robert Hutchings Goddard: 'God Pity a One-Dream Man.'" *American Heritage* 31, no. 4 (June–July 1980): 25–32.
[Rice, Franklin P., ed.] *The Worcester of Eighteen Hundred and Ninety-Eight: Fifty Years a City*. Worcester, Mass.: F. S. Blanchard, 1899.
Robillard, G. "Explorer Rocket Research Program." *American Rocket Society Journal*, July 1959, 492–96.

Rosen, Milton W. "Big Rockets." *International Science and Technology*, December 1962, 66–71, 87.

———. *The Viking Rocket Story*. New York: Harper, 1955.

Rosenzweig, Roy. *Eight Hours for What We Will: Workers and Leisure in an Industrial City, 1870–1920*. Cambridge: Cambridge University Press, 1983.

Rotta, Julius C. *Die Aerodynamische Versuchsanstalt in Göttingen, ein Werk Ludwig Prandtls: Ihre Geschichte von den Anfängen bis 1925*. Göttingen: Vandenhoeck und Ruprecht, 1990.

Ruland, Bernd. *Wernher von Braun: Mein Leben für die Raumfahrt*. Offenburg: Burda, 1969.

Rynin, N. A. *Interplanetary Flight and Communication*. Vol. 3, no. 8, *Theory of Space Flight*. Translated by R. Hardin. NASA TT F-647. Jerusalem: Israel Program for Scientific Translations, 1971.

Sänger-Bredt, Irene. "The Silver Bird Story: A Memoir." In R. C. Hall, *Rocketry and Astronautics* (3rd–6th Symposia), 1: 195–201.

Sänger-Bredt, Irene, and Rolf Engel. "The Development of Regeneratively Cooled Liquid Rocket Engines in Austria and Germany, 1926–1942." In Durant and James, *First Steps Toward Space*, 217–46.

Sapolsky, Harvey M. *The Polaris System Development: Bureaucratic and Programmatic Success in Government*. Cambridge, Mass.: Harvard University Press, 1972.

Schilling, Martin. "The Development of the V-2 Rocket Engine." In Benecke and Quick, *German Guided Missile Development*, 281–96.

Schoettle, Enid Curtis Bok. "The Establishment of NASA." In *Knowledge and Power*, edited by Sanford A. Lakoff, 162–270. New York: Free Press, 1966.

"Searchlights to Space Systems: The Story of Arma Division, 1918–'61." *Arma Engineering* 4, no. 3 (Fall 1961): 4–12.

Seely, Bruce. "Research, Engineering, and Science in American Engineering Colleges: 1900–1960." *Technology and Culture* 34 (1993): 344–67.

Sharpe, Mitchell R., and Bettye R. Burkhalter. "Mercury-Redstone: The First American Man-Rated Space Launch Vehicle." In Becklake, *Rocketry and Astronautics* (22nd–23rd Symposia), 341–88.

Siddiqi, Asif A. *Challenge to Apollo: The Soviet Union and the Space Race, 1945–1974*. SP-2000-4408. Washington, D.C.: NASA, 2000. Later published in two volumes as *Sputnik and the Soviet Space Challenge* and *The Soviet Space Race with Apollo* (Gainsville: University Press of Florida, 2003).

Simmons, John M. "The Navaho Lineage." *Threshold: An Engineering Journal of Power Technology* (Rocketdyne), no. 2 (December 1987): 16–23.

Slater, Alan E., ed. "Research and Development at the Jet Propulsion Laboratory, GALCIT." *Journal of the British Interplanetary Society* 6 (September 1946): 41–54.

Sloop, John L., ed. *History of Rocketry and Astronautics: Proceedings of the Seventeenth History Symposium of the International Academy of Astronautics*. AAS History Series 12. San Diego: Univelt, 1991.

———. *Liquid Hydrogen as a Propulsion Fuel, 1945–1959*. SP-4404. Washington, D.C.: NASA, 1978.

Smith, Robert A., and Henry M. Minami. "History of Atlas Engines." In Doyle, *Liquid Rocket Engine Development*, 53–66.

Snyder, Conway W. "Caltech's *Other* Rocket Project: Personal Recollections." *Engineering and Science*, Spring 1991, 2–13.

Sparks, J. F., and M. P. Friedlander III. "Fifty Years of Solid Propellant Technical Achievements at Atlantic Research Corporation." Paper AIAA-99-2932, presented at the 35th AIAA/ASME/SAE/ASEE Joint Propulsion Conference and Exhibit, Los Angeles, June 20–24, 1999.

Spinardi, Graham. *From Polaris to Trident: The Development of US Fleet Ballistic Missile Technology*. Cambridge: Cambridge University Press, 1994.

"Stage III for USAF's Minuteman III." *Aerojet General Booster*, September 1967.

Starr, S. "The Launch of Bumper 8 from the Cape: The End of an Era and the Beginning of Another." Paper IAAA-01-IAA.2.3.05 presented at the 52nd Congress of the International Astronautical Federation, Toulouse, October 1–5, 2001.

Stine, G. Harry. *ICBM: The Making of the Weapon that Changed the World*. New York: Orion, 1991.

Stone, Irving. "Aerojet Second Stage Must Withstand Heaviest Stresses." *Aviation Week and Space Technology*, September 3, 1962, 68–71.

———. "Hercules Stage 3 Uses Glass Fiber Case." *Aviation Week and Space Technology*, September 10, 1962, 162–67.

———. "Minuteman Propulsion—Part I: Minuteman ICBM Solid Motor Stages Enter Production Phase." *Aviation Week and Space Technology*, August 27, 1962, 54–62.

Stuart, John. "New Rocket Engine Is in 'Final Stages': Power for the X-2 Now Being Developed at Caldwell, N.J., Plant of Curtiss-Wright." *New York Times*, April 7, 1949, 59.

Stuhlinger, Ernst. "Army Activities in Space—A History." *IRE Transactions on Military Electronics* (Institute of Radio Engineers), vol. MIL-4, nos. 2–3 (April–July 1960): 64–69. Seen in "Juno Launch Vehicles," folder 012072, NHRC.

Stuhlinger, Ernst, et al., eds. *Astronautical Engineering and Science: From Peenemünde to Planetary Space; Honoring the Fiftieth Birthday of Wernher von Braun*. New York: McGraw-Hill, [1963].

———. *From Peenemünde to Outer Space: Commemorating the Fiftieth Birthday of Wernher von Braun, March 23, 1962*. Huntsville, Ala.: NASA Marshall Space Flight Center, 1962.

Stuhlinger, Ernst, and Michael J. Neufeld. "Wernher von Braun and Concentration Camp Labor: An Exchange." *German Studies Review* 26, no. 1 (February 2003): 121–26.

Stuhlinger, Ernst, and Frederick I. Ordway III. *Wernher von Braun: Aufbruch in den Weltraum*. Munich: Bechtle, 1992.

———. *Wernher von Braun, Crusader for Space: A Biographical Memoir*. Malabar, Fla.: Krieger, 1994.

Stumpf, David K. *Regulus, the Forgotten Weapon: A Complete Guide to Chance Vought's Regulus I and II*. Paducah, Ky.: Turner, 1996.

———. *Titan II: A History of a Cold War Missile Program.* Fayetteville: University of Arkansas Press, 2000.

Sutton, E[rnie] S. "From Polymers to Propellants to Rockets—A History of Thiokol." Paper AIAA-99-2929, presented at the 35th AIAA/ASME/SAE/ASEE Joint Propulsion Conference and Exhibit, Los Angeles, June 20–24, 1999.

———. "From Polysulfides to CTPB Binders—A Major Transition in Solid Propellant Binder Chemistry." Paper AIAA-84-1236, presented at the AIAA/SAE/ASME 20th Joint Propulsion Conference, Cincinnati, June 11–13, 1984.

———. *How a Tiny Laboratory in Kansas City Grew into a Giant Corporation: A History of Thiokol and Rockets, 1926–1996.* Chadds Ford, Pa.: privately printed, 1997.

Sutton, George P. *History of Liquid Propellant Rocket Engines.* Reston, Va.: AIAA, 2006.

Swenson, Loyd S., Jr., James M. Grimwood, and Charles C. Alexander. *This New Ocean: A History of Project Mercury.* SP-4201. Washington, D.C.: NASA, 1998.

Tatarewicz, Joseph N. "The Hubble Space Telescope Servicing Mission." In Mack, *From Engineering Science to Big Science,* 365–96.

Teutsch, Fr[iedrich]. *Die Siebenbürger Sachsen in Vergangenheit und Gegenwart.* 2nd ed. Hermannstadt: W. Krafft, 1924.

Texas A&M University Undergraduate Catalog. "History and Development." <http://www.tamu.edu/admissions/catalogs/06-07_UG_Catalog/gen_info/history_development/history.htm>.

Thiokol Chemical Corporation. *Rocket Propulsion Data.* 3rd ed. Bristol, Pa.: Thiokol, 1961.

Thomas, Shirley. *Men of Space: Profiles of the Leaders in Space Research, Development, and Exploration.* 8 vols. Philadelphia: Chilton, 1960–68.

———. "Theodore von Kármán's Caltech Students." Paper IAA-92-191 presented at the 43rd Congress of the International Astronautical Federation, Washington, D.C., August 28–September 5, 1992.

Tomayko, James E. "Helmut Hoelzer's Analog Computer." *Annals of the History of Computing* 7, no. 3 (July 1985): 227–40.

"Trevor Gardner, Ex-Air Force Aide: Research Chief Who Quit in Dispute in 1956 Dies at 48." *New York Times,* September 29, 1963.

Truax, Robert C. "Annapolis Rocket Motor Development." In Durant and James, *First Steps Toward Space,* 295–301.

———. "Liquid Propellant Rocket Development by the U.S. Navy during World War II: A Memoir." In Sloop, *Rocketry and Astronautics* (17th Symposium), 57–67.

———. "The Pioneer Rocket Project of the U.S. Navy." *Journal of the American Rocket Society,* no. 74 (June 1948): 62–65.

Umholz, Philip D. "The History of Solid Rocket Propulsion and Aerojet." Paper AIAA-99-2927, presented at the 35th AIAA/ASME/SAE/ASEE Joint Propulsion Conference, Los Angeles, June 20–24, 1999.

"Upper Air Rocket Summary, V-2 No. 19." ouray.cudenver.edu/~wrbeggs/uars19.html, accessed February 11, 2002.

U.S. Army Ordnance Missile Command, Public Information Office. *History of Redstone Arsenal*. Redstone Arsenal, Ala.: Ordnance Missile Command, 1962.

U.S. Congress, House of Representatives. "Organization and Management of Missile Programs." 11th Report by the Committee on Government Operations. 86th Cong., 1st sess., September 2, 1959. H. Rep. 1121.

———. "Organization and Management of Missile Programs." Hearings before a Subcommittee of the Committee on Government Operations. 86th Cong., 1st sess., February 4–March 20, 1959.

Van Dyke, Gretchen J. "Sputnik: A Political Symbol and Tool in 1960 Campaign Politics." In Launius, Logsdon, and Smith, *Reconsidering Sputnik*, 363–99.

Van Riper, A. Bowdoin. *Rockets and Missiles: The Life Story of a Technology*. Westport, Conn.: Greenwood Press, 2004.

Vincenti, Walter G. "The Air Propeller Tests of W. F. Durand and E. P. Lesley: A Case Study in Technological Methodology." *Technology and Culture* 20 (1979): 714–49.

———. *What Engineers Know and How They Know It: Analytical Studies from Aeronautical History*. Baltimore: Johns Hopkins University Press, 1990.

von Braun, Magnus. *Weg durch vier Zeitepochen*. Limburg an der Lahn: C. A. Stark, 1965.

von Braun, Wernher. "Konstruktive, theoretische und experimentelle Beiträge zu den Problem der Flüssigkeitsrakete." Ph.D. diss., Friedrich-Wilhelms-Universität zu Berlin, 1934.

———. "A Minimum Satellite Vehicle Based on Components Available from Missile Developments of the Army Ordnance Corps." In *Exploring the Unknown: Selected Documents in the History of the U.S. Civil Space Program*, edited by John M. Logsdon with Linda J. Lear, 1: 274–81. SP-4218. Washington, D.C.: NASA, 1995.

———. "The Prophet of Space Travel—Hermann Oberth." <www.meaus.com/spaceProphet.html>.

———. "The Redstone, Jupiter, and Juno." In Emme, *Rocket Technology*, 107–21.

———. "Reminiscences of German Rocketry." *Journal of the British Interplanetary Society* 15, no. 3 (May–June 1956): 125–45.

———. "Rundown on Jupiter-C." *Astronautics*, October 1958, 32–33, 80–84.

von Braun, Wernher, and Frederick I. Ordway III. *Space Travel: A History*. 4th ed. Revised by Dave Dooling. New York: Harper and Row, 1975.

von Kármán, Theodore. *The Wind and Beyond*. With Lee Edson. Boston: Little, Brown, 1967.

von Kármán, Theodore, and Frank J. Malina. "Characteristics of the Ideal Solid Propellant Rocket Motor." JPL report 1-4, 1940. In *Collected Works of Theodore von Kármán*, 4: 94–106. London: Butterworths, 1956.

Walker, Chuck. *Atlas: The Ultimate Weapon*. With Joel Powell. Burlington, Ont.: Apogee, 2005.

Walker, Mark. *German National Socialism and the Quest for Nuclear Power, 1939–1949*. Cambridge: Cambridge University Press, 1989.

Wandycz, Piotr S. *The Price of Freedom: A History of East Central Europe from the Middle Ages to the Present*. London: Routledge, 1992.

Warren, Francis A. *Rocket Propellants*. New York: Reinhold, 1958.

Warren, Francis A., and Charles H. Herty III. "U.S. Double-Base Solid Propellant Tactical Rockets of the 1950–1955 Era." Draft of paper presented at AIAA 9th Annual Meeting and Technical Display, Washington, D.C., January 8–10, 1973.

[Watson, Clement Hayes.] *Adventures in Partnership: The Story of Polaris*. Danbury, Conn.: Danbury Printing and Litho, [1963].

Weart, Spencer R. "The Physics Business in America, 1919–1940: A Statistical Reconnaissance." In *The Sciences in the American Context: New Perspectives*, edited by Nathan Reingold, 295–358. Washington, D.C.: Smithsonian Institution Press, 1979.

Wegener, Peter P. *The Peenemünde Wind Tunnels: A Memoir*. New Haven: Yale University Press, 1996.

Werrell, Kenneth P. *The Evolution of the Cruise Missile*. Maxwell AFB, Ala.: Air University Press, 1985.

Who's Who in World Aviation. Washington, D.C.: American Aviation Publications, 1955.

Wiech, Raymond E., Jr., and Robert F. Strauss. *Fundamentals of Rocket Propulsion*. New York: Reinhold, 1960.

Wiggins, Joseph W. "The Earliest Large Solid Rocket Motor—The Hermes." Paper presented at the AIAA 9th Annual Meeting and Technical Display, Washington, D.C., 8–10 January 1973.

———. "Hermes: Milestone in U.S. Aerospace Progress." *Aerospace Historian* 21, no. 1 (March 1974): 34–40.

Wills, Albert Potter. "Webster, Arthur Gordon." In *Dictionary of American Biography*, 10: 584–85. New York: Scribner, 1927–36.

Wilson, Andrew. "Jupiter/Juno—America's First Satellite Launcher." *Spaceflight* 23 (January 1981): 12–17.

Wilson, Brian A. "The History of Composite Motor Case Design." Paper AIAA-93-1782, presented at the AIAA/SAE/ASME/ASEE 29th Joint Propulsion Conference and Exhibit, Monterey, Calif., June 28–30, 1993.

Winter, Frank H. "Birth of the VfR: The Start of Modern Astronautics." *Spaceflight* 19 (July–August 1977): 243–56.

———. "'Black Betsy': The 6000C-4 Rocket Engine, 1945–1989, Part 1." In Becklake, *Rocketry and Astronautics* (22nd–23rd Symposia), 221–52.

———. "'Black Betsy': The 6000C-4 Rocket Engine, 1945–1989, Part 2." In Hunley, *Rocketry and Astronautics* (24th Symposium), 237–58.

———. "The East Parking Lot Rocket Experiments of North American Aviation, Inc., 1946–1949." IAA paper 99-IAA.2.2.07, presented at the 50th Congress of the International Astronautical Federation, Amsterdam, October 4–8, 1999.

———. *The First Golden Age of Rocketry*. Washington, D.C.: Smithsonian Institution Press, 1990.

———. "Harry Bull, American Rocket Pioneer." In *History of Rocketry and Astronautics: Proceedings of the Ninth, Tenth, and Eleventh History Symposia of the International Academy of Astronautics*, edited by Frederick I. Ordway III, 291–309. AAS History Series 9. San Diego: Univelt, 1989.

———. *Prelude to the Space Age: The Rocket Societies, 1924–1940*. Washington, D.C.: Smithsonian Institution Press, 1983.

———. "Reaction Motors Division of Thiokol Chemical Corporation: A Project History, 1958–1972 (Part III)." In Sloop, *Rocketry and Astronautics* (17th Symposium), 175–201.

———. "Rocketdyne—A Giant Pioneer in Rocket Technology: The Earliest Years, 1945–1955." Paper 97-IAA.2.2.08, presented at the 48th Congress of the International Astronautical Federation, Turin, October 6–10, 1997.

———. *Rockets into Space*. Cambridge, Mass.: Harvard University Press, 1990.

Winter, Frank H., and George S. James. "Highlights of 50 Years of Aerojet, A Pioneering American Rocket Company, 1942–1992." Paper IAA-93-679 presented at the 44th Congress of the International Astronautical Federation, Graz, Austria, October 16–22, 1993.

Winter, Frank H., and Frederick I. Ordway III. "Reaction Motors, Inc., 1941–1958, Part 2: Research and Development Efforts." In Launius, *Rocketry and Astronautics* (15th–16th Symposia), 101–27.

W.T.G. "Thor: A Study of a Great Weapons System." *Flight* 74 (December 5, 1958): 862–72.

Wuerth, J. M. "The Evolution of Minuteman Guidance and Control." *Journal of the Institute of Navigation* 23 (Spring 1976): 64–75.

Wyld, James H. "The Liquid-Propellant Rocket Motor." *Mechanical Engineering* 69, no. 6 (June 1947): 457–62.

Yaffee, Michael. "Ablation Wins Missile Performance Gain." *Aviation Week*, July 18, 1960, 54–55, 57, 59–61, 65.

———. "Pyrolitic Graphite Studied for Re-entry." *Aviation Week*, July 25, 1960, 26–28.

Yang, Vigor, and William E. Anderson, eds. *Liquid Rocket Engine Combustion Instability*. Washington, D.C.: AIAA, 1995.

York, Herbert. *Race to Oblivion: A Participant's View of the Arms Race*. New York: Simon and Schuster, 1970.

Zaloga, Steven J. *The Kremlin's Nuclear Sword: The Rise and Fall of Russia's Strategic Nuclear Forces, 1945–2000*. Washington, D.C.: Smithsonian Institution Press, 2002.

Index

Page numbers for illustrations and their captions are italicized. The letter t *following a page number denotes a table.*

J. D. Hunley is a retired U.S. Air Force and NASA historian. He has written "Minuteman and the Development of Solid Rocket Launch Technology," chapter 6 in *To Reach the High Frontier: A History of U.S. Launch Vehicles*, edited by Roger D. Launius and Dennis R. Jenkins (2002), and numerous articles in the field of aerospace history. During the ten years he served as a NASA historian, which included six years as chief historian for the NASA Dryden Flight Research Center, he edited more than a dozen books and monographs on aviation and space history, including *The Birth of NASA: The Diary of T. Keith Glennan* (1993).